Background and Recent Developments of

METRIC FIXED POINT THEORY

Background and Recent Developments of METRIC FIXED POINT THEORY

EDITED BY

DHANANJAY GOPAL

POOM KUMAM

MUJAHID ABBAS

CRC Press
Taylor & Francis Group
Boca Raton London New York

CRC Press is an imprint of the
Taylor & Francis Group, an **informa** business
A CHAPMAN & HALL BOOK

CRC Press
Taylor & Francis Group
6000 Broken Sound Parkway NW, Suite 300
Boca Raton, FL 33487-2742

Printed on acid-free paper
Version Date: 20171010

International Standard Book Number-13: 978-0-8153-6945-5 (Hardback)

Visit the Taylor & Francis Web site at
http://www.taylorandfrancis.com

and the CRC Press Web site at
http://www.crcpress.com

Contents

Foreword

This book, "Background and Recent Developments of Metric Fixed Point Theory," is an attempt to present an attractive research area of metric fixed point theory in a simple way with focus on clarity of arguments in the proofs of classical and recent results in this area. A collection of examples illustrates a variety of essential concepts. Fixed point theory makes this book an easily accessible source of knowledge. This book touches on several research directions within the fixed point theory and opens new avenues of investigation to extend and explore it further. Written by well known fixed point theorists with vast teaching experience, this book is particularly suitable for young mathematicians who want to study fixed point theory and to pursue their research careers in that area.

The presentation of Banach fixed point theorem and its generalizations provides a handy account of development and progress in metric fixed point theory dealing with single valued mapping. The book also includes interesting common fixed point results for families of commuting mappings. A thorough discussion on recently introduced concepts of α-admissible mappings and related results invites active researchers to contribute and to extend this theory further.

A fixed point theory discussed in the framework of fuzzy and soft metric will not only provide an adequate background but also will describe advanced results in this direction. Best proximity point theory has attracted the attention of several mathematicians working in optimization theory. This motivates us to include a chapter on cyclic mappings satisfying certain contractive conditions and best proximity point results in the setup of metric spaces. The last chapter of this book highlights the significance of fixed point theory in connection with numerical analysis, economics, integral equations, boundary value problems and difference equations. We believe that this effort will inspire young mathematicians to explore more applications of metric fixed point theory and extend the boundaries of this theory further.

Sompong Dhompongsa
Chiang Mai University
Chiang Mai, Thailand

Yeol Je Cho
Gyeongsang National University
Jinju, Korea

Preface

Functional analysis is an important branch of mathematics divided mainly into linear and nonlinear categories. We live in a nonlinear world in which infinitesimal inputs may result in macroscopic outputs and vice versa. For this reason, functional analysis has evolved into a separate and practical area of mathematics.

Functional analysis is an important branch of mathematics which can mainly be divided into two categories: Linear and Nonlinear. As, we are living in a nonlinear world where infinitesimal inputs may result in macroscopic outputs or vice versa, so nonlinear functional analysis has become an independent and a useful subject.

The theory of fixed points deals with the conditions which guarantee the existence of points x of a set X which solve an operator equation $x = Tx$, where T is a transformation defined on a set X. The solution set of such a problem can be empty, a finite set, a countable or uncountable infinite set. Fixed point theory provides essential tools for solving problems arising in various branches of mathematical analysis. Split feasibility problems, variational inequality problems, nonlinear optimization problems, equilibrium problems, complementarity problems, selection and matching problems, and proving solutions of integral and differential equations are closely related with fixed point theory. A long list of problems fall into the category of solving a fixed point problem. In particular, the solutions have deep roots in nonlinear functional analysis.

Research in fixed point theory generally includes (a) the investigation of less restrictive conditions on mappings that guarantees the existence of a fixed point, (b) study of conditions which assure uniqueness of a fixed point, (c) modification, enrichment and extension of the structures of domains of definition to obtain more general spaces, (d) identification of characterization, (e) constructions or approximations of fixed points and (f) the study of structure of the set of fixed points of mapping under consideration.

The Banach contraction principle proved in 1922 lies at the heart of metric fixed point theory and has played a fundamental role in many aspects of nonlinear functional analysis. The principle arises from the use of successive approximations (an idea initiated by Picard in 1890) to establish the existence and uniqueness of the solution of an operator equation $T(x) = x$. This principle serves as a powerful tool with a wide range of applications; in particular it can be employed to prove the existence of solutions of differential or integral equations. Due to its applications in mathematics and other related disciplines, it has been generalized in many directions.

Extensions of Banach contraction principle have been obtained either by generalizing the domain of the mappings or by extending the contractive condition on the mappings. The literature abounds with numerous extensions and variants that led to far-reaching applications.

The aim of this monograph is to present the recent results dealing with existence and uniqueness of fixed points of certain mappings and those aspects of investigations which fall within the scope of metric fixed point theory.

This monograph is divided into seven chapters. Each chapter contributed by different authors contains an "Introduction" which summarizes the material needed to read the chapter independent of others and contains a necessary background, several examples, and comprehensive literature to explain the concepts presented therein. This could be helpful for those who want to pursue research careers in fixed point theory and related areas.

Chapter 1 is motivational in nature and focuses on contractive conditions on mappings to guarantee the existence and uniqueness of fixed points. It gives a brief introduction of Banach fixed point principle and trace some well known extensions and generalizations of this principle.

Chapter 2 deals with comprehensive survey and motivation of coincidence and common fixed point theory. Several examples are presented to compare various weaker forms of commuting mappings which help the reader better understand the contents presented in this chapter.

Chapter 3 presents admissible mappings which provide a necessary machinery to obtaining numerous extensions of contractions in many situations. Although it is not possible to include all the available results on α-admissibility, an attempt has been made to introduce some fundamental results involving α-admissibility concepts and references of related works have been indicated.

Chapter 4 deals with a concise study of fixed point theorems for fuzzy contractive type mappings in fuzzy metric spaces as well as fuzzy metric-like spaces. The results presented in detail were selected to illustrated the direction of research over the last four decades up to most recent contributions.

Chapter 5 presents the famous Banach contraction principle and some related results in the setup of soft metric spaces and fixed point theory. This deals with the restrictions and limitations of certain concepts of soft metric spaces which offer challenges to establishing a bridge between metric and soft metric fixed point theory. Some interesting examples are provided to highlight the problems of extensions of metric fixed point theory.

Chapter 6 includes the cyclic representation of mappings for existence of best proximity points, common best proximity points for proximity com- muting mappings and also contains the necessary properties to achieve interesting results.

The last chapter is devoted to elementary applications of fixed point results in finding the roots of functional equations, establishing the existence and uniqueness of solutions of systems of linear algebraic equation, and studying

Markov processes, integral equations, initial value and boundary value problems, periodic and impulsive boundary value problems, equilibrium of difference equations and periodic solutions of cyclic systems.

We greatly admire and are deeply indebted to our friends and colleagues working in fixed point theory for their encouragement and support, especially Professor Sompong Dhompongsa, Chiang Mai University, Thailand and Professor Yeol-Je Cho, Gyeongsang National University, Jinju, Korea. In particular, we wish to express our deepest thanks to our colleagues who contributed their recent research work in the form of chapters for inclusion in this book.

Our deepest gratitude and thanks are also due to our family members who always encouraged us and refreshed our energies with their sweet words while we were busy completing this project.

The authors are very thankful to Aastha Sharma, Shikha Garg and the staff from the CRC Press for their unfailing support, cooperation and patience in publishing this book.

Dhananjay Gopal
Poom Kumam
Mujahid Abbas

Editors and Authors

Mujahid Abbas received a doctorate in mathematics from Universitat Politecnica, Valencia, Spain, and completed post-doctorate programs at Indiana University, Bloomington, Indiana, U.S., and University of Birmingham, U.K. Dr. Abbas is a professor of mathematics at Government College University, Lahore, Pakistan. He also serves as an extraordinary or adjunct professor at the University of Pretoria in South Africa and at King Abdulaziz University in Jeddah, Saudi Arabia. He participated in various research programs and activities at the University of Ton Duc Thang University in Ho Chi Minh City, Vietnam and at King Saud University in Riyadh, Saudi Arabia. His research focuses on fixed point theory, fuzzy logic, and optimization theory and he has published several papers in leading journals.

Ravindra K. Bisht earned a Ph.D. at Kumaun University in Nainital, India. He is an assistant professor of mathematics at the National Defence Academy in Pune, India. His research interests include metric fixed point theory and discrete dynamical systems. Dr. Bisht wrote more than 30 publications and serves as a reviewer for major analysis journals such as *Mathematical Review* and *Zentralblatt Math.*

Dhauanjay Gopal has a doctorate in mathematics from Guru Ghasidas University, Bilaspur, India and is currently an assistant professor of applied mathematics at the Sardar Vallabhbhai National Institute of Technology, Surat, India. He is the author or co-author of several papers published in journals and conference proceedings and pursues research on the theories of nonlinear analysis and fuzzy metric fixed points.

Poom Kumam earned a Ph.D. in mathematics from Naresuan University, Phitsanulok, Thailand. He is the associate dean for research and networking on the Faculty of Science of King Mongkut's University of Technology, Thonburi, Thailand. He also leads the university's Theoretical and Computational Science Center and directs its Computational and Applied Science for Smart Innovation Cluster. Dr. Kumam serves on the editorial boards of several international journals and wrote more than 400 papers for the Scopus and Web of Science databases. His research focuses on fixed point theory and optimization, mainly related to solving problems in pure and applied sciences.

Chirasak Mongkolkeha earned a Ph.D. from King Mongkut's University of Technology in Thonburi, Thailand. He is a lecturer in the Department of Mathematics, Statistics and Computers at Kasetsart University, Bangkok, Thailand. Dr. Mongkolkeha's areas of research interest include fixed point theory and its applications, geometric properties of sequences and Banach spaces. Several of his works have been published in international journals of pure and applied mathematics and he has served as a referee for several international journals specializing in those subjects.

Ghulam Murtaza is an assistant professor in the Department of Mathematics at Government College University, Faisalabad, Pakistan. His areas of interest include topology, fuzzy mathematics, fixed point theory, and their applications. Several of Dr. Murtaza's research articles have been published in well known mathematics journals.

Deepesh Kumar Patel received a Ph.D. in mathematics from Sardar Vallabhbhai National Institute of Technology, Surat, India. Dr. Patel is an assistant professor at the Visvesvaraya National Institute of Technology, Nagpur, India. His areas of research interest include fixed point theory and its applications in nonlinear analysis. He has written several research articles and served as a reviewer for international journals focusing on fixed point theory.

Salvador Romaguera is a full professor in the Department of Applied Mathematics at Universitat Politecnica, Valencia, Spain. His research focuses on topology, fuzzy mathematics, fixed point theory, and their applications. Dr. Romaguera's work has been published in several journals in the areas of general and applied mathematics and computer science. He also serves as an editor and referee for several mathematics journals.

Satish Shukla earned a Ph.D. at Barkatullah University, Bhopal, India in 2000 and is currently an assistant professor of applied mathematics at the Shri Vaishnav Institute of Technology and Science, Indore, India. He has written several papers published in journals and conference proceedings and continues his research in theories of nonlinear analysis and metric fixed point.

Wutiphol Sintunavarat was awarded a Ph.D. in applied mathematics from King Mongkut's University of Technology Thonburi, Thailand. He is an assistant professor on the Faculty of Science and Technology at Thammasat University in Rangsit, Thailand and served as an invited researcher at Kyungnurn University and Gyeongsang National University, both in South Korea, and at Sakarya University, Adapazam, Turkey. Dr. Sintunaarat studies fixed point theory and its applications in nonlinear functional analysis. He has published more than 115 articles in international mathematical journals.

Symbol Descriptions

$\mathbb{R}$	Set of all real numbers
$\mathbb{R}^+$	Set of all positive real numbers
$\mathbb{N}$	Set of all natural numbers
$\mathbb{Q}$	Set of all rational numbers
$(a_{ij})_{n\times m}$	Matrix of order $n \times m$ with the entry a_{ij} in i^{th} row and j^{th} column
$\mathbb{R}^n$	Set of all n-tuples with real coordinates
$C(I, \mathbb{R})$	Class of all real-valued functions defined on interval I
$C^k(I, \mathbb{R})$	Class of all real-valued functions defined on interval I having continuous k^{th} order derivative
U	Universal set
E or A	Set of parameters
(F, A)	Soft set over U
$\tilde{U}$	Absolute soft set
$\mathbb{R}(A)^*$	Set of non-negative soft real numbers
$\overline{0}$	Null soft number
$\overline{1}$	Unit soft number
$\overline{c}$	Constant soft real number
F^x_λ	Soft point
$SP(F, A)$	Collection of all soft points of (F, A)
$(\tilde{U}, d)$	Soft metric space
$\{U^x_{\lambda,n}\}_n$	Sequence of soft points

Chapter 1

Banach Fixed Point Theorem and Its Generalizations

Dhananjay Gopal, Deepesh Kumar Patel and Satish Shukla

1.1 Introduction

The term *metric fixed point theory* refers to those fixed point theoretical results in which geometric conditions on the underlying spaces and/or mappings play a crucial role. The first ever fixed point theorem in metric space appeared in explicit form in Banach's thesis [2], known as the "Banach Contraction Principle"(BCP), used to established the existence of a solution to an integral equation. Due to its simplicity and elegant proof, it is perhaps the most widely applied fixed point theorem in many branches of mathematics.

The BCP has been generalized in different directions. In this chapter, we attempt to present some well known extensions and generalizations of BCP which play an important role in the development of metric fixed point theory.

1.2 Banach fixed point theorem

Definition 1.1 Let (X, d) be a metric space and let $f : X \to X$ be a mapping.

(a) A point $x \in X$ is called a fixed point of f if $x = fx$.

(b) f is called a contraction if there exists a fixed constant $\alpha < 1$ such that

$$d(fx, fy) \leq \alpha d(x, y) \quad \text{for all } x, y \in X. \tag{1.1}$$

A contraction mapping is also known as a Banach contraction. If (1.1) is replaced by strict inequality and it holds for $\alpha = 1$ and $x \neq y$, then f is called contractive (or strict contractive). If (1.1) holds for $\alpha = 1$, then f is called nonexpansive, and if (1.1) holds for a fixed $\alpha < +\infty$, then f is called Lipschitz continuous. Clearly, for the mapping f, the following obvious implications hold:

Contraction $\Rightarrow$ contractive $\Rightarrow$ nonexpansive $\Rightarrow$ Lipschitz continuous.

Any contraction mapping is clearly continuous. The following theorem shows that a more generalized fact holds for such mappings.

Theorem 1.1 *Any contraction mapping is uniformly continuous.*

Proof. Let $f : X \to X$ be a contraction mapping. Choose $\varepsilon > 0$ arbitrarily and take $0 < \delta < \dfrac{\varepsilon}{\alpha}$. Then

$$d(fx, fy) \leq \alpha d(x, y) < \alpha\delta \quad \text{if} \quad d(x, y) < \delta.$$

So $d(fx, fy) < \varepsilon$ whenever $d(x, y) < \delta$, $x, y \in X$. Therefore f is uniformly continuous.

Note. We may note here that a function $f : X \to Y$, where (X, d) and (Y, d') are metric spaces, is said to be uniformly continuous on X if for every $\varepsilon > 0$ there exists a $\delta > 0$ (depending only on ε) such that $d'(fx', fx'') < \varepsilon$ whenever $d(x', x'') < \delta$ for an arbitrary pair of points x', x'' of the space X. Any uniformly continuous function is clearly continuous but there exist continuous functions which are not uniformly continuous.

Let $fx = x^2$ for all $x \in \mathbb{R}$. Then f is continuous on $\mathbb{R}$. We show that it is not uniformly continuous on $\mathbb{R}$. If possible, suppose that f is uniformly continuous on $\mathbb{R}$. Then corresponding to $\varepsilon = 1$ there exists a $\delta > 0$ such that $|fx' - fx''| < 1$ whenever $|x' - x''| < \delta$. Then

$$\frac{\delta}{2}\left|2x + \frac{\delta}{2}\right| = \left|f\left(x + \frac{\delta}{2}\right) - fx\right| < 1$$

for all $x \in \mathbb{R}$. But clearly, this inequality does not hold for all sufficiently large values of x. So f cannot be uniformly continuous on $\mathbb{R}$.

Next, we consider an example which shows that a contraction mapping may not have a fixed point.

Example 1.1 Let $X = (0, \frac{1}{4})$ be a metric space with the usual metric on X. Let $f : X \to X$ be given by $fx = x^2$. Then

$$|fx - fy| = |x^2 - y^2| = (x + y)|x - y| < \frac{1}{2}|x - y| \quad \text{for } x, y \in X.$$

So f is a contraction mapping. If x_0 is a fixed point of f then we must have $fx_0 = x_0$, i.e., $x_0^2 = x_0$ and $x_0 = 0$ or 1. Neither 0 nor 1 lies in $X = (0, \frac{1}{4})$. So f has no fixed point in X.

In the above example we have shown that a contraction mapping may not have a fixed point. It may be noted that the space X is not complete here, and we will see that it is not a coincidence (see Theorem 1.2).

Definition 1.2 Let $f : X \to X$ where X is a metric space. Let $x_0 \in X$. We determine successively the elements of a sequence starting from x_0 as follows:

$$x_1 = fx_0,\ x_2 = fx_1,\ x_3 = fx_2,\ \ldots,\ x_{n+1} = fx_n,\ \ldots.$$

This procedure of constructing a sequence of elements from an element is called iteration.

Clearly,

$$\begin{aligned} x_1 &= fx_0, \\ x_2 &= fx_1 = ffx_0 = f^2 x_0, \end{aligned}$$

in general,

$$x_n = f^n x_0. \tag{1.2}$$

The following theorem is known as Banach's fixed point theorem. Sometimes it is also called the Banach contraction principle (BCP).

Theorem 1.2 (Banach) *Let (X, d) be a complete metric space and $f : X \to X$ be a contraction, i.e., f satisfies* (1.1). *Then there exists a unique fixed point of f in X.*

Proof. Let $x_0 \in X$ be an arbitrary element. Starting from x_0 we form the iteration as (1.2), i.e., $x_n = f^n x_0 = fx_{n-1}$ for all $n \in \mathbb{N}$. Now we verify that $\{x_n\}$ is a Cauchy sequence. We have

$$\begin{aligned} d(x_1, x_2) &= d(fx_0, fx_1) \le \alpha d(x_0, x_1) = \alpha d(x_0, fx_0), \\ d(x_2, x_3) &= d(fx_1, fx_2) \le \alpha d(x_1, x_2) \le \alpha^2 d(x_0, fx_0), \\ d(x_3, x_4) &= d(fx_2, fx_3) \le \alpha d(x_2, x_3) \le \alpha^3 d(x_0, fx_0). \end{aligned}$$

In general, for any positive integer n,

$$d(x_n, x_{n+1}) \le \alpha^n d(x_0, fx_0).$$

Also, for any positive integer p,

$$\begin{aligned} d(x_n, x_{n+p}) &\le d(x_n, x_{n+1}) + d(x_{n+1}, x_{n+2}) + \cdots + d(x_{n+p-1}, x_{n+p}) \\ &\le \alpha^n d(x, fx) + \alpha^{n+1} d(x, fx) + \cdots + \alpha^{n+p-1} d(x_0, fx_0) \\ &= (\alpha^n + \alpha^{n+1} + \cdots + \alpha^{n+p-1}) d(x_0, fx_0) \\ &\le \frac{\alpha^n - \alpha^{n+p}}{1 - \alpha} d(x_0, fx_0) \qquad (1.3) \\ &\le \frac{\alpha^n}{1 - \alpha} d(x_0, fx_0), \quad \text{since } 0 < \alpha < 1. \qquad (1.4) \end{aligned}$$

Because $\alpha < 1$, the relation (1.4) shows that $d(x_n, x_{n+p}) \to 0$ as $n \to \infty$. Therefore $\{x_n\}$ is a Cauchy sequence. Since, by hypothesis, the space X is complete, the sequence $\{x_n\}$ is convergent. Suppose

$$\lim_{n\to\infty} x_n = x^* \in X. \tag{1.5}$$

We show that x^* is a fixed point of the mapping f, i.e., $fx^* = x^*$. We have

$$\begin{aligned} d(x^*, fx^*) &\leq d(x^*, x_n) + d(x_n, fx^*) \\ &= d(x^*, x_n) + d(fx_{n-1}, fx^*) \\ &\leq d(x^*, x_n) + \alpha d(x_{n-1}, x^*). \end{aligned}$$

This implies, $d(x^*, fx^*) = 0$ because of (1.5). So $fx^* = x^*$. Therefore, x^* is a fixed point of f. We now verify that there exists only one fixed point of f. Suppose $y^* \in X$ is such that $fy^* = y^* \neq x^*$. Then

$$d(x^*, y^*) = d(fx^*, fy^*) \leq \alpha d(x^*, y^*). \tag{1.6}$$

As $d(x^*, y^*) > 0$, from (1.6) we obtain $\alpha \geq 1$, which is a contradiction to $0 < \alpha < 1$. Hence, $x^* = y^*$, and so f has a unique fixed point in X. This proves the theorem.

Note. We consider the inequality (1.3)

$$d(x_n, x_{n+p}) \leq \frac{\alpha^n - \alpha^{n+p}}{1-\alpha} d(x_0, fx_0).$$

Fix $n \in \mathbb{N}$, then as $p \to \infty$, the right hand side tends to $\dfrac{\alpha^n}{1-\alpha} d(x_0, fx_0)$ (since $\alpha < 1$) and the left hand side tends to $d(x_n, x^*)$ because $x_{n+p} \to x^*$. So

$$d(x_n, x^*) \leq \frac{\alpha^n}{1-\alpha} d(x_0, fx_0) \text{ for all } n \in \mathbb{N}. \tag{1.7}$$

The relation (1.7) gives an estimation for the error of the nth approximation.

Remark 1.1 It is important that the contractive constant α should be strictly less than 1 in the proof of BCP. It gives us control over the rate of convergence of $\{x_n\} = \{f^n x_0\}$ to the fixed point since $\alpha^n \to 0$ as $n \to \infty$. If we assume f is a contractive mapping instead of a contraction, we lose that control over the convergence, and the fixed point of f need not exist.

Example 1.2 *[21]* Consider the usual metric space $(\mathbb{R}, d)$. Define $f : \mathbb{R} \to \mathbb{R}$ as:

$$fx = x + \frac{\pi}{2} - \tan^{-1} x.$$

Then for $x, y \in \mathbb{R}$, $x < y$, we have

$$\begin{aligned} fy - fx &= (y - x) - (\tan^{-1} y - \tan^{-1} x) \\ &= y - x - \frac{y - x}{1 + z^2} \end{aligned}$$

where $x < z < y$, by Lagrange's mean value theorem. Therefore

$$fy - fx = (y - x)\left(1 - \frac{1}{1+z^2}\right).$$

So

$$|fx - fy| = |y - x|\left|1 - \frac{1}{1+z^2}\right|.$$

If $|fx - fy| \geq |y - x|$, then the above implies

$$\left|1 - \frac{1}{1+z^2}\right| \geq 1, \text{ i.e. } \frac{z^2}{1+z^2} \geq 1,$$

which is definitely not true for any value of z. Hence, we must have

$$|fx - fy| < |y - x|,$$

i.e., $|fx - fy| \leq |y - x|$ for all $x, y \in \mathbb{R}$, $x \neq y$. Therefore, the mapping f is contractive and f has no fixed point. For, if there exists $p \in \mathbb{R}$ such that $fp = p$, then

$$p + \frac{\pi}{2} - \tan^{-1} p = p$$

$$\text{i.e. } \tan^{-1} p = \frac{\pi}{2},$$

which is not true because $\tan^{-1} x < \frac{\pi}{2}$ for every $x \in X$.

Note. The mapping f in Example 1.2 is contractive but is not a contraction because, if it is a contraction, as $\mathbb{R}$ is a complete metric space, f would have a fixed point.

We have seen that a contractive mapping need not have a fixed point. But the following theorem due to Edelstein [12] shows that if the space is considerably restricted (compact) then a unique fixed point of a contractive mapping is ensured.

Theorem 1.3 *Let (X, d) be a metric space and $f : X \to X$ be such that*

$$d(fx, fy) < d(x, y), \quad x, y \in X, \ x \neq y. \tag{1.8}$$

Suppose that there exists a point $x_0 \in X$ such that the sequence of iterates $\{f^n x_0\}$ has a subsequence converging to $\xi \in X$. Then ξ is the unique fixed point of f.

Similarly, a nonexpansive mapping on a complete metric space need not have a fixed point. For instance, consider the translation operator by a nonzero vector in a Banach space, which is clearly a nonexpansive, fixed point free mapping. On the other hand, a fixed point of a nonexpansive map need not be unique. For instance, consider an identity operator which is obviously nonexpansive and rich with fixed points. Thus, the fixed point theory of nonexpansive mappings is fundamentally different from that of the contraction mapping, and thus we shall not discuss it in this chapter.

After the Banach and Edelstein theorems, an open question for the researchers is whether there exists any other non-contractive type condition which does not force the map to be continuous and still has a unique fixed point.

Kannan [17] affirmatively found the answer and proved a fixed point theorem in complete metric spaces using an iterative method for a self mapping which is neither contraction nor contractive in nature.

Theorem 1.4 (Kannan) *Let $f : X \to X$, where (X, d) is a complete metric space and f satisfies the condition*

$$d(fx, fy) \leq \beta[d(x, fx) + d(y, fy)], \tag{1.9}$$

where $0 < \beta < \frac{1}{2}$ and $x, y \in X$. Then f has unique fixed point in X.

Proof. Let $x_0 \in X$ and the iterative sequence be $x_n = f^n x_0 = fx_{n-1}$ for all $n \in \mathbb{N}$. Then

$$d(x_1, x_2) = d(fx_0, fx_1) \leq \beta[d(x_0, fx_0) + d(x_1, fx_1)] = \beta[d(x_0, x_1) + d(x_1, x_2)]$$

and so $d(x_1, x_2) \leq \frac{\beta}{1-\beta} d(x_0, fx_0)$. Also,

$$d(x_2, x_3) = d(fx_1, fx_2) \leq \beta[d(x_1, fx_1) + d(x_2, fx_2)] = \beta[d(x_1, x_2) + d(x_2, x_3)]$$

and so $d(x_2, x_3) \leq \frac{\beta}{1-\beta} d(x_1, x_2) \leq \left(\frac{\beta}{1-\beta}\right)^2 d(x_0, fx_0)$.
Similarly, one can obtain

$$d(x_3, x_4) \leq \left(\frac{\beta}{1-\beta}\right)^3 d(x_0, fx_0).$$

In general, if n is any positive integer, then

$$d(x_n, x_{n+1}) \leq \left(\frac{\beta}{1-\beta}\right)^n d(x_0, fx_0).$$

If p is any positive integer, then

$$\begin{aligned} d(x_n, x_{n+p}) &\leq d(x_n, x_{n+1}) + d(x_{n+1}, x_{n+2}) + \cdots + d(x_{n+p-1}, x_{n+p}) \\ &\leq (r^n + r^{n+1} + \cdots + r^{n+p-1}) d(x_0, fx_0), \text{ where } r = \frac{\beta}{1-\beta} \\ &< \frac{r^n}{1-r} d(x_0, fx_0). \end{aligned}$$

Since $0 < \beta < \dfrac{1}{2}$, we have $0 < r < 1$, and so $d(x_n, x_{n+p}) \to 0$ as $n \to \infty$. Therefore $\{x_n\}$ is a Cauchy sequence. Because the space X is complete, we have

$$\lim_{n\to\infty} x_n = x^* \in X. \tag{1.10}$$

We show that x^* is a fixed point of f. We have

$$\begin{aligned} d(x^*, fx^*) &\leq d(x^*, x_n) + d(x_n, fx^*) \\ &= d(x^*, x_n) + d(fx_{n-1}, fx^*) \\ &\leq d(x^*, x_n) + \beta[d(x_{n-1}, fx_{n-1}) + d(x^*, fx^*)] \end{aligned}$$

or $(1-\beta)d(x^*, fx^*) \leq d(x^*, x_n) + \beta d(x_{n-1}, x_n) \to 0$ as $n \to \infty$ because of (1.10).
So $fx^* = x^*$ and x^* is a fixed point of f. We now prove that x^* is the only fixed point of f. Suppose $fy^* = y^*$ and $y^* \neq x^*$. Then

$$d(x^*, y^*) = d(fx^*, fy^*) \leq \beta[d(x^*, fx^*) + d(y^*, fy^*)] = 0.$$

This contradiction shows that $x^* = y^*$ and uniqueness follows.

Note. If we compare the conditions (1.1) and (1.9) we see that the condition (1.1) implies the continuity of the mapping while (1.9) does not. Moreover, the following examples show that the conditions (1.1) and (1.9) are independent.

Example 1.3 Let $X = [0,1]$, $fx = \dfrac{x}{4}$ for $x \in \left[0, \frac{1}{2}\right)$, $fx = \dfrac{x}{5}$ for $x \in \left[\frac{1}{2}, 1\right]$. Here, f is discontinuous at $x = \dfrac{1}{2}$ and so (1.1) is not satisfied. But it is easily seen that the condition (1.9) is satisfied if $\beta = \dfrac{4}{9}$. Let $X = [0,1]$, $fx = \dfrac{x}{3}$ for $x \in [0,1]$. Here the condition (1.1) is satisfied it may be verified that (1.9) is not satisfied for $x = \dfrac{1}{3}, y = 0$.

D.N. Sarkhel [34] gave an alternative proof of Kannan's fixed point theorem using Banach's fixed point theorem. In fact he showed that if the mapping f satisfies (1.9) in a complete metric space (X, d), the same mapping f satisfies (1.1) in a complete metric space (X, d') for a suitable metric d'. His proof runs as follows:

Proof. Suppose that (X, d) is a complete metric space and $f : X \to X$ satisfies (1.9). For all $x, y \in X$, let

$$d'(x,x) = 0 \quad \text{and} \quad d'(x,y) = d(x, fx) + d(y, fy) \quad \text{if } x \neq y.$$

Now $d(x,y) \leq d(x,fx) + d(fx,fy) + d(fy,y)$ and also $0 < \beta < \frac{1}{2}$. This implies that for all $x, y \in X$, $d(x,y) \leq 2d'(x,y)$. So every Cauchy sequence in (X, d') is also a Cauchy sequence in (X, d).

We now show that (X, d') is complete. Let $\{x_n\}$ be a Cauchy sequence in (X, d'), and so it is a Cauchy sequence in (X, d). Since (X, d) is complete, $x \in X$ such that $d(x_n, x) \to 0$ as $n \to \infty$. If $x_n \neq x$ for some n, then $n' > n$ such that $x_n \neq x$, so we have

$$\begin{aligned} d'(x, x_n) &= d(x, fx) + d(x_n, fx_n) \\ &\leq d(x, x_n) + 2d(x_n, fx_n) + d(fx_n, fx) \\ &\leq d(x, x_n) + 2d'(x_n, x_{n'}) + \beta d'(x_n, x), \end{aligned}$$

and $(1 - \beta)d'(x, x_n) \leq d(x, x_n) + 2d'(x_n, x_{n'})$. This shows that $d'(x, x_n) \to 0$ as $n \to \infty$ and the space (X, d') becomes complete. Since

$$d(fx, f^2x) + d(fy, f^2y) \leq \beta[d(x, fx) + d(fx, f^2x)] + \beta[d(y, fy) + d(fy, f^2y)],$$

putting $\alpha = \dfrac{\beta}{1-\beta}$, we see that $\alpha < 1$ and for all $x, y \in X$,

$$d'(fx, fy) \leq \alpha d'(x, y),$$

i.e., f satisfies (1.1) in the complete metric space (X, d') and by Banach's fixed point theorem, f has unique fixed point in X.

Kannan [17] generated a widespread interest in the study of fixed points of generalized contractive mappings and soon these were followed by a number of papers involving contractive definitions. Most of them did not require continuity of the mappings; see, for example, [5, 40].

In our opinion, the Kannan theorem is also very important because Subrahmanyam [36] proved that Kannan theorem characterizes the metric completeness, that is, a metric space X is complete if and only if every Kannan mapping on X has a fixed point. Contractions (in the sense of Banach) do not have this property; Connell [9] gave an example of a metric space X such that X is not complete and every contraction on X has a fixed point.

On the other hand, it is also important to note that the contraction of Kannan type with constant $\beta = \frac{1}{2}$ in complete metric space does not guarantee the existence of fixed points of f (see [31]). Moreover, the assumption of continuity of the mapping and the compactness condition on metric space are required for the existence of a fixed point for a strict type Kannan contraction as shown in the following result:

Theorem 1.5 *[15] (see also [10]) Let (X, d) be a compact metric space and let $f : X \to X$ be a continuous mapping that satisfies the condition*

$$d(fx, fy) < \frac{1}{2}[d(x, fx) + d(y, fy)], \tag{1.11}$$

for all $x, y \in X$ with $x \neq y$. Then f has unique fixed point in X.

We are concluding this section with the following open problem:

Does there exist a complete but noncompact metric space (X, d) and a continuous mapping $f : X \to X$ such that

$$d(fx, fy) < \frac{1}{2}[d(x, fx) + d(y, fy)], \tag{1.12}$$

for all $x, y \in X$ with $x \neq y$, and f is fixed point free?

1.3 Some other generalizations of BCP

There have been numerous extensions of a milder form of BCP. In this section we present some of these. The first generalization in this direction of importance is the following result of Rakotch [30].

Theorem 1.6 *Let (X, d) be a complete metric space, and suppose that $f : X \to X$ satisfies*

$$d(fx, fy) \leq \eta(d(x, y))d(x, y) \quad \text{for all } x, y \in X, \tag{1.13}$$

where $\eta : \mathbb{R}^+ \to [0, 1)$ is a decreasing function. Then f has a unique fixed point.

A variant of Rakotch's theorem has been given by Geraghty [13], in which the function η satisfies a simpler condition that $\eta(t_n) \to 1 \implies t_n \to 0$.

In [4], Boyd and Wong obtained a more general result as follows:

Theorem 1.7 (Boyd and Wong) *Let (X, d) be a complete metric space, and suppose that $f : X \to X$ satisfies*

$$d(fx, fy) \leq \psi(d(x, y)) \quad \text{for all } x, y \in X, \tag{1.14}$$

where $\psi : \mathbb{R}^+ \to [0, \infty)$ is upper semicontinuous from the right, i.e., for any sequence $t_n \downarrow t \geq 0 \implies \limsup_{n\to\infty} \psi(t_n) \leq \psi(t)$ and satisfies $0 \leq \psi(t) < t$ for $t > 0$. Then f has a unique fixed point.

Proof. Let x_0 be an arbitrary but fixed element in X. Define a sequence of iterates $\{x_n\}$ in X by

$$x_n = fx_{n-1} \ (= f^n x_0) \quad \text{for all } n \geq 1.$$

Set $a_n = d(x_{n-1}, x_n)$. Note that the sequence $\{a_n\}$ is monotonically decreasing and bounded below and thus is convergent. We get $\lim_{n\to\infty} a_n = a$. If $a > 0$, we have

$$a_{n+1} \leq \psi(a_n)$$

so that $a \leq \limsup_{t \to a^+} \psi(t) \leq \psi(a)$, which is a contradiction. We now show that $\{x_n\}$ is a Cauchy sequence. Suppose that is not so. Then there exists $\varepsilon > 0$ and sequences of integers $\{m(k)\}, \{n(k)\}$, with $m(k) \geq n(k) \geq k$ and such that

$$d_k = d(x_{m_k}, x_{n_k}) \geq \varepsilon \text{ for } k = 1, 2, \cdots. \tag{1.15}$$

We may assume that $d(x_{m_k - 1}, x_{n_k}) < \varepsilon$ by choosing $m(k)$ to be the smallest number exceeding $n(k)$ for which (1.15) holds. Now

$$\varepsilon \leq d_k \leq d(x_{m_k}, x_{m_k - 1}) + d(x_{m_k - 1}, x_{n_k}) < a_{m_k} + \varepsilon \leq a_k + \varepsilon$$

which implies that $d_k \to \varepsilon^+$ as $k \to \infty$. But now

$$\begin{aligned} d_k &\leq d(x_{m_k}, x_{n_k}) \leq d(x_{m_k}, x_{m_k+1}) + d(x_{m_k+1}, x_{n_k+1}) + d(x_{n_k+1}, x_{n_k}) \\ &\leq 2a_k + \psi(d_k). \end{aligned}$$

Thus, as $k \to \infty$ we obtain $\varepsilon \leq \psi(\varepsilon)$, which is a contradiction for $\varepsilon > 0$. Hence, $\{x_n\}$ is a Cauchy sequence in X. Since X is complete, $x_n \to z \in X$. Thus, by the continuity of f, we get $fz = z$. Uniqueness of z follows from the contractive condition.

Example 1.4 Let $X = [0, 1) \cup \{1, 2, 3, \cdots\}$. For the metric, let

$$d(x, y) = \begin{cases} |x - y| & \text{if } x, y \in [0, 1], \\ x + y & \text{if one of } x, y \notin [0, 1]. \end{cases}$$

Then (X, d) is a complete metric space. Now define a mapping $f : X \to X$ by

$$fx = \begin{cases} x - \frac{1}{2}x^2 & \text{if } x, y \in [0, 1], \\ x - 1 & \text{if } x = 2, 3, 4, \cdots. \end{cases}$$

Then for $x, y \in [0, 1]$ with $x - y = t > 0$, we have

$$d(fx, fy) = (x - y)(1 - \frac{1}{2}(x + y)) \leq t(1 - \frac{1}{2}t)$$

and if $x \in \{2, 3, 4, \cdots\}$ with $x > y$, then

$$d(fx, fy) = fx + fy < x - 1 + y = d(x, y) - 1.$$

Thus, if we define ψ by

$$\psi(t) = \begin{cases} t - \frac{1}{2}t^2, & 0 \leq t \leq 1, \\ t - 1, & 1 < t < \infty \end{cases}$$

then ψ is upper semicontinuous from the right on $[0, \infty)$, $\psi(t) < t$ for all $t > 0$ and (1.14) holds. However as $n \to \infty$, $\frac{d(fx,0)}{d(x,0)} \to 1$, there can be no decreasing function η with $\eta(t) < 1$ for $t > 0$ and (1.13) holds.

Remark 1.2 In [4], Boyd and Wong observed that the upper semicontinuity of ψ can be dropped if the space X is metrically convex. While in [23], Matkowski further extended this result by assuming that ψ is continuous at 0 and that there exists a sequence $t_n \downarrow 0$ for which $\psi(t_n) < t_n$.

The following variant of the previous result is given by Matkowski [22] in which the continuity of ψ is replaced with another suitable assumption.

Theorem 1.8 (Matkowski) *Let (X, d) be a complete metric space, and suppose that $f : X \to X$ satisfies*

$$d(fx, fy) \leq \psi(d(x, y)) \quad \textit{for all } x, y \in X,$$

where $\psi : \mathbb{R}^+ \to \mathbb{R}^+$ is monotonically nondecreasing and satisfies $\lim_{n\to\infty} \psi^n(t) = 0$ for all $t > 0$. Then f has a unique fixed point.

Proof. Let x_0 be an arbitrary but fixed element in X. Define a sequence of iterates $\{x_n\}$ in X by

$$x_n = fx_{n-1} \; (= f^n x_0) \quad \text{for all } n \geq 1.$$

Set $a_n = d(x_{n+1}, x_n)$. Note that

$$0 \leq \limsup_{n\to\infty} a_n \leq \limsup_{n\to\infty} \psi^n(d(x_1, x_0)) = 0.$$

Thus, $\lim_{n\to\infty} a_n = 0$. Now, we show that $\{x_n\}$ is a Cauchy sequence. Also, note that for any $\varepsilon > 0$, $\psi(\varepsilon) < \varepsilon$. Since $\lim_{n\to\infty} a_n = 0$, so for $\varepsilon > 0$, we can choose $n \in \mathbb{N}$ such that $a_n \leq \varepsilon - \psi(\varepsilon)$. Now, define

$$M = \{x \in X : d(x, x_n) \leq \varepsilon\}.$$

Then for any $y \in M$, we have

$$\begin{aligned} d(fy, x_n) &\leq d(fy, fx_n) + d(fx_n, x_n) \\ &\leq \psi(d(y, x_n)) + d(x_{n+1}, x_n) \\ &\leq \psi(\varepsilon) + \varepsilon - \psi(\varepsilon) = \varepsilon. \end{aligned}$$

Thus, $fy \in M$, that is, $f(M) \subseteq M$. It follows that $d(x_m, x_n) \leq \varepsilon$ for all $m \geq n$. We obtain the conclusion by following the rest of the proof as in Theorem 1.7.

Using somewhat different approach Meir and Keeler [24] extended Theorem 1.7 as follows.

Theorem 1.9 (Meir and Keeler) *Let (X, d) be a complete metric space, and suppose that $f : X \to X$ satisfies the condition: for each $\varepsilon > 0$, there exists $\delta > 0$ such that for all $x, y \in X$,*

$$\varepsilon \leq d(x, y) \leq \varepsilon + \delta \implies d(fx, fy) \leq \varepsilon. \tag{1.16}$$

Then f has a unique fixed point.

Clearly, the condition (1.16) implies that the mapping f is contractive. Thus, f is continuous and has a unique fixed point if it exists. Further, the condition (1.16) implies $d(x_n, x_{n+1})$ decreasing to zero. Finally, it is easy to show that $\{x_n\}$ is a Cauchy sequence in a complete metric space X by using the contrary technique.

The following example shows that (1.14) may be violated while the hypothesis of Theorem (1.9) is fulfilled.

Example 1.5 Let $X = [0,1] \cup \{3,4,6,7,\cdots,3n,3n+1,\cdots\}$ with the Euclidean distance, and let $f(x)$ be defined as follows

$$f(x) = \begin{cases} \frac{x}{2} & \text{if } 0 \le x \le 1, \\ 0 & \text{if } x = 3n, \\ 1 - \frac{1}{n+2} & \text{if } x = 3n+1. \end{cases}$$

Then f satisfies the condition (1.16) whereas (1.14) cannot; to see this consider $x = 3n$ and $y = 3n+1$. Then we get

$$d(x,y) = 1 = d(fx, fy) \quad \text{as } n \to \infty$$

and so $\psi(1)$ would have to be 1, a contradiction to the property of ψ.

In [7], Ćirić has obtained the following generalization of BCP.

Theorem 1.10 *Let (X,d) be a complete metric space, and let $f : X \to X$ be a quasi-contraction, that is, for a fixed constant $\alpha < 1$*

$$d(fx, fy) \le \alpha \max\{d(x,y), d(x,fx), d(y,fy), d(x,fy), d(y,fx)\} \tag{1.17}$$

Then f has a unique fixed point.

It has been observed in [32] that the Theorem 1.10 is also true if we replace (1.17) with the following equivalent contractive condition.

$$d(fx, fy) \le \alpha \max\left\{d(x,y), d(x,fx), d(y,fy), \frac{1}{2}[d(x,fy) + d(y,fx)]\right\}. \tag{1.18}$$

Several such contractive conditions have been studied by Rhoades [32], Jachymaski [14] and Ćirić [8].

In 1965, Prešić [28, 29] extended the BCP into the product spaces, and used his extended version in obtaining the convergence of some particular sequences. Later, several researchers obtained various generalizations and applications of this interesting result.

Let (X,d) be a metric space and $k \ge 1$ be an integer. A point $x \in X$ is called a fixed point of the mapping $f\colon X^k \to X$ if $f(x,x,\ldots,x) = x$.

Theorem 1.11 (Prešić) *Let (X, d) be a complete metric space, k a positive integer and $f : X^k \to X$ be a mapping satisfying*

$$d(f(x_1, x_2, \ldots, x_k), f(x_2, x_3, \ldots, x_{k+1})) \leq \sum_{i=1}^{k} \alpha_i d(x_i, x_{i+1}), \tag{1.19}$$

for every $x_1, x_2, \ldots, x_{k+1} \in X$, where $\alpha_1, \alpha_2, \ldots, \alpha_k$ are non-negative constants such that $\sum\limits_{i=1}^{k} \alpha_i < 1$. Then there exists a unique point $u \in X$ such that $f(u, u, \ldots, u) = u$. Moreover, if $x_1, x_2, \ldots, x_k$ are arbitrary points in X, then the sequence $\{x_n\}$ generated by

$$x_{n+k} = f(x_n, x_{n+1}, \ldots, x_{n+k-1}) \tag{1.20}$$

is convergent and $\lim x_n = f(\lim x_n, \lim x_n, \ldots, \lim x_n)$.

Proof. Let $x_1, x_2, \ldots, x_k \in X$. We define a sequence $\{x_n\}$ in X by

$$x_{n+k} = f(x_n, x_{n+1}, \ldots, x_{n+k-1}) \text{ for all } n \in \mathbb{N}.$$

We shall show that the sequence $\{x_n\}$ is a Cauchy sequence.

For simplicity, set $d_n = d(x_n, x_{n+1})$. We shall prove by induction that

$$d_n \leq \mu\theta^n, \tag{1.21}$$

is true for each $n \in \mathbb{N}$, where $\theta = \left[\sum\limits_{i=1}^{k} \alpha_i\right]^{1/k}$ and $\mu = \max\left\{\dfrac{d_1}{\theta}, \dfrac{d_2}{\theta^2}, \ldots, \dfrac{d_k}{\theta^k}\right\}$.
By definition of μ, (1.21) is obviously true for $n \in \{1, 2, \ldots, k\}$. Suppose that the following k inequalities

$$d_n \leq \mu\theta^n,\ d_{n+1} \leq \mu\theta^{n+1}, \ldots,\ d_{n+k-1} \leq \mu\theta^{n+k-1}$$

hold. We obtain from (1.19) that

$$\begin{aligned}
d_{n+k} &= d(x_{n+k}, x_{n+k+1}) \\
&= d(f(x_n, x_{n+1}, \ldots, x_{n+k-1}), f(x_{n+1}, x_{n+2}, \ldots, x_{n+k})) \\
&\leq \alpha_1 d(x_n, x_{n+1}) + \alpha_2 d(x_{n+1}, x_{n+2}) + \cdots + \alpha_k d(x_{n+k-1}, x_{n+k}) \\
&= \alpha_1 d_n + \alpha_2 d_{n+1} + \cdots + \alpha_k d_{n+k-1} \\
&\leq \alpha_1 \mu\theta^n + \alpha_2 \mu\theta^{n+1} + \cdots + \alpha_k \mu\theta^{n+k-1} \\
&\leq \alpha_1 \mu\theta^n + \alpha_2 \mu\theta^n + \cdots + \alpha_k \mu\theta^n \qquad \left(\text{as } \theta = \left[\textstyle\sum_{i=1}^{k} \alpha_i\right]^{1/k} < 1\right) \\
&= [\alpha_1 + \alpha_2 + \cdots + \alpha_k]\mu\theta^n \\
&= \mu\theta^{n+k}.
\end{aligned}$$

Hence, by induction, (1.21) is true for each $n \in \mathbb{N}$.
Now, for $n, m \in \mathbb{N}$ with $m > n$, we obtain from (1.21) that

$$\begin{aligned} d(x_n, x_m) &\le d(x_n, x_{n+1}) + d(x_{n+1}, x_{n+2}) + \cdots + d(x_{m-1}, x_m) \\ &= d_n + d_{n+1} + \cdots + d_{m-1} \\ &\le \mu\theta^n + \mu\theta^{n+1} + \mu\theta^{n+2} + \cdots + \mu\theta^{m-1} \\ &\le \mu\theta^n [1 + \theta + \theta^2 + \cdots] \\ &= \frac{\mu}{1-\theta}\theta^n. \end{aligned}$$

As $\theta < 1$, we conclude that $\{x_n\}$ is a Cauchy sequence in X. Since X is a complete metric space, $u \in X$ such that $\lim_{n\to\infty} x_n = u$. For any $n \in \mathbb{N}$ we have

$$\begin{aligned} d(f(u,u,\ldots,u),u) &\le d(f(u,u,\ldots,u), x_{n+k}) + d(x_{n+k}, u) \\ &= d(f(u,u,\ldots,u), f(x_n, x_{n+1}, \ldots, x_{n+k-1})) + d(x_{n+k}, u) \\ &\le d(f(u,u,\ldots,u), f(u,\ldots,u,x_n)) \\ &\quad + d(f(u,\ldots,u,x_n), f(u,\ldots,u,x_n,x_{n+1})) \\ &\quad\quad + \cdots + d(f(u,x_n,\ldots,x_{n+k-2}), f(x_n,x_{n+1},\ldots,x_{n+k-1})) \\ &\quad\quad\quad + d(x_{n+k}, u). \end{aligned}$$

Using (1.19) in the previous inequality we obtain

$$\begin{aligned} d(f(u,u,\ldots,u),u) \le \alpha_k d(u,x_n) &+ [\alpha_{k-1} d(u,x_n) + \alpha_k d(x_n,x_{n+1})] \\ &+ \cdots + [\alpha_1 d(u,x_n) + \alpha_2 d(x_n,x_{n+1}) \\ &\quad + \cdots + \alpha_k d(x_{n+k-2}, x_{n+k-1})] + d(x_{n+k}, u). \end{aligned}$$

Since $\lim_{n\to\infty} x_n = u$, it follows from previous the inequality that $d(f(u,u,\ldots,u),u) = 0$, i.e., $f(u,u,\ldots,u) = u$. Hence, u is a fixed point of f.

For uniqueness of fixed point, suppose u and v are two fixed points of f and $u \ne v$. Then, we obtain from (1.19) that

$$\begin{aligned} d(u,v) &= d(f(u,u,\ldots,u), f(v,v,\ldots,v)) \\ &\le d(f(u,u,\ldots,u), f(u,\ldots,u,v)) + d(f(u,\ldots,u,v), f(u,\ldots,u,v,v)) \\ &\qquad + \cdots + d(f(u,v,\ldots,v), f(v,v,\ldots,v)) \\ &\le \alpha_k d(u,v) + \alpha_{k-1} d(u,v) + \cdots + \alpha_1 d(u,v) \\ &= \sum_{i=1}^{k} \alpha_i d(u,v) < d(u,v). \end{aligned}$$

This contradiction shows that $d(u,v) = 0$, that is, $u = v$.

Remark 1.3 In the above theorem, if we put $k = 1$, we obtain the BCP. Therefore, the above theorem is an extension of the BCP into product spaces.

Example 1.6 Consider the convergence of the following sequence of non-negative numbers:

$$\{x_n\}\colon \frac{8}{3}x_{n+2} = \frac{1}{1+x_n} + \frac{1}{1+x_{n+1}}, \quad n = 1, 2, \ldots \tag{1.22}$$

where $x_n \geq 0$ for all $n \in \mathbb{N}$ and $x_1, x_2 \geq 0$ are arbitrary. Let $X = [0, \infty)$ and $d(x, y) = |x - y|$ for all $x, y \in X$, so that (X, d) is a complete metric space. Define $f\colon X^2 \to X$ by

$$f(x, y) = \frac{3}{8}\left(\frac{1}{1+x} + \frac{1}{1+y}\right) \quad \text{for all } x, y \in X.$$

Then, we have

$$\begin{aligned} d(f(x_1, x_2), f(x_2, x_3)) &= \frac{3}{8}\left|\left(\frac{1}{1+x_1} + \frac{1}{1+x_2}\right) - \left(\frac{1}{1+x_2} + \frac{1}{1+x_3}\right)\right| \\ &= \frac{3}{8}\left|\frac{1}{1+x_1} - \frac{1}{1+x_3}\right| \\ &\leq \frac{3}{8} \cdot \frac{|x_3 - x_2| + |x_2 - x_1|}{(1+x_1)(1+x_3)} \\ &\leq \frac{3}{8}\,[d(x_1, x_2) + d(x_2, x_3)] \end{aligned}$$

for all $x_1, x_2, x_3 \in X$. Therefore, from Theorem 1.11 the sequence $\{x_n\}$ converges to the unique fixed point of f in X which can be obtained by solving $x = f(x, x)$, i.e., $\frac{8}{3}x = \frac{2}{1+x}$. This equation gives two values of x, namely, $x = \frac{1}{2}$ and $-\frac{3}{2}$ which are the fixed points of f. Notice that only the fixed point $x = \frac{1}{2} \in X$ is admissible, and it is the unique limit of the given sequence for any arbitrary $x_1, x_2 \in [0, \infty)$.

In [1], Alber *et al.* suggested a generalization of BCP in the setting of Hilbert spaces and subsequently Rhoades [33] extended and improved their results to metric spaces.

Theorem 1.12 *Let (X, d) be a complete metric space, and suppose that $f : X \to X$ satisfies the following inequality*

$$d(fx, fy) \leq d(x, y) - \psi(d(x, y)) \ \textit{for all } x, y \in X, \tag{1.23}$$

where $\psi : [0, \infty) \to [0, \infty)$ is a continuous and nondecreasing function such that $\psi(t) = 0$ if and only if $t = 0$. Then f has unique fixed point.

Note that if one takes $\psi(t) = (1-\alpha)t$ where $0 < \alpha < 1$, then the inequality (1.23) reduces to the inequality (1.1).

Remark 1.4 A mapping $f : X \to X$ with inequality (1.23) is called a weakly contractive map. The function ψ involved in Theorem 1.12 known as altering distance (also called control function) was initially used in metric fixed point theory by Khan *et al.* [18]. This function and its generalization have been used in fixed point problems in metric and probabilistic metric spaces; see, for example, [25, 26, 35].

In [11], the following generalization of previous theorem appears.

Theorem 1.13 *Let (X, d) be a complete metric space, and suppose that $f : X \to X$ satisfies the following inequality*

$$\phi(d(fx, fy)) \leq \phi(d(x, y)) - \psi(d(x, y)) \text{ for all } x, y \in X, \tag{1.24}$$

where both the functions $\phi, \psi : [0, \infty) \to [0, \infty)$ are continuous and nondecreasing such that $\psi(t) = 0 = \phi(t)$ if and only if $t = 0$. Then f has a unique fixed point.

Recently, more general results in this direction have appeared. For example, one of the results in [6] follows.

Theorem 1.14 *Let (X, d) be a complete metric space, and suppose that $f : X \to X$ satisfies the following inequality*

$$\phi(d(fx, fy)) \leq \phi(m(x, y)) - \psi(\max\{d(x, y), d(y, fy)\}), \tag{1.25}$$

where

$$m(x, y) = max\left\{ d(x, y), d(x, fx), d(y, fy), \frac{1}{2}[d(x, fy) + d(y, fx)] \right\},$$

for all $x, y \in X$, and $\phi, \psi : [0, \infty) \to [0, \infty)$ are functions such that ϕ is alternating distance and ψ is continuous with $\psi(t) = 0$ if and only if $t = 0$. Then f has a unique fixed point.

In 2004, Berinde [3] introduced the notion of weak contraction (also known as almost contraction) and generalised BCP.

Definition 1.3 Let (X, d) be a metric space. A map $f : X \to X$ is called a weak contraction if there exists a constant $\delta \in (0, 1)$ and some $L \geq 0$ such that

$$d(fx, fy) \leq \delta \cdot d(x, y) + Ld(y, fx) \text{ for all } x, y \in X. \tag{1.26}$$

Remark 1.5 Due to symmetry of the distance, the weak contraction condition (1.26) implicitly includes the following dual condition:

$$d(fx, fy) \leq \delta \cdot d(x, y) + Ld(x, fy) \text{ for all } x, y \in X, \tag{1.27}$$

obtained from (1.26) by formally replacing $d(fx, fy)$ and $d(x, y)$ by $d(fy, fx)$ and $d(y, x)$, respectively, and then interchanging x and y. Consequently, to check the weak contractiveness of f, it is necessary to check both (1.26) and (1.27). Obviously, any strict contraction satisfies (1.26) with $\delta = a$ and $L = 0$, and hence is a weak contraction (that possesses a unique fixed point).

Theorem 1.15 *Let (X, d) be a complete metric space and $f : X \to X$ a weak contraction. Then f has a fixed point.*

It is possible to force the uniqueness of the fixed point of a weak contraction by imposing an additional contractive condition as follows:

Theorem 1.16 *Let (X, d) be a complete metric space and $f : X \to X$ a weak contraction for which there exists $\theta \in (0, 1)$ and some $L_1 \geq 0$ such that*

$$d(fx, fy) \leq \theta \cdot d(x, y) + L_1.d(x, fx), \quad \text{for all } x, y \in X. \tag{1.28}$$

Then f has a unique fixed point.

Remark 1.6 The main merit of weak contractions is that they unify large classes of contractive type operators including quasi-contractions whose fixed points can be obtained by means of the Picard iteration and for which both a priori and posteriori estimates are also available.

On the other hand, Suzuki [37] established a new type of generalization of the BCP and characterized the metric completeness of the underlying space.

Theorem 1.17 (Suzuki) *Let (X, d) be a complete metric space, and suppose that $f : X \to X$. Define a nonincreasing function $\theta : [0, 1) \to (1/2, 1]$ by*

$$\theta(r) = \begin{cases} 1 & \text{if } 0 \leq r \leq \frac{\sqrt{5}-1}{2}, \\ (1-r)r^{-2} & \text{if } \frac{\sqrt{5}-1}{2} \leq r \leq 2^{-1/2}, \\ (1+r)^{-1} & \text{if } 2^{-1/2} \leq r < 1. \end{cases}$$

Assume that there exists $r \in [0, 1)$ such that

$$\theta(r)d(x, fx) \leq d(x, y) \implies d(fx, fy) \leq rd(x, y) \text{ for all } x, y \in X.$$

Then f has a unique fixed point.

Proof. By the definition of $\theta(r)$, it is clear that $\theta(r) \leq 1$. So

$$\theta(r)d(x, fx) \leq d(x, fx)$$

holds for every $x \in X$. By hypothesis of the theorem,

$$d(fx, f^2x) \leq rd(x, fx) \tag{1.29}$$

for all $x \in X$. We now fix $u \in X$ and define a sequence $\{u_n\}$ in X by $u_n = f^n u$. Then by repeated application of (1.29) we get,

$$d(u_n, u_{n+1}) \leq r^n d(u, fu).$$

Since $r \in [0,1)$, we have $r^n \to 0$ as $n \to \infty$. So, $\sum_{n=1}^{\infty} d(u_n, u_{n+1}) < \infty$ and thus $\{u_n\}$ is a Cauchy sequence. Since X is complete, $\{u_n\}$ converges to some point $z \in X$. We next show

$$d(fx, z) \leq rd(x, z) \tag{1.30}$$

for all $x \in X \setminus \{z\}$. For $x \in X \setminus \{z\}$, there exists $\nu \in \mathbb{N}$ such that $d(u_n, z) \leq \frac{d(x,z)}{3}$ for all $n \in \mathbb{N}$ with $n \geq \nu$. Then we have

$$\begin{aligned} \theta(r)d(u_n, fu_n) &\leq d(u_n, fu_n) = d(u_n, u_{n+1}) \\ &\leq d(u_n, z) + d(u_{n+1}, z) \\ &\leq \frac{2}{3}d(x, z) = d(x, z) - \frac{d(x, z)}{3} \\ &\leq d(x, z) - d(u_n, z) \leq d(u_n, x). \end{aligned}$$

Hence and by hypothesis, $d(u_{n+1}, fx) \leq rd(u_n, x)$ for $n \geq \nu$. Letting n tend to ∞, we get $d(fx, z) \leq rd(x, z)$. That is, we have shown (1.30). Arguing by contradiction, we assume that $f^j z \neq z$ for all $j \in \mathbb{N}$. Then (1.30) yields

$$d(f^{j+1}z, z) \leq r^j d(fz, z) \quad \text{for all } j \in \mathbb{N}. \tag{1.31}$$

Consider the following three cases:

- $0 \leq r \leq \frac{\sqrt{5}-1}{2}$
- $\frac{\sqrt{5}-1}{2} < r < 2^{-1/2}$
- $2^{-1/2} \leq r < 1$

In the case where $0 \leq r \leq \frac{\sqrt{5}-1}{2}$, we note $r^2 + r - 1 \leq 0$ and $2r^2 < 1$. If we assume $d(f^2z, z) < d(f^2z, f^3z)$, then we have

$$\begin{aligned} d(z, fz) &\leq d(z, f^2z) + d(fz, f^2z) \\ &< d(f^2z, f^3z) + d(fz, f^2z) \\ &\leq r^2 d(z, fz) + rd(z, fz) \\ &\leq d(z, fz). \end{aligned}$$

This is a contradiction. So we have

$$d(f^2z, z) \geq d(f^2z, f^3z) = \theta(r)d(f^2z, f \circ f^2z).$$

By hypothesis and (1.31), we have

$$\begin{aligned} d(z, fz) &\leq d(z, f^3z) + d(f^3z, fz) \\ &\leq r^2 d(z, fz) + rd(f^2z, z) \\ &\leq r^2 d(z, fz) + r^2 d(fz, z) = 2r^2 d(z, fz) \\ &\leq d(z, fz). \end{aligned}$$

This is a contradiction. If $\frac{\sqrt{5}-1}{2} < r < 2^{-1/2}$, we note that $2r^2 < 1$. If we assume $d(f^2z, z) < \theta(r)d(f^2z, f^3z)$, in the view of (1.29),

$$\begin{aligned} d(z, fz) &\leq d(z, f^2z) + d(fz, f^2z) \\ &< \theta(r)d(f^2z, f^3z) + d(fz, f^2z) \\ &\leq \theta(r)r^2d(z, fz) + rd(z, fz) = d(z, fz). \end{aligned}$$

This is a contradiction. Hence, $d(f^2z, z) \geq \theta(r)d(f^2z, fof^2z)$. As in the previous case, we can prove

$$d(z, fz) \leq 2r^2d(z, fz) < d(z, fz).$$

This is a contradiction. In the third case, where $2^{-1/2} \leq r < 1$, we note that for $x, y \in X$, either

$$\theta(r)d(x, fx) \leq d(x, y) \quad \text{or} \quad \theta(r)d(fx, f^2x) \leq d(fx, y)$$

holds. Indeed, if

$$\theta(r)d(x, fx) > d(x, y) \quad \text{and} \quad \theta(r)d(fx, f^2x) > d(fx, y),$$

then we have

$$\begin{aligned} d(x, fx) &\leq d(x, y) + d(fx, y) \\ &< \theta(r)\big(d(x, fx) + rd(fx, f^2x)\big) \\ &\leq \theta(r)\big(d(x, fx) + rd(x, fx)\big) \\ &= d(x, fx). \end{aligned}$$

This is a contradiction. Since either

$$\theta(r)d(u_{2n}, u_{2n+1}) \leq d(u_{2n}, z) \quad \text{or} \quad \theta(r)d(u_{2n+1}, u_{2n+2}) \leq d(u_{2n+1}, z)$$

holds for every $n \in \mathbb{N}$, either

$$d(u_{2n+1}, fz) \leq rd(u_{2n}, z) \quad \text{or} \quad d(u_{2n+2}, fz) \leq rd(u_{2n+1}, z)$$

holds for every $n \in \mathbb{N}$. Since $\{u_n\}$ converges to z, the above inequalities imply a subsequence of $\{u_n\}$ which converges to fz. This implies $fz = z$. This is a contradiction. Therefore in all the cases, there exists $j \in \mathbb{N}$ such that $f^jz = z$. Since $\{f^nz\}$ is a Cauchy sequence, we obtain $fz = z$. That is, z is a fixed point of f. The uniqueness of a fixed point follows easily from (1.30).

Obviously, the class of contraction mappings given in Theorem 1.17 contains the class of usual contractions. However, it has been observed in [37] that Suzuki's and Kannan's contractions are independent but both types characterize the metric completeness of the underlying spaces.

Very recently, the following natural generalization of BCP appeared in the literature of metric fixed point theory.

(a) F-contraction

(b) $\mathcal{Z}$-contraction

(c) JS-contraction

The notion of F-contraction was introduced by Wardowski [38].

Definition 1.4 A mapping $f : X \to X$ is said to be an F-contraction if there exists $\tau > 0$ such that for every $x, y \in X$

$$d(fx, fy) > 0 \implies \tau + F(d(fx, fy)) \leq F(d(x, y)), \tag{1.32}$$

where the function $F : \mathbb{R}^+ \to \mathbb{R}$ satisfie the following conditions:

(F1) F is strictly increasing, i.e., for all $\alpha, \beta \in \mathbb{R}^+$ such that $\alpha < \beta$, $F(\alpha) < F(\beta)$;

(F2) for each sequence $\{\alpha_n\}_{n\in\mathbb{N}}$ of positive numbers $\lim\limits_{n\to\infty} \alpha_n = 0$ if and only if $\lim\limits_{n\to\infty} F(\alpha_n) = -\infty$;

(F3) there exists $k \in (0, 1)$ such that $\lim\limits_{\alpha\to 0^+} \alpha^k F(\alpha) = 0$.

When we consider various types of the mapping F in (1.32), we obtain a variety of contractions, some of a type known in the literature.

Example 1.7 Let $F : \mathbb{R}^+ \to \mathbb{R}$ be given by $F(\alpha) = \ln \alpha$; then F satisfies conditions $(F1), (F2)$ and $(F3)$ of Definition 1.4. Each mapping $f : X \to X$ satisfying (1.32) is an F-contraction such that

$$d(fx, fy) \leq e^{-\tau} d(x, y) \quad \text{for all } x, y \in X,\ fx \neq fy. \tag{1.33}$$

It is clear that if $fx = fy$ for $x, y \in X$, the above inequality holds trivially, i.e., f is a Banach contraction.

Example 1.8 Let $F : \mathbb{R}^+ \to \mathbb{R}$ be given by $F(\alpha) = \ln \alpha + \alpha$, then F satisfies conditions $(F1)$ through $(F3)$ of Definition 1.4 and the condition (1.32) is of the form

$$d(fx, fy) e^{d(fx,fy)-d(x,y)} \leq d(x, y) e^{-\tau} \quad \text{for all } x, y \in X,\ fx \neq fy. \tag{1.34}$$

Example 1.9 Consider $F(\alpha) = -1/\sqrt{\alpha}$, $\alpha > 0$. Then F satisfies $(F1)$ through $(F3)$, $((F3)$ for $k \in (\frac{1}{2}, 1))$. In this case each F-contraction f satisfies

$$d(fx, fy) \leq \frac{1}{(1 + \tau\sqrt{d(x, y)})^2} d(x, y) \quad \text{for all } x, y \in X,\ fx \neq fy.$$

Here, we obtain a special case of nonlinear contraction of the type $d(fx, fy) \leq \alpha(d(x, y)) d(x, y)$ as mentioned in [4, 30].

Example 1.10 Let $F(\alpha) = \ln(\alpha^2 + \alpha)$, $\alpha > 0$. Obviously F satisfies $(F1)$ through $(F3)$ and for F-contraction f, the following condition holds:

$$\frac{d(fx, fy)(d(fx, fy) + 1)}{d(x, y)(d(x, y) + 1)} \leq e^{-\tau} \quad \text{for all } x, y \in X, \;\; fx \neq fy.$$

In the above Examples 1.7 through 1.10, the contractive conditions are satisfied for $x, y \in X$ such that $fx = fy$.

Remark 1.7 If F_1 and F_2 are two mappings satisfying conditions $(F1)$ through $(F3)$ of Definition 1.4, $F_1(\alpha) \leq F_2(\alpha)$ for all $\alpha > 0$ and $G = F_2 - F_1$ is nondecreasing, every F_1-contraction f is an F_2-contraction. Indeed, from $(F1)$ and (1.32) every F-contraction f is a contractive mapping, i.e., $d(fx, fy) < d(x, y)$ and we have $G(d(fx, fy)) \leq G(d(x, y))$ for all $x, y \in X$, $fx \neq fy$. So

$$\begin{aligned} \tau + F_2(d(fx, fy)) &= \tau + F_1(d(fx, fy)) + G(d(fx, fy)) \\ &\leq F_1(d(x, y)) + G(d(x, y)) \\ &= F_2(d(x, y)). \end{aligned}$$

Theorem 1.18 (Wardowski) *Let (X, d) be a complete metric space and $f : X \to X$ be an F-contraction. Then f has a unique fixed point $x^* \in X$, and for every $x_0 \in X$, a sequence $\{f^n x_0\}_{n\in\mathbb{N}}$ is convergent to x^*.*

Proof. First, let us observe that f has at most one fixed point. Indeed, if $x_1^*, x_2^* \in X$, $fx_1^* = x_1^* \neq x_2^* = fx_2^*$, we get $\tau \leq F(d(x_1^*, x_2^*)) - F(d(fx_1^*, fx_2^*)) = 0$, which is a contradiction.

In order to show that f has a fixed point, let $x_0 \in X$ be arbitrary and fixed. We define a sequence $\{x_n\}_{n\in\mathbb{N}} \subset X$, $x_{n+1} = fx_n$, $n = 0, 1, \ldots$. Denote $\gamma_n = d(x_{n+1}, x_n), n = 0, 1, \ldots$. If there exists $n_0 \in \mathbb{N}$ for which $x_{n_0+1} = x_{n_0}$, then $fx_{n_0} = x_{n_0}$ and the proof is finished.

Suppose now that $x_{n+1} \neq x_n$, for every $n \in \mathbb{N}$. Then $\gamma_n > 0$ for all $n \in \mathbb{N}$ and using (1.32), the following holds for every $n \in \mathbb{N}$:

$$F(\gamma_n) \leq F(\gamma_{n-1}) - \tau \leq F(\gamma_{n-2}) - 2\tau \leq F(\gamma_0) - n\tau. \tag{1.35}$$

From (1.35), we obtain $\lim\limits_{n\to\infty} F(\gamma_n) = -\infty$ that together with $(F2)$ gives

$$\lim_{n\to\infty} \gamma_n = 0. \tag{1.36}$$

From $(F3)$ there exists $k \in (0, 1)$ such that

$$\lim_{n\to\infty} \gamma^n F(\gamma_n) = 0. \tag{1.37}$$

By (1.35), the following holds for all $n \in \mathbb{N}$:

$$\gamma_n^k F(\gamma_n) - \gamma_n^k F(\gamma_0) \leq \gamma_n^k (F(\gamma_0) - n\tau) - \gamma_n^k F(\gamma_0) = -\gamma_n^k n\tau \leq 0. \tag{1.38}$$

Letting $n \to \infty$ in (1.38), and using (1.36) and (1.37), we obtain

$$\lim_{n\to\infty} n\gamma_n^k = 0. \tag{1.39}$$

Now, let us observe that from (1.39) there exists $n_1 \in \mathbb{N}$ such that $n\gamma_n^k \leq 1$ for all $n \geq n_1$. Consequently we have

$$\gamma_n \leq \frac{1}{n^{1/k}}, \text{ for all } n \geq n_1. \tag{1.40}$$

In order to show that $\{x_n\}_{n\in\mathbb{N}}$ is a Cauchy sequence, consider $m, n \in \mathbb{N}$ such that $m > n > n_1$. From the definition of the metric and from (1.40), we get

$$d(x_m, x_n) \leq \gamma_{m-1} + \gamma_{m-2} + \cdots + \gamma_n < \sum_{i=n}^{\infty} \gamma_i \leq \sum_{i=n}^{\infty} \frac{1}{i^{1/k}}.$$

From the above and from the convergence of the series $\sum_{i=n}^{\infty} \frac{1}{i^{1/k}}$, it follows that $\{x_n\}_{n\in\mathbb{N}}$ is a Cauchy sequence.

Now, from the completeness of X, there exists $x^* \in X$ such that $\lim_{n\to\infty} x_n = x^*$. Finally, the continuity of f yields $d(fx^*, x^*) = \lim_{n\to\infty} d(fx_n, x_n) = \lim_{n\to\infty} d(x_{n+1}, x_n) = 0$, which completes the proof.

The next example shows a mapping f on certain metric space which is an F_2-contraction but not an F_1-contraction mentioned in Remark 1.7.

Example 1.11 Let $X = \{S_n : n = 1, 2, 3, \ldots\}$ where $S_n = \frac{n(n+1)}{2}$. Define metric d on X as $d(x, y) = |x - y|$. Then (X, d) is a complete metric space. Let $f : X \to X$ be given by

$$f(S_n) = \begin{cases} S_{n-1} & \text{for } n > 1, \\ S_1 & \text{for } n = 1. \end{cases}$$

Then S_1 is a fixed point of f.
Let $F_1 : \mathbb{R}^+ \to \mathbb{R}$ be given by $F(\alpha) = \ln \alpha$; then f is not an F_1-contraction since we know that every F_1-contraction f satisfies $d(fx, fy) \leq e^{-\tau} d(x, y)$ for all $x, y \in X$, $fx \neq fy$ for some $\tau > 0$, i.e., f is a Banach contraction.
Here

$$\lim_{n\to\infty} \frac{d(f(S_n), f(S_1))}{d(S_n, S_1)} = \lim_{n\to\infty} \frac{S_{n-1} - S_1}{S_n - 1} = 1.$$

Let $F_2 : \mathbb{R}^+ \to \mathbb{R}$ be given by $F(\alpha) = \ln \alpha + \alpha$.

Now for $m \in \mathbb{N}$, $m > 2$ we have

$$\begin{aligned}
\frac{d(f(S_m), f(S_1))}{d(S_m, S_1)} e^{d(f(S_m), f(S_1)) - d(S_m, S_1)} &= \frac{S_{m-1} - 1}{S_m - 1} e^{S_{m-1} - S_m} \\
&= \frac{m^2 - m - 2}{m^2 + m - 2} e^{-m} \\
&< e^{-m} < e^{-1},
\end{aligned}$$

and for every $m, n \in \mathbb{N}$; $m > n > 1$ we have

$$\begin{aligned}
\frac{d(f(S_m), f(S_n))}{d(S_m, S_n)} e^{d(f(S_m), f(S_n)) - d(S_m, S_n)} &= \frac{S_{m-1} - S_{n-1}}{S_m - S_n} e^{S_n - S_{n-1} + S_{m-1} - S_m} \\
&= \frac{m + n - 1}{m + n + 1} e^{n-m} \\
&< e^{n-m} \leq e^{-1}.
\end{aligned}$$

Thus, f is an F_2-contraction.

Using a simulation function Khojasteh *et al.* [19] defined a new type of contraction which generalizes the Banach contraction principle and unifies several known types of contractions involving $d(fx, fy)$ and $d(x, y)$.

Definition 1.5 Let $\zeta : [0, \infty) \times [0, \infty) \to \mathbb{R}$ be a mapping; then ζ is called a simulation function if it satisfies the following conditions:

(ζ1) $\zeta(0, 0) = 0$;

(ζ2) $\zeta(t, s) < s - t$ for all $t, s > 0$;

(ζ3) if $\{t_n\}, \{s_n\}$ are sequences in $[0, \infty)$ such that $\lim_{n\to\infty} t_n = \lim_{n\to\infty} s_n > 0$, then

$$\limsup_{n\to\infty} \zeta(t_n, s_n) < 0.$$

We denote the set of all simulation functions by $\mathcal{Z}$.

Example 1.12 The following functions $\zeta_i : [0, \infty) \times [0, \infty) \to \mathbb{R}$, $i = 1, 2$ belong to $\mathcal{Z}$:

1. $\zeta_1(t, s) = \frac{s}{s+1} - t$ for all $t, s \in [0, \infty)$.
2. $\zeta_2(t, s) = \lambda s - t$ for all $t, s \in [0, \infty)$.

Definition 1.6 Let (X, d) be a metric space, $f : X \to X$ a mapping and $\zeta \in \mathcal{Z}$. Then f is called a $\mathcal{Z}$-contraction with respect to ζ if the following condition is satisfied:

$$\zeta(d(fx, fy), d(x, y)) \geq 0 \text{ for all } x, y \in X. \tag{1.41}$$

A simple example of a $\mathcal{Z}$-contraction is the Banach contraction which can be obtained by taking $\lambda \in [0, 1)$ and $\zeta(t, s) = \lambda s - t$ for all $s, t \in [0, \infty)$ in the above definition.

Remark 1.8 It is clear from the definition of simulation function that $\zeta(t, s) < 0$ for all $t \geq s > 0$. Therefore, if f is a $\mathcal{Z}$-contraction with respect to $\zeta \in \mathcal{Z}$,

$$d(fx, fy) < d(x, y) \text{ for all distinct } x, y \in X.$$

This shows that every $\mathcal{Z}$-contraction is contractive mapping and is therefore continuous.

The next theorem is the existence theorem for a fixed point of $\mathcal{Z}$-contraction.

Theorem 1.19 *Let (X, d) be a complete metric space and $f : X \to X$ be a $\mathcal{Z}$-contraction with respect to ζ. Then f has a unique fixed point u in X and for every $x_0 \in X$ the Picard sequence $\{x_n\}$, where $x_n = fx_{n-1}$ for all $n \in \mathbb{N}$ converges to the fixed point of f.*

The following example shows that the above theorem is a proper generalization of the Banach contraction principle.

Example 1.13 Let $X = [0, 1]$ and $d : X \times X \to \mathbb{R}$ be defined by $d(x, y) = |x - y|$. Then (X, d) is a complete metric space. Define a mapping $f : X \to X$ as $fx = \frac{x}{x+1}$ for all $x \in X$. Then f is a $\mathcal{Z}$-contraction with respect to $\zeta \in \mathcal{Z}$, where $\zeta(t, s) = \frac{s}{s+1} - t$ for all $t, s \in [0, \infty)$. Indeed, if $x, y \in X$, then

$$\begin{aligned}
\zeta(d(fx, fy), d(x, y)) &= \frac{d(x, y)}{1 + d(x, y)} - d(fx, fy) \\
&= \frac{|x - y|}{1 + |x - y|} - \left| \frac{x}{x + 1} - \frac{y}{y + 1} \right| \\
&= \frac{|x - y|}{1 + |x - y|} - \left| \frac{|x - y|}{(x + 1)(y + 1)} \right| \geq 0.
\end{aligned}$$

But f is not a Banach contraction. Also note that all the conditions of Theorem 1.19 are satisfied and f has a unique fixed point $u = 0 \in X$.

Using a new type of control function Jleli and Samet [16] generalised the Banach contraction principle by introducing the following.

Definition 1.7 A mapping $f : (X, d) \to (X, d)$ is said to be a JS-contraction if there exists $\theta \in \Theta$ and $\lambda \in (0, 1)$ such that

$$\theta(d(fx, fy)) \leq \theta(d(x, y))^{\lambda} \text{ for all } x, y \in X \text{ with } fx \neq fy, \tag{1.42}$$

where Θ denotes the set of functions $\theta : \mathbb{R}^+ \to (1, \infty)$ satisfying the following conditions:

θ_1 : θ is nondecreasing;

θ_2 : for each sequence $\{t_n\} \subset \mathbb{R}^+$, $\lim_{n\to\infty} \theta(t_n) = 1$ iff $\lim_{n\to\infty} t_n = 0^+$;

θ_3 : there exists $r \in (0,1)$ and $l \in (0,\infty]$ such that $\lim_{t\to 0^+} \frac{\theta(t)-1}{t^r} = l$.

Theorem 1.20 *[see [16], Corollary 2.1] : Let (X,d) be a complete metric space and $f : X \to X$ be a JS-contraction. Then f has a unique fixed point in X.*

Remark 1.9 The Banach contraction principle follows immediately from Theorem 1.20. Indeed, let $f : X \to X$ be a JS-contraction. Then, if we choose $\theta(t) = e^{\sqrt{t}} \in \Theta$ and $\lambda = \sqrt{k}$ in (1.42), we get $\sqrt{d(fx,fy)} \leq k\sqrt{d(x,y)}$, i.e., $d(fx,fy) \leq kd(x,y)$ for all $x,y \in X$, which means f is a Banach contraction.

Example 1.14 Let X be the set defined by $X = \{\tau_n \ : \ n \in \mathbb{N}\}$ where $\tau_n = \frac{n(n+1)}{2}$, for all $n \in \mathbb{N}$. We endow X with the metric d given by $d(x,y) = |x-y|$ for all $x,y \in X$. Then (X,d) is a complete metric space. Let $f : X \to X$ be the map defined by $f\tau_1 = \tau_1, f\tau_n = \tau_{n-1}$ for all $n \geq 2$. Clearly, the Banach contraction is not satisfied. In fact, we can check easily that $\lim_{n\to\infty} \frac{d(f\tau_n,f\tau_1)}{d(\tau_n,\tau_1)} = 1$. Now consider the function $\theta : \mathbb{R}^+ \to (1,\infty)$ defined by $\theta(t) = e^{\sqrt{te^t}}$. Then $\theta \in \Theta$ and f satisfies all the hypotheses of Theorem 1.20 with $\lambda = e^{-\frac{1}{2}}$ and thus f has a unique fixed point $u = \tau_1 \in X$.

Remark 1.10 The above new classes of contractions have generated much interest amongst fixed point theorists. It is worthwhile to note that Ćirić, Suzuki and Berinde type F-contractions have been investigated in [39, 27]; whereas to the best of our knowledge only Suzuki type $\mathcal{Z}$-contractions and Ćirić type JS-contractions have been formulated in [20, 41]. It would be interesting if one could formulate the Ćirić, Berinde type $\mathcal{Z}$-contraction as well as Suzuki, Berinde type JS-contraction and associated fixed point theorems.

References

[1] Alber Y.I., Guerre-Delabriere S., Principles of weakly contractive maps in Hilbert spaces, in: I. Gohberg, Yu. Lyubich (Eds.), New Results in Operator Theory, Birkhäuser, Basel, 1997; 98:7-22.

[2] Banach S., Surles operations dansies ensembles abstraits et leur application aux quations integrales. Fund. Math. 1922; 3:133-181.

[3] Berinde V., Approximating fixed points of weak contractions using the Picard iteration. Nonlinear Anal. Forum. 2004; 9(1):43-53.

[4] Boyd D.W., Wong J.S.W., Nonlinear contractions. Proc. Am.Math. Soc. 1969; 20:458-464.

[5] Chatterjea S.K., Fixed-point theorems. C.R. Acad. Bulgare Sci. 1972; 25:727-730.

[6] Choudhury B.S., Konarb P., Rhoades B.E., Metiya N., Fixed point theorems for generalized weakly contractive mappings. Nonlinear Anal. 2011; 74:2116-2126.

[7] Ćirić L.B., A generalization of Banach's contraction principle. Proc. Am. Math. Soc. 1974; 45:267-273.

[8] Ćirić L.B., Fixed Point Theory: Contraction Mapping Principle. Faculty of Mechanical Engneering, University of Belgrade, Serbia; 2003.

[9] Connell E.H., Properties of fixed point spaces. Proc. Am. Math. Soc. 1959; 10:974-979.

[10] Delefina R., Elisabetta M., Contractive Kannan maps in compact spaces. Riv. Mat. Univ. Parma. 1979; 4(5):141-145.

[11] Dutta P.N., Choudhury B.S., A generalisation of contraction principle in metric spaces. Fixed Point Theor. Appl. 2008; Article ID 406368.

[12] Edelstein M., On fixed and periodic points under contractive mappings. J. Lond. Math. Soc. 1962; 37:74-79.

[13] Geraghty M., On contraction mappings. Proc. Am. Math. Soc. 1973; 40:604-608.

[14] Jachymski J.R., Equivalence of some contractivity properties over metrical structure. Proc. Am. Math. Soc. 1997; 125:2327-2335.

[15] Jaroslaw G., Fixed point theorems for Kannan type mappings. J. Fixed Point Theory Appl. 2017; DOI 10.1007/s11784-017-0402-8.

[16] Jleli M. Samet B., A new generalization of the Banach contraction principle. J. Inequal. Appl. 2014; 38.

[17] Kannan R., Some results on fixed point, Bull. Calcutta Math. Soc. 1968; 60:71-76.

[18] Khan M.S., Sessa S., Fixed points theorems by altering distances between the points. Bull. Aust. Math. Soc. 1984; 30:1-9.

[19] Khojasteh F., Shukla S., Radenovic S., A new approach to the study of fixed point theory for simulation function. Filomat. 2015; 29(6): 1189-1194.

[20] Kumam P., Gopal D., Budhia L., A new fixed point theorem under Suzuki type $\mathcal{Z}$-contraction mapping. J. Math. Anal., 2017; 8(1):113-119.

[21] Lahiri B.K, Elements of Functional Analysis. Kolkata: World Press; 2009.

[22] Matkowski J., Integrable solutions of functional equations. Diss. Math. 1975; 127.

[23] Matkowski J., Nonlinear contractions in metrically convex spaces. Publ. Math. Debrecen. 1994; 45:103-114.

[24] Meir A., Keeler E., A theorem on contraction mappings. J. Math. Anal. Appl. 1969; 28:326-329.

[25] Mihet D., Altering distances in probabilistic Menger spaces. Nonlinear Anal. 2009; 71:2734-2738.

[26] Naidu S.V.R., Some fixed point theorems in metric spaces by altering distances. Czech. Math. J. 2003; 53:205-212.

[27] Piri H., Kumam P., Some fixed point theorems concerning F-contraction in complete metric spaces. Fixed Point Theory Appl., 2014; 210.

[28] Prešić S. B., Sur la convergence des suites. Comptes Rendus de l'Acad. des Sci. de Paris. 1965; 260:3828-3830.

[29] Prešić S. B., Sur une classe d'inéquations aux diffćerences finite et sur la convergence de certaines suites. Publ. de L'Inst. Math. Belgrade. 1965; 19(5):75-78.

[30] Rakotch E., A note on contractive mappings. Proc. Am. Math. Soc. 1962; 13:459-465.

[31] Reich S., Kannan's fixed point theorem, Boll. Un. Mat. Ital. 1971; 4:1-11.

[32] Rhoades B.E., A comparison of various definitions of contractive mappings. Proc. Am. Math. Soc. 1977; 226:257-290.

[33] Rhoades B.E., Some theorems on weakly contractive maps. Nonlinear Anal. 2001; 47:2683-2693.

[34] Sarkhel D.N., Banach's fixed point theorem implies Kannan's. Bull. Cal. Math. Soc. 1999; 91(2):143-144.

[35] Sastry K.P.R., Babu G.V.R., Some fixed point theorems by altering distances between the points. Indian J. Pure Appl. Math. 1999; 30:641-647.

[36] Subrahmanyam V., Completeness and fixed points. Monatsh. Math. 1975; 80:325-330.

[37] Suzuki T., A generalized Banach contraction principle that characterizes metric completeness. Proc. Am. Math. Soc. 2008; 136:1861-1869.

[38] Wardowski D., Fixed points of a new type of contractive mappings in complete metric spaces. Fixed Point Theory Appl. 2012; Article ID 94(2012).

[39] Wardowski D., Dung N. Van, Fixed points of F-weak contractions on complete metric spaces. Demonst. Math. 2014; 47(1):146-155.

[40] Zamfirescu T., Fixed point theorems in metric spaces. Arch. Math., 1972; 23:292-298.

[41] Zhilong L., Shujun J., Fixed point theorems of JS-quasi-contractions. Fixed Point Theory Appl. 2016; 2016:40 DOI: 10.1186/s13663-016-0526-3.

Chapter 2

Metrical Common Fixed Points and Commuting Type Mappings

Dhananjay Gopal and Ravindra K Bisht

2.1 Introduction

The well-known Banach contraction principle tells us to let f be a self mapping of a complete metric space (X, d) satisfying the condition

$$d(fx, fy) \leq kd(x, y), \quad 0 \leq k < 1,$$

for each $x, y \in X$. Then f has a unique fixed point, that is, there exists a unique $z \in X$ such that $f(z) = z$.

In 1975, Pfeller [69] observed interdependence between the commuting mappings and the fixed point concepts. He proves that an involution σ of a circle has a fixed point if it commutes with a fixed point free involution different from σ. (A mapping $\sigma : S^1 \to S^1$ is an involution if $\sigma^2 = \sigma$.)

A fixed point of a self mapping of a metric space X can also be perceived to be a common fixed point of f with the identity mapping on X. A common point theorem ensures the existence of a common fixed point of a pair of mappings under suitable assumptions on the space and on the mappings. Those assumptions are sufficient and include conditions of commutativity, containment of ranges of mappings. continuity of at least one mapping or weaker notion, contractivity, and all substantial common fixed point theorem attempts to obtain or soften required values of one of more such conditions. In addition to ensuring existence of a common point, it may be necessary to prove its uniqueness. From a computational view, a constructive algorithm to calculate the value of a common fixed point is desirable. Such algorithms often require iterates of the given mappings.

The existence, uniqueness and approximations of common fixed points are important features of the common point theorem. The Banach contraction principle for single mapping and Jungck's common fixed point theorem (Theorem 2.1 below) for a pair of mappings cover all three features convincingly.

Jungck [31] described the interdependence of common fixed points and commuting mappings. His result in the setting of complete metric spaces yields an abstraction of the Banach contraction principle and partially answers the historical open question [14, 27, 40]: for a pair of commuting self mappings on the [0, 1] interval, what additional conditions guarantee that f and g have a common fixed point?

Theorem 2.1 *[31] A self mapping g of a complete metric space (X, d) has a fixed point if and only if $0 < k < 1$ and $f : X \to X$ commute with g such that:*

(a) $fX \subseteq gX$;

(b) $d(fx, fy) \le kd(gx, gy)$, *for all* $x, y \in X$.

Moreover, f and g have a common fixed point if conditions (i) and (ii) hold.

Jungck's results allow us to investigate common fixed point theorems for commuting and noncommuting pairs of mappings satisfying contractive conditions. Machuca [38] developed a constructive technique from Jungck's result and other researches established common fixed point theorems for three, four, and sequence mappings. Condition (*b*) of Theorem 2.1 is known as the Machuca-Jungck (MJ) contraction condition.

Generalizations of the Jungck's contraction condition have been extensively used to study common fixed points of contractive mappings. If f and g are two self mappings of a metric space (X, d), general contractive conditions assume the forms below and extend the theorem:

(c) A ϕ-type contractive condition (in the sense of Boyd and Wong [15]):

$$d(fx, fy) \le \phi(d(gx, gy)),$$

where $\phi : \mathbb{R}^+ (:= [0, \infty)) \to \mathbb{R}^+$ is such that ϕ is upper semicontinuous from the right and $\phi(t) < t$ for each $t > 0$. In [39] Matkowski assumed ϕ to be increasing and $\lim_{n \to \infty} \phi^n(t) = 0$ for every $t \in [0, \infty)$.

(d) Given $\epsilon > 0$ there exists a $\delta > 0$ such that

$$\epsilon \le d(gx, gy) < \epsilon + \delta \Rightarrow d(fx, fy) < \epsilon.$$

Condition (d) is referred us a Meir-Keeler type (ϵ, δ) contractive condition. It can easily be seen that if f and g satisfy (d) then f and g also satisfy the contractive condition

$$d(fx, fy) < d(gx, gy).$$

In [39] the contractive condition (d) has been replaced by a slightly weaker contractive condition of the following form

(e) *Given* $\epsilon > 0$, *there exists a* $\delta > 0$ *such that*

$$\epsilon < d(gx, gy) < \epsilon + \delta \Rightarrow d(fx, fy) \leq \epsilon.$$

It is well-known that; $b \Longrightarrow c \Longrightarrow d \Longrightarrow e$, but the converse is not true.

Proving a common fixed point theorem for mappings satisfying the above contractive conditions involves the following steps. The first is the most crucial part of common fixed point theorems consisting of constructive procedures yielding a Cauchy sequence which converges to a point in X (where X is complete). The second step is to show the coincidence point by assuming suitable weaker forms of commutating and/or continuity type conditions; and the third step automatically gives rise to the coincidence point, which is simply a unique common fixed point due to contractive condition. Observing carefully the second step, one can find the existence of the coincidence point by suitable conditions including choice of weaker forms of commuting conditions. Many authors have considered generalizations of commuting mappings; see Table 2.1 in Section 2.3. Now, it has been shown that weak compatibility is the minimal noncommuting condition for the existence of common fixed points of contractive type mapping pairs.

However, in the case of nonunique common fixed point theorems, conditionally absorbing mappings appeared as the most suitable conditions (see [51]). In recent works several authors introduced some weaker noncommuting notions and showed that their notions yielded weak compatibility as a proper subclass. This is, however, of no use when searching for common fixed points. In fact, most of the generalized commutativity notions fall in the subclass of weak compatibility in the setting of a unique common fixed point (or unique point of coincidence). These generalizations are significant in the sense that they are defined over fewer assumptions on the definition of a metric space but for their actual applications we must look to noncontractive conditions since they allow more than one point of coincidence or common fixed point.

In 2010, Haghi *et al.* [26] presented the following lemma, and showed that some coincidence points and common fixed point generalizations in fixed point theory are not real generalizations as they could easily be assumed from the corresponding fixed point theorems. However, the consequence of their proved lemma is limited to only a few coincidence or common fixed point theorems cited in the literature. Therefore, coincidence or common fixed point theorems are indeed real generalizations of fixed point theorems.

Lemma 2.1 *[26] Let* X *be a nonempty set and* $f : X \to X$ *a function. Then there exists a subset* $E \subset X$ *such that* $f(E) = f(X)$ *and* $f : E \to X$ *is one-to-one.*

2.2 Comparison of weaker forms of commuting mappings

Let f and g be two mappings of a set X. For a pair of self mappings $f, g : X \to X$, we will use fg to denote the composition $f \circ g$. If $w = fx = gx$ for some x in X, then x is called a coincidence point of f and g, and w is called a point of coincidence (POC) of f and g. The set of coincidence points (CP) of f and g will be denoted $C(f, g)$. Let $PC(f, g)$ represent the set of points of coincidence of f and g; $x \in X$ is a common fixed point of f and g if $x = fx = gx$. The set of all common fixed points of f and g is denoted by $F(f, g)$.

Two self mappings f and g of a metric space (X, d) are said to be commuting iff $fgx = gfx$ for all x in X.

The study of common fixed points of a pair of self mappings satisfying contractive type conditions becomes interesting in view of the fact that even commuting continuous mappings on such well behaved entities as compact convex sets may fail to have a coincidence or common fixed point. When we extend such studies to the class of noncommuting contractive mapping pairs, the analysis becomes still more interesting [44].

In 1982, Sessa introduced weak commutativity and relaxed the commutativity condition of mappings to a smaller subset of the domain of mappings.

Definition 2.1 (Sessa [73]) Two self mappings f and g of a metric space (X, d) are called weakly commuting iff $d(fgx, gfx) \leq d(fx, gx)$ for all x in X.

Notice that commuting mappings are obviously weakly commuting. However, weakly commuting mappings need not be commuting.

Example 2.1 Let $X = [0, 1]$ be equipped with the the usual metric d on X. Define constant mappings f and $g : X \to X$ by

$$fx = b \text{ and } gx = c, b \neq c.$$

Then f and g are weakly commuting but not commuting since $d(fgx, gfx) = |b - c| = d(fx, gx)$.

In order to enlarge the domain of noncommuting mappings, Pathak [55, 57] obtained several new classes of noncommuting notions, namely, weak* commuting and weak** commuting mappings.

Definition 2.2 (Pathak [55]) Two self mappings f and g of a metric space (X, d) are called weak* commuting iff $d(fgx, gfx) \leq d(f^2x, g^2x)$ for all x in X.

Definition 2.3 (Pathak [57]) Two self mappings f and g of a metric space (X, d) are called weak** commuting iff $fX \subset gX$ and for any $x \in X$,

$$\begin{aligned} d(f^2g^2x, g^2f^2x) &\le d(f^2gx, gf^2x) \le d(fg^2x, g^2fx) \\ &\le d(fgx, gfx) \le d(f^2x, g^2x). \end{aligned}$$

It is easy to check that commuting mappings are weak* commuting and weak** commuting. The following example shows that the reverse implication does not hold.

Example 2.2 *[55]* Consider $X = [0, 1]$ with the usual metric d on X. Define $f, g : X \to X$ by

$$fx = \frac{x}{x+3} \text{ for all } x \text{ and } gx = \frac{x}{3} \text{ for all } x.$$

Then f and g are weak* commuting and weak** commuting but f and g are not commuting mappings.

Remark 2.1 Notice that if $f^2 = f$ and $g^2 = g$, weak* commutativity or weak** commutativity reduces to weak commutativity.

Definition 2.4 (Pathak [60]) Two self mappings f and g of a metric space (X, d) are called weakly uniformly contraction mappings iff $d(fgx, ggx) \le d(fx, gx)$ and $d(ffx, gfx) \le d(fx, gx)$ for all x in X.

In view of Example 2.2, we remark that commuting mappings are weakly uniformly contractions. However, weakly uniformly contraction mappings need not be weakly commuting.

In 1986, Jungck generalized the concept of weak commutativity by introducing compatible mappings [32], also called asymptotically commuting mappings by Tivari and Singh [81] in an independent work.

Definition 2.5 (Jungck [32], Tivari and Singh [81]) Two self mappings f and g of a metric space (X, d) are called compatible or asymptotically commuting iff $\lim_{n\to\infty} d(fgx_n, gfx_n) = 0$, whenever $\{x_n\}$ is a sequence in X such that $\lim_{n\to\infty} fx_n = \lim_{n\to\infty} gx_n = t$ for some t in X.

Clearly, weakly commuting mappings are compatible, but in view of the following example the converse does not hold [32].

Example 2.3 Let $X = [0, \infty)$ and d be the usual metric on X. Define $f, g : X \to X$ by

$$fx = x^3 \text{ for all } x \text{ and } gx = ax^3 \text{ for all } x, a \ge 1.$$

Then $d(fgx, gfx) > d(fx, gx)$. Therefore f and g are not weakly commuting mappings. However, f and g are compatible mappings.

Remark 2.2 It is important to note that weak commutativity is essentially a point property while compatibility uses the machinery of sequences.

Remark 2.3 In Math. Rev. 89 h : 54030 (see also [44]) Singh remarked that for a pair of weakly commuting mappings on a metric space (X, d), there may not exist even a single sequence $\{x_n\}$ in X for which $\lim_n fx_n = \lim_n gx_n = t$ for some t in X (see also [44]). In this case the mappings f and g are still compatible but only vacuously. The following example shows that in this situation they can be weakly commuting.

Example 2.4 Let $X = [0, \infty)$ and d be the usual metric on X. Define $f, g : X \to X$ by

$$fx = x \text{ for all } x \text{ and } gx = x + a \text{ for all } x, a \geq 1.$$

Then f and g are weakly commuting mappings but there does not exist even a single sequence for which the condition of compatibility is satisfied. However, $fgx = gfx = a + x$, i.e., $d(fgx, gfx) = 0$.

Singh [76] has also shown by an example that weak commutativity of a pair of self mappings f and g on a metric space (X, d) depends on the choice of the metric. This is true for compatibility as well.

Example 2.5 *[76]* Let $X = [0, \infty)$ be endowed with the usual metric. Define $f, g : X \to X$ by

$$fx = 1 + x \text{ and } gx = 2 + x^2.$$

Then $d(fgx, gfx) = 2x$ and $d(fx, gx) = (x^2 - x + 1)$. One may observe that f and g are not weakly commuting on X with respect to the usual metric. But if X is endowed with the discrete metric d, then $d(fgx, gfx) = 1 = d(fx, gx)$ for $x > 0$. So f and g are weakly commuting on X when endowed with a discrete metric.

A large body of literature of common fixed points has grown around compatible mappings and their weaker forms and became an area of vigorous research activity. However, fixed point theory for noncompatible mappings is equally interesting and Pant [42] performed work along these lines. This nonconstructive but powerful approach provides a new tool to deal with common fixed points of Lipschitz type mapping pairs. The best examples of noncompatible mappings are found among pairs which are necessarily discontinuous at their common fixed points [42]. It may be observed that the mappings f and g are said to be noncompatible if there exists a sequence $\{x_n\}$ in X such that $\lim_n fx_n = \lim_n gx_n = t$ for some t in X but $\lim_n d(fgx_n, gfx_n)$ is either nonzero or nonexistent.

Definition 2.6 *[1]* Two self mappings f and g of a metric space (X, d) are said to satisfy the (E.A.) property if there exists a sequence $\{x_n\}$ in X such that

$$\lim_n fx_n = \lim_n gx_n = t, \qquad \text{for some } t \in X.$$

It may be observed that the (E.A.) property is equivalent to the previously known notion of tangential mappings introduced by Shastri *et al.* [72]. Noncompatible mappings satisfy the (E.A.) property but the converse is not true in general. Sintunavarat and Kumam [78] introduced an interesting property, namely a common limit in the range property (in short CLR_g) which completely accommodates the condition of closedness of the ranges of the involved mappings and has an edge over the (E.A.) property.

Definition 2.7 *[78]* Two self mappings f and g of a metric space (X, d) are said to satisfy the common limit in the range of g property (in short CLR_g) if there exists a sequence $\{x_n\}$ in X such that

$$\lim_n fx_n = \lim_n gx_n = gx, \qquad \text{for some } x \in X.$$

The notions of (E.A.) property and (CLR_g) property circumvent the most crucial parts of common fixed point theorems consisting of constructive procedures yielding a Cauchy sequence. It is important to note that in the setting of metric spaces, there is no general method for the study of common fixed points of nonexpansive or Lipschitz type mappings. The notions of noncompatibility, (E.A.) property and (CLR_g) property are well suited for studying common fixed points of strict contractive conditions, nonexpansive type mapping pairs and Lipschitz type mapping pairs in ordinary metric spaces, which are not even complete.

Remark 2.4 Notice that noncompatibilty is a consequence of the definition of compatibility but the (E.A.) property presupposes the existence of a sequence $\{x_n\}$ in X satisfyng

$$\lim_n fx_n = \lim_n gx_n = t, \qquad \text{for some } t \in X.$$

The same is true for the (CLR_g) property.

Definition 2.8 (Singh and Mishra [75]) If for $x_0 \in X$ there exists a sequence $\{x_n\}$ in X such that $fx_{n+1} = gx_n, n = 0, 1, 2, ...$, then $O(g, f; x_0) = \{fx_n : n = 0, 1, 2, ...\}$ is an orbit for g and f. Mappings f and g are weakly x_0 preorbitally commuting iff there exists a positive integer N such that $d(fgx_n, gfx_n) \leq d(fx_n, gx_n)$ for every $\{x_n\}$ $(n \geq N)$ occurring in $O(g, f; x_0)$.

Weakly commuting mappings are preorbitally commuting but the converse is not true in general.

Example 2.6 *[17]* Let $X = [0, \infty)$ and d be the usual metric on X. Define $f, g : X \to X$ by

$$fx = 3x^3 \text{ for all } x \text{ and } gx = x^3 \text{ for all } x.$$

Then f and g are not weakly commuting mappings but f and g are preorbitally commuting mappings (e.g., $\{x_n\} = 0$).

Definition 2.9 Two self mappings f and g of a metric space (X, d) are called:

(i) compatible of type (A) (Jungck *et al.* [34]) iff

$$\lim_{n\to\infty} d(ffx_n, gfx_n) = 0 \text{ and } \lim_{n\to\infty} d(fgx_n, ggx_n) = 0,$$

(ii) compatible of type (B) (Pathak *et al.* [53]) iff

$$\lim_{n\to\infty} d(fgx_n, ggx_n) \leq \tfrac{1}{2}\left[\lim_{n\to\infty} d(fgx_n, ft) + \lim_{n\to\infty} d(ft, ffx_n)\right]$$

and

$$\lim_{n\to\infty} d(gfx_n, ffx_n) \leq \tfrac{1}{2}\left[\lim_{n\to\infty} d(gfx_n, gt) + \lim_{n\to\infty} d(gt, ggx_n)\right],$$

(iii) compatible of type (C) (Pathak *et al.* [54]) iff

$$\lim_{n\to\infty} d(fgx_n, ggx_n) \leq$$
$$\tfrac{1}{3}\left[\lim_{n\to\infty} d(fgx_n, ft) + \lim_{n\to\infty} d(ft, ffx_n) + \lim_{n\to\infty} d(ft, ggx_n)\right],$$

and

$$\lim_{n\to\infty} d(gfx_n, ffx_n) \leq$$
$$\tfrac{1}{3}\left[\lim_{n\to\infty} d(gfx_n, gt) + \lim_{n\to\infty} d(gt, ggx_n) + \lim_{n\to\infty} d(gt, ffx_n)\right],$$

whenever $\{x_n\}$ is a sequence in X such that $\lim_{n\to\infty} fx_n = \lim_{n\to\infty} gx_n = t$ for some t in X.

Proposition 2.1 *[12] Let f and g be continuous mappings from a metric space (X, d) into itself. Then the following are equivalent:*

(i) f and g are compatible of type (A),

(ii) f and g are compatible of type (B),

(iii) f and g are compatible of type (C),

(iv) f and g are compatible.

It is clear that compatible mappings of type (A) $\Longrightarrow$ compatible of type (B) $\Longrightarrow$ compatible of type (C), but the converse is not true in general.

Example 2.7 *[12]* Let $X = [1, 20]$ and d be the usual metric on X. Define $f, g : X \to X$ as

$$f(x) = \begin{cases} 1 & \text{if } x = 1, \\ 3 & \text{if } 1 < x \leq 7, \\ x - 6 & \text{if } 7 < x \leq 20, \end{cases} \quad \text{and} \quad g(x) = \begin{cases} 1 & \text{if } x \in \{1\} \cup (7, 20], \\ 2 & \text{if } 1 < x \leq 7. \end{cases}$$

It may be observed that f and g are compatible of type (C) but neither compatible of type (A) nor compatible of type (B) (consider the sequence $\{x_n\}$ given by $x_n = 7 + \frac{1}{n} : n > 0$).

Example 2.8 *[49]* Let $X = [2, 12]$ and d be the usual metric on X. Define $f, g : X \to X$ as

$$f(x) = \begin{cases} 2 & \text{if } x = 2 \text{ or } x > 5, \\ 12 & \text{if } 2 < x \le 5, \end{cases} \quad \text{and} \quad g(x) = \begin{cases} 2 & \text{if } x = 2, \\ 12 & \text{if } 2 < x \le 5, \\ \frac{x+1}{3} & \text{if } x > 5. \end{cases}$$

It may be observed that f and g are compatible mappings of type (A), but neither commuting nor compatible mappings. To see this let us consider the sequence $\{x_n\}$ given by $x_n = 5 + \frac{1}{n} : n > 0$. Then $fx_n \to 2$, $gx_n \to 2$, $fgx_n = ggx_n \to 12$, $gfx_n = ffx_n \to 2$, $\lim_{n\to\infty} d(fgx_n, ggx_n) = \lim_{n\to\infty} d(ffx_n, gfx_n) = 0$ and $\lim_{n\to\infty} d(fgx_n, gfx_n) = \lim_{n\to\infty} d(ffx_n, ggx_n) \neq 0$.

Examples 2.8 and 2.9 (below) show that compatible mappings and compatible mappings of type (A) are independent.

Example 2.9 *[66]* Let X = $\mathbb{R}$ equipped with the usual metric d. Define self mappings f and g as follows:

$$f(x) = x \quad \text{and} \quad g(x) = \begin{cases} 0 & \text{if } x \text{ is an integer}, \\ 1 & \text{if } x \text{ is not an integer}. \end{cases}$$

Then for the sequence $\{x_n\} = 1 + \frac{1}{n+1}$ we get $\lim_n fx_{n\to\infty} = \lim_{n\to\infty} gx_n = 1$, $\lim_n d(fgx_n, gfx_n) = 0$, but $\lim_{n\to\infty} d(ffx_n, gfx_n) \neq 0$, $\lim_{n\to\infty} d(gfx_n, ggx_n) \neq 0$ and $\lim_{n\to\infty} d(ffx_n, ggx_n) \neq 0$. Therefore f and g are compatible but not compatible of type (A).

Definition 2.10 (Pant [44]) Two self mappings f and g of a metric space (X, d) are called R-weakly commuting iff there exists some positive real number R such that $d(fgx, gfx) \le Rd(fx, gx)$ for all x in X.

Notice that weak commutativity of a pair of self mappings implies their R-weak commutativity and the converse is true only when $R \le 1$.

Example 2.10 *[44]* Let $X = [0, \infty)$ be endowed with the usual metric. Define $f, g : X \to X$ by

$$fx = 2x - 1 \text{ and } gx = x^2 \text{ for all } x \in X.$$

Then $d(fgx, gfx) = 2d(fx, gx)$. Thus f and g are R-weakly commuting ($R = 2$) but are not weakly commuting.

Definition 2.11 (Pathak *et al.* [61]) Two self mappings f and g of a metric space (X, d) are called

(i) R-weakly commuting of type (A_f) iff there exists some positive real number R such that $d(fgx, ggx) \le Rd(fx, gx)$ for all x in X.

(ii) R-weakly commuting of type (A_g) iff there exists some positive real number R such that $d(ffx, gfx) \leq Rd(fx, gx)$ for all x in X.

It may be noticed that definition (ii) can be obtained from definition (i) by interchanging the roles of f and g. Further, R-weakly commuting pair of self mappings are independent of R-weakly commuting of type (A_f) or type (A_g). Example 2.10 shows that $d(fgx, ggx) > Rd(fx, gx)$ for each $x > 1$ and some $R > 0$ (e.g., say $R = 4$). Thus f and g are R-weakly commuting but not R-weakly commuting of type (A_f).

The next example shows that f and g are R-weakly commuting of type (A_f) but not R-weakly commuting mappings.

Example 2.11 *[61]* Let $X = [-1, 2]$ be equipped with the usual metric d. Define self mappings f and g on X as follows:

$$f(x) = \begin{cases} 1 & \text{if } x \in [-1, 1], \\ \frac{3}{4} & \text{if } x \in (1, \frac{5}{4}), \\ 1 + \frac{1}{32}x^2 & \text{if } x \in [\frac{5}{4}, 2], \end{cases} \quad \text{and } g(x) = \begin{cases} 1 + \frac{1}{4}x^2 & \text{if } x \in [-1, 1), \\ 1 & \text{if } x = 1, \\ 2 & \text{if } x \in (1, \frac{5}{4}), \\ 1 - \frac{1}{8}x^2 & \text{if } x \in [\frac{5}{4}, 2]. \end{cases}$$

In this example f and g are an R-weakly commuting pair (A_f) for $R = 6$ but not R-weakly commuting mappings. Thus R-weakly commuting mappings and R-weakly commuting of (A_f) or (A_g) mappings are independent.

It may be noted that both compatible and noncompatible mappings can be R-weakly commuting of type (A_g) or (A_f).

Example 2.12 *[49]* Let $X = [0, \infty)$ be endowed with the usual metric. Define $f, g : X \to X$ by

$$fx = 2x - 1 \text{ and } gx = 3x - 2 \text{ for all } x \in X.$$

Then it can be verified in this example that f and g are compatible. Furthermore, f and g are R-weakly commuting of type (A_f) with $R = 3$ and R-weakly commuting of type (A_g) with $R = 2$.

Example 2.13 *[49]* Let $X = [2, 20]$ and d be the usual metric on X. Define $f, g : X \to X$ as follows:

$$f(x) = \begin{cases} 2 & \text{if } x = 2 \text{ or } x > 5, \\ 12 & \text{if } 2 < x \leq 5, \end{cases} \quad \text{and } g(x) = \begin{cases} 2 & \text{if } x = 2, \\ \frac{x+31}{3} & \text{if } 2 < x \leq 5, \\ \frac{x+1}{3} & \text{if } x > 5. \end{cases}$$

It may be observed that f and g are R weakly commuting of type (A_g) since $d(gfx, ggx) \leq d(fx, gx)$ for all $x \in X$. To see that f and g are noncompatible, let us consider a sequence $\{x_n\}$ given by $x_n = 5 + \frac{1}{n} : n > 0$. Then $fx_n \to 2$, $gx_n \to 2$, $fgx_n \to 12$, $gfx_n \to 2$, and $\lim_{n \to \infty} d(fgx_n, gfx_n) \neq 0$.

Definition 2.12 *[64]* Two self mappings f and g of a metric space (X, d) are called R-weakly commuting of type (P) iff there exists some positive real number R such that $d(ffx, ggx) \leq Rd(fx, gx)$ for all x in X.

The next example shows when two mappings representing parallel straight lines on the real plane shall be commuting, weakly commuting, R-weakly commuting or analogous definitions of R-weakly commuting mappings.

Example 2.14 *[45]* Let X = $\mathbb{R}$ be equipped with the usual metric d. Define self mappings f and g as follows:

$$f(x) = mx + a \text{ and } g(x) = mx + b,\ a \neq b.$$

Then $d(fgx, gfx) = d(ffx, ggx) = |m - 1|.|a - b|$; $d(ffx, gfx) = d(fgx, ggx) = |a - b|$ and $d(fx, gx) = |a - b|$. Thus f and g will be:

(i) commuting if $m = 1$,

(ii) weakly commuting if $0 \leq m \leq 2$,

(iii) R-weakly commuting or R-weakly commuting of type (P) if $|m-1| \geq 1$.

Remark 2.5 If f and g are R-weakly commuting or R-weakly commuting (A_f) or R-weakly commuting of type (A_g) or R-weakly commuting (P) and if z is their coincidence point, i.e., $fz = gz$, we get $ffz = fgz = gfz = ggz$. Thus at a coincidence point, all the analogous notions of R-weak commutativity including R-weak commutativity are equivalent and imply their commutativity.

Definition 2.13 (Jungck and Pathak [35]) Two self mappings f and g of a metric space (X, d) are called f-biased iff

$$\alpha \lim_{n\to\infty} d(fx_n, fgx_n) \leq \alpha \lim_{n\to\infty} d(gx_n, gfx_n),$$

where $\alpha \lim_n$ stays for $\limsup_n$ or $\liminf_n$, whenever $\{x_n\}$ is a sequence in X such that $\lim_{n\to\infty} fx_n = \lim_{n\to\infty} gx_n = t$ for some t in X.

Similarly, the definition of g-biased can be obtained from the definition of f-biased by interchanging the roles of f and g.

Jungck and Pathak [35] have shown that if f and g are compatible, they are both f-biased and g-biased, but the converse is not true.

Example 2.15 Let $X = [0, 1]$ and d be the usual metric on X. Define $f, g : X \to X$ by

$$fx = 1 - 2x,\ gx = 2x \text{ for } x \in [0, \tfrac{1}{2}],$$
$$fx = 0, gx = 1 \text{ for } x \in (\tfrac{1}{2}, 1].$$

Then f and g are both f-biased and g-biased but not compatible.

Definition 2.14 (Jungck and Pathak [35]) Two self mappings f and g of a metric space (X, d) are called weakly f-biased iff

$$fx = gx \text{ implies } d(fgx, fx) \leq d(gfx, gx).$$

Clearly, all biased mappings are weakly biased (see Proposition 1.1 in [35]) but the converse is false in general.

Sahu *et al.* [71] have shown that intimate mappings are more general than compatible mappings.

Definition 2.15 (Sahu *et al.* [71]) Two self mappings f and g of a metric space (X, d) are called f-intimate iff

$$\alpha \lim_{n\to\infty} d(fx_n, fgx_n) \leq \alpha \lim_{n\to\infty} d(gx_n, ggx_n),$$

where $\alpha \lim_{n\to\infty}$ stays for $\limsup_n$ or $\liminf_n$, whenever $\{x_n\}$ is a sequence in X such that $\lim_{n\to\infty} fx_n = \lim_{n\to\infty} gx_n = t$ for some t in X.

Example 2.16 [71] shows that intimate mappings need not be compatible.

Example 2.16 Let $X = [0, 1]$ and d be the usual metric on X. Define $f, g : X \to X$ as follows:

$$fx = \frac{2}{x+2} \text{ and } gx = \frac{1}{1+x} \text{ for } x \in [0, 1].$$

For this, consider the sequence $\{x_n\}$ given by $x_n = \frac{1}{n} : n > 0$. Then $fx_n \to 1$, $gx_n \to 1$ and $\lim_{n\to\infty} d(fgx_n, fx_n) < \lim_{n\to\infty} d(ggx_n, gx_n)$, i.e., f and g are f-intimate but $\lim_{n\to\infty} d(fgx_n, gfx_n) \neq 0$.

Definition 2.16 (Cho *et al.* [18]) Two self mappings f and g of a metric space (X, d) are called semi-compatible iff

(i) $ft = gt$ implies $fgt = gft$;

(ii) $\lim_n fx_n = \lim_n gx_n = t$ for some t in X implies $\lim_n d(fgx_n, gt) = 0$.

It may be noted that semi-compatible mappings need not be compatible..

Example 2.17 *[45]* Let $X = [2, 6]$ and d be the usual metric on X. Define $f, g : X \to X$ as follows:

$$f(x) = \begin{cases} 2 & \text{if } x < 3, \\ 4 & \text{if } x = 3, \\ \frac{x+21}{12} & \text{if } 3 < x \leq 6. \end{cases} \quad \text{and} \quad g(x) = \begin{cases} 2 & \text{if } x = 2, \\ 2x & \text{if } 2 < x \leq 3, \\ \frac{2x}{3} & \text{if } 3 < x \leq 6. \end{cases}$$

Then f and g are semi-compatible but noncompatible. To see this, let us consider a decreasing sequence $\{x_n\}$ given by $3 < x_n < 4$ and $\lim_n x_n = 3$. Then $fx_n \to 2$, $gx_n \to 2$, $fgx_n \to 2 = g2$, $gfx_n \to 4$, $\lim_{n\to\infty} d(fgx_n, gt) = 0$, $fg2 = gf2$ and $\lim_{n\to\infty} d(fgx_n, gfx_n) \neq 0$.

In 1997, Pathak *et al.* [56] weakened the compatibility of type (A) by splitting it into two parts, namely f-compatible and g-compatible.

Definition 2.17 Two self mappings f and g of a metric space (X, d) are called f-compatible (Pathak *et al.* [56]) iff

$$\lim_{n\to\infty} d(fgx_n, ggx_n) = 0,$$

whenever $\{x_n\}$ is a sequence in X such that $\lim_{n\to\infty} fx_n = \lim_{n\to\infty} gx_n = t$ for some t in X.

The definition of g-compatible can be obtained from the definition of f-compatible by interchanging the roles of f and g.

The following propositions have been proved [56].

Proposition 2.2 *[56] Let $f, g : (X, d) \to (X, d)$ be mappings with g continuous. Then f and g are compatible iff they are f-compatible.*

Proposition 2.3 *[56] Let f and g be continuous mappings from a metric space (X, d) into itself. Then the following are equivalent:*

(i) f and g are compatible,

(ii) f and g are f-compatible,

(iii) f and g are g-compatible.

Remark 2.6 If f and g are discontinuous mappings, the concepts compatible, f-compatible, g-compatible are independent.

Example 2.18 *[56]* Let $X = \mathbb{R}$ and d be the usual metric on X. Define $f, g : X \to X$ as follows:

$$fx = \tfrac{1}{x^3} \text{ if } x \neq 0,\ fx = 2 \text{ if } x = 0,$$
$$gx = \tfrac{1}{x^2} \text{ if } x \neq 0,\ gx = 1 \text{ if } x = 0.$$

Then f and g are compatible but neither f-compatible nor g-compatible. To see that, let us consider a sequence $\{x_n\}$ given by $x_n = n$. Then $fx_n \to 0$, $gx_n \to 0$, $\lim_{n\to\infty} d(fgx_n, gfx_n) = 0$ but $\lim_{n\to\infty} d(fgx_n, ggx_n) = \lim_{n\to\infty} d(ffx_n, gfx_n) = \infty$ as $n \to \infty$.

Example 2.19 *[56]* Let X $= [0, 1]$ be equipped with the usual metric d. Define self mappings f and g as follows:

$$f(x) = \begin{cases} \frac{1}{x^3} & \text{if } x \neq 0; \\ 2 & \text{if } x = 0. \end{cases} \qquad g(x) = \begin{cases} \frac{1}{x^2} & \text{if } x \neq 0; \\ 1 & \text{if } x = 0. \end{cases}$$

Then f and g are compatible and g-compatible but not f-compatible.

Example 2.20 *[56]* Let $X = [0, \infty)$ and d be the usual metric on X. Define $f, g : X \to X$ as follows:

$$fx = x \text{ if } 0 \leq x < 1, \; fx = 2 \text{ if } 1 \leq x < \infty,$$
$$gx = 2 - x \text{ if } 0 \leq x < 1, \; fx = 2 \text{ if } 1 \leq x < \infty.$$

Then f and g are both f-compatible and g-compatible but not compatible.

Definition 2.18 Two self mappings f and g of a metric space (X, d) are called:

(i) compatible of type (P) (Pathak *et al.* [62]) iff

$$\lim_{n\to\infty} d(ffx_n, ggx_n) = 0,$$

(ii) compatible of type (C) (Singh [77]) iff

$$\lim_{n\to\infty} d(ffx_n, ggx_n) = 0 \quad \text{and} \quad \lim_{n\to\infty} d(fgx_n, gfx_n) = 0,$$

whenever $\{x_n\}$ is a sequence in X such that $\lim_{n\to\infty} fx_n = \lim_{n\to\infty} gx_n = t$ for some t in X.

Proposition 2.4 *[2] Let $f, g : (X, d) \to (X, d)$ be mappings. If f and g are either compatible or compatible of type (A) or f-compatible, or g-compatible or compatible of type (P) or compatible of type (C) and $fz = gz$ for some $z \in X$, then $ffz = fgz = gfz = ggz$.*

Proof. Let $\{x_n\}$ be a sequence in X defined by $x_n = z, n \in \mathbb{N}$ and $fz = gz$ for some $z \in X$. Then we have $\lim_n fx_n = \lim_n gx_n = fz = gz$. Since f and g are either compatible or compatible of type (A) or f-compatible, or g-compatible or compatible of type (P) or compatible of type (C), we have

$$d(fgz, gfz) = \lim_{n\to\infty} d(fgx_n, gfx_n) = 0$$

or $d(ffz, gfz) = \lim_{n\to\infty} d(ffx_n, ggfx_n) = 0 = d(ggz, gfz) = \lim_{n\to\infty} d(ggx_n, gfx_n)$

$$\text{or} \quad d(ffz, gfz) = \lim_{n\to\infty} d(ffx_n, ggx_n) = 0.$$

Therefore, $ffz = fgz = gfz = ggz$.

Definition 2.19 Two self mappings f and g of a metric space (X, d) are called compatible mappings of type (f) (Pathak *et al.* [58]) iff

$$\lim_{n\to\infty} d(gfx_n, gx_n) + \lim_{n\to\infty} d(fgx_n, gfx_n) = \lim_{n\to\infty} d(fgx_n, fx_n),$$

whenever $\{x_n\}$ is a sequence in X such that $\lim_{n\to\infty} fx_n = \lim_{n\to\infty} gx_n = t$ for some t in X.

In 1996, Jungck generalized the notion of compatible mappings.

Definition 2.20 (Jungck [33], Shastry *et al.* [72] and Dhage[22], Shrivastava *et al.* [74]) Two self mappings f and g of a metric space (X, d) are called weakly compatible (partially commuting or coincidentally commuting, compatible type (N)) iff f and g commute on the set of coincidence points.

In 1998, Pant investigated existence of common fixed points for noncompatible mappings and pointwise R-weak commutativity, which he defined in 1994 [44] without giving the concept a name.

Definition 2.21 (Pant [43]) Two self mappings f and g are called pointwise R-weakly commuting on X iff given x in X there exists $R > 0$ such that $d(fgx, gfx) \leq Rd(fx, gx)$.

It is obvious from the definition that f and g can fail to be pointwise R-weakly commuting only if there exists some x in X such that $fx = gx$ while $fgx \neq gfx$, i.e., only if they possess coincidence points at which they do not commute.

Remark 2.7 *[45]* Compatible mappings are pointwise R-weakly commuting. To see this, let $fx = gx$. Consider the constant sequence $\{x_n : x_n = x\}$. Then $\lim_n fx_n = fx = gx = \lim_n gx_n = t$. Compatibility of f and g implies that $\lim_n d(fgx_n, gfx_n) = 0$, i.e., $d(gfx, fgx) = 0$ or $fgx = gfx$. However, pointwise R-weakly commuting mappings need not be compatible (see Example 2.21).

Remark 2.8 *[45]* Pointwise R-weak commutativity is necessary for the existence of common fixed points of contractive type mapping pairs. Suppose f and g constitute a contractive pair of self mappings of a metric space (X, d) having a common fixed point, say z; then $z = fz = gz$ and $fgz = gfz = fz = gz = z$. Suppose that f and g are not pointwise R-weakly commuting. Then there exists a point w in X such that $fw = gw$ while $fgw \neq gfw$. We thus have $fw = gw$ and $fz = gz$ with $fw \neq fz$. This is not possible in view of contractive conditions. For example, if f and g satisfy the contractive condition

$$d(fx, fy) < \max\{d(gx, gy), d(fx, gx), d(fy, gy), d(fx, gy), d(fy, gx)\},$$

which is one of the most general contractive conditions, we get

$$\begin{aligned} d(fw, fz) &< \max\{d(gw, gz), d(fw, gw), d(fz, gz), d(fw, gz), d(fz, gw)\} \\ &= d(fw, fz), \end{aligned}$$

a contradiction. This shows that a common fixed point satisying contractive conditions implies pointwise R-weakly commuting.

(It is also well known that pointwise R-weak commutativity is equivalent to commutativity at coincidence points and in the setting of metric spaces this notion is equivalent to weak compatibility.)

If f and g are compatible or f-compatible or g-compatible or compatible of type (A), they are obviously weakly compatible but as shown in the following example the converse is not true.

Example 2.21 *[2]* Let X $=\mathbb{R}$ be equipped with the usual metric d. Define self mappings f and g as

$$f\,(x) = [x] \quad \text{and} \quad g(x) = \begin{cases} -1 & \text{if } x \leq 0, \\ 0 & \text{if } 0 < x < 2, \\ 1 & \text{if } x \geq 2, \end{cases}$$

where [x] denotes the integral part of x. Then for the sequence $\{x_n\} = \frac{1}{n}$ we get $\lim_{n\to\infty} fx_n = \lim_{n\to\infty} gx_n = 0$ and $\lim_{n\to\infty} d(fgx_n, gfx_n) \neq 0$, $\lim_{n\to\infty} d(ffx_n, gfx_n) \neq 0$ and $\lim_{n\to\infty} d(gfx_n, ggx_n) \neq 0$. Therefore f and g are neither compatible nor f-compatible nor g-compatible nor compatible of type (A), type (P) nor type (C) but are weakly compatible as they commute at their coincidence points $x = -1, 1$.

In order to obtain new common fixed point theorems, one should be careful to use nontrivial noncommuting conditions. For example see the following result.

Corollary 2.1 *[19] Let f and g be weakly compatible self mappings of a complete metric space (X, d) satisfying*

(i) $fX \subseteq gX$,

(ii) $d(fx, fy) \leq kd(gx, gy)$, for all $x, y \in X$ and some $0 \leq k < 1$.

Then f and g have a unique common fixed point in X.

In some cases the condition of completeness mentioned in the above corollary may be replaced by (E.A) property along with some condition on the ranges of the involved mappings [1, 72].

It may be observed that the conclusion of above corollary is not true. This can be seen from the following counter example [49].

Example 2.22 *[49]* Let $X = [2, 20]$ and d be the usual metric on X. Define self mappings f and g on X as follows:

$$f(x) = \begin{cases} 6 & \text{if } 2 \leq x \leq 5, \\ \dfrac{x+5}{5} & \text{if } x > 5, \end{cases} \qquad g(x) = \begin{cases} 12 & \text{if } 2 \leq x \leq 5, \\ \dfrac{x+1}{3} & \text{if } x > 5. \end{cases}$$

Then f and g satisfy the following conditions of Corollary 2.1, but do not have a common fixed point.

(i) $fX = (2, 5] \bigcup \{6\}, gX = (2, 7] \bigcup \{12\}$ and $fX \subseteq gX$;

(ii) f and g satisfy a particular contraction condition $d(fx, fy) \leq \frac{3}{5} d(gx, gy)$;

(iii) f and g are trivially weakly compatible;

(iv) f and g are also tangential mappings. To see this let $\{x_n\}$ be the sequence in X given by $x_n = 5 + \epsilon_n$ where $\epsilon_n \to 0$ as $n \to \infty$. Then, $\lim_{n\to\infty} fx_n = \lim_{n\to\infty} gx_n = 2$.

Example 2.23 *[80]* Let $X = [0, 2]$ and $\mathbb{Q}$ be the set of rational numbers equipped with the usual metric d on X. Define self mappings f and g on X as follows:

$$f(x) = \begin{cases} \sqrt{3} & \text{if } x \in X \bigcap \mathbb{Q}, \\ x & \text{if } X \setminus \mathbb{Q}, \end{cases} \qquad g(x) = \frac{1}{2} fx.$$

Then f and g satisfy all the conditions of Corollary 2.1, but do not have a common fixed point.

One can see in the above examples that both the mappings f and g satisfy weak compatibility conditions vacuously, yet f and g are common fixed point free mappings. We can redefine weakly compatible mappings in the following way.

Definition 2.22 *[2]* Two self mappings f and g of a metric space (X, d) are called nontrivially weakly compatible if they commute on the set of coincidence points whenever the set of their coincidences is nonempty.

Definition 2.23 (Pathak *et al.* [59]) Two self mappings f and g of a metric space (X, d) are called compatible of type (I) iff

$$d(t, gt) \leq \limsup d(t, fgx_n)$$

whenever $\{x_n\}$ is a sequence in X such that $\lim_{n\to\infty} fx_n = \lim_{n\to\infty} gx_n = t$ for some t in X.

The following examples show that weakly compatible mappings and compatible mappings of type (I) are independent.

Example 2.24 *[59]* Let $X = [0, \infty)$ and d be the usual metric on X. Define $f, g : X \to X$ by

$$f(x) = \begin{cases} \cos x & \text{if } x \neq 1, \\ 0 & \text{if } x = 1, \end{cases} \qquad g(x) = \begin{cases} e^x & \text{if } x \neq 1, \\ 0 & \text{if } x = 1. \end{cases}$$

Then $fx = gx$ iff $x = 0$ and $x = 1$. Also at these points $fgx = gfx$. It means the mappings f and g are weakly compatible. It can also be noted that f and g not compatible of type (I). To see this, let $\{x_n\}$ be a sequence in X such that $fx_n \to t, gx_n \to t$ as $n \to \infty$. Then for $t = 1$, $d(t, gt) > d(t, fgx_n)$.

Example 2.25 *[59]* Let $X = [0, 1)$ be endowed with the usual metric. Define $f, g : X \to X$ by

$$fx = 2x + 1 \text{ and } gx = x^2 + 1 \text{ for all } x \in X.$$

Then at $x = 0$, $fx = gx$. But $fgx \neq gfx$, which shows that f and g are not weakly compatible but compatible of type (I). To see this, let $\{x_n\}$ be a sequence in X such that $fx_n \to t, gx_n \to t$. Then for $t = 1$, $d(t, gt) < d(t, fgx_n)$.

Definition 2.24 (Pathak *et al.* [65]) Two self mappings f and g of a metric space (X, d) are called g-biased of type (A) iff

$$\alpha \lim_{n\to\infty} d(gx_n, ggx_n) \leq \alpha \lim_{n\to\infty} d(fx_n, fgx_n),$$

where $\alpha \lim_n$ stays for $\limsup_n$ or $\liminf_n$. whenever $\{x_n\}$ is a sequence in X such that $\lim_{n\to\infty} fx_n = \lim_{n\to\infty} gx_n = t$ for some t in X.

Definition 2.25 (Pathak *et al.* [63]) Two self mappings f and g of a metric space (X, d) are said to be weakly compatible mappings of type (f) with index p at a point x in X iff $p > 0$ such that $fx = gx$ implies

$$d(fgx, fx)^p + d(fgx, gfx)^p \leq d(gfx, gx)^p.$$

Definition 2.26 (Fisher and Murthy [24]) Two self mappings f and g of a metric space (X, d) are called biased mappings of type (A_f) iff

$$fx = gx \text{ implies } d(gfx, fx) \leq d(ffx, gx) \text{ for some } x \text{ in } X.$$

Definition 2.27 (Pathak and Tiwari [67]) Let f and g be mappings from a metric space (X, d) into itself. The pair of mappings (f, g) is said to be ϕ-weakly compatible of type (f, g) at $x \in X$, iff for every $p > 0$, $fx = gx$ implies

$$\phi\left(\frac{d(fgx, fx)^{p+1} + d(fgx, gfx)^{p+1}}{d(fgx, fx) + d(fgx, gfx)}\right) \leq d(gfx, gx)^p,$$

where $\phi : [0, 1) \to [0, 1)$ is upper semicontinuous, nondecreasing and $\phi(t) < t$ for all $t > 0$, and $d(fgx, fx) + d(fgx, gfx) \neq 0$.

If $\phi(t) = ht$, where $0 < h < 1$, the pair of mappings (f, g) is said to be h-weakly compatible of type (f, g).

Example 2.26 *[67]* Consider $X = [0, 1]$ with the Euclidean metric d on X and $\phi(t) = \frac{1}{2}t$. Define $f, g : X \to X$ by

$$fx = 1 + x \text{ and } gx = 1 + 2x.$$

Here for $x = 0$ and $p > 0$, (f, g) is ϕ-weakly compatible of type (f, g) but (f, g) is not ϕ-weakly compatible of type (g, f) for $p > 1$. Moreover the pair (f, g) is neither weakly commuting nor weakly compatible.

In 2008, Gopal *et al.* [25] attempted to generalize Pant's [43] results by introducing a new notion of absorbing maps.

Definition 2.28 *[25]* Let f and g $(f \neq g)$ be two self mappings of a metric space (X, d). Then f is called g-absorbing if there exists some real number $R > 0$ such that $d(gx, gfx) \leq Rd(fx, gfx)$ for all x in X. Similarly, g is called f-absorbing if there exists some real number $R > 0$ such that $d(fx, fgx) \leq Rd(fx, gx)$ for all x in X.

The mapping f will be called pointwise g-absorbing if given x in X, $R > 0$ such that $d(gx, gfx) \leq Rd(fx, gx)$. On similar lines we can define pointwise f-absorbing mapping. If we take $g = Ix$ (identity mapping), f can be considered a trivially I-absorbing mapping.

The following examples show that absorbing mappings are neither subclasses of compatible mappings nor subclasses of noncompatible mappings.

Example 2.27 *[25]* Let $X = [0, 1]$ and d be the usual metric on X and define $f, g : X \to X$ by $fx = 1$ for $x \neq 1$, $f1 = 0$ and $gx = 1$ for all x in X. Then the mapping f is g-absorbing for any $R > 1$ but the pair of mappings (f, g) does not commute at their coincidence point $x = 0$.

Example 2.28 *[25]* Let $X = [0, 1]$ and d be the usual metric on X and define $f, g : X \to X$ by $fx = 1 - x$ and $gx = (1 - x)^2$ for all x in X. Then the pair of mappings (f, g) is R-weakly commuting and hence compatible but f is neither a g-absorbing nor f-absorbing mapping.

Example 2.29 *[25]* Let $X = [0, 1]$ and d be the usual metric on X and define $f, g : X \to X$ as

$$fx = \frac{x}{6} \text{ and } gx = 1 - \frac{x}{3} \text{ for all } x \text{ in } X.$$

In this example $f(X) \cap g(X) = \phi$ and hence the pair of mappings (f, g) are vacuously compatible. It can also be seen that for $R \geq 2$, f is g-absorbing and g is f-absorbing.

Al-Thagafi and Shahzad [6] introduced occasionally weakly compatible (OWC) mappings as generalizations of nontrivially weakly compatible mappings. Jungck and Rhoades [36] utilized the concept of OWC and proved several results under different contractive conditions. In view of the paper of Bisht and Pant [9], under contractive conditions, proving the existence of common fixed points by assuming OWC as defined in [36] is equivalent to proving the existence of common fixed points by assuming the existence of common fixed points.

Moreover, it was shown by Dorić *et al.* in [23] that in the presence of a unique point of coincidence, OWC reduces to weak compatibility. Hence, several generalizations obtained by using OWC and similar conditions are not real (see [4, 5, 8, 37]).

Definition 2.29 (Al-Thagafi and Shahzad[6]) Two self mappings f and g of a metric space (X, d) are said to be occasionally weakly compatible (OWC) if there exists at least one coincidence point at which f and g commute, i.e., if $fx = gx$ for some $x \in X$, then $fgx = gfx$.

Example 2.30 Let $X = [10, 20]$ be equipped with the usual metric d on X. Define mappings f and $g : X \to X$ by

$$f10 = 10,\ fx = \frac{(80-x)}{4} \text{ if } 10 < x \leq 20,$$
$$g10 = 10,\ gx = \frac{(x+55)}{5} \text{ if } 10 < x \leq 20.$$

Then f and g have a common fixed point at $x = 10$. Also, f and g have a coincidence point $x = 20$ at which they do not commute.

In [7] Al-Thagafi and Shahzad have shown that if (X, d) is a discrete metric space and $C(f, g) \neq \emptyset$, then f and g are weakly compatible iff they are weakly commuting. Moreover, in order to use occasional weak compatibility, it is necessary to assume that X is nondiscrete and that $C(f, g) \neq \emptyset$.

Remark 2.9 Besides commutativity of the mappings, OWC requires mappings to have coincidence points and therefore, imposes a very strong condition on the mappings. By assuming the existence of a coincidence point, the OWC circumvents the most crucial parts of fixed point theorems consisting of constructive procedures yielding coincidence points. Conditions or constructive procedures yielding coincidence points are important parts of fixed point theorems and strong assumptions like OWC do not and should not obviate the need for constructive procedures.

Definition 2.30 (Chen and Li [16]) Two self mappings f and g of metric space (X, d) are said to be Banach operator pairs iff the set $F(g)$ is f-invariant, namely $f(F(g)) \subset F(g)$.

It is easy to check that the commuting pair (f, g) is a Banach operator pair but the converse is not true in general.

Example 2.31 Let $X = [1, \infty)$ and d be the usual metric on X. Define $f, g : X \to X$ by

$$fx = x^3 \quad \text{for all } x \quad \text{and} \quad gx = 3x - 2 \text{ for all } x.$$

Then $F(g) = \{1\}$. Here (f, g) is a Banach operator pair but f and g are not commuting.

Definition 2.31 (Pathak and Hussain [52]) Two self mappings f and g of a metric space (X, d) are said to be P-operators iff $x \in X$ such that

$$x \in C(f, g) \text{ and } d(x, fx) \leq \delta(C(f, g)),$$
$$\text{where } \delta(A) = \sup\{\max\{d(x, y), d(y, x)\} : x, y \in A\} \text{ for } A \subset X.$$

Pathak and Hussain [52] have shown by an example that occasionally weakly compatible mappings are P-operators. If the self mappings f and g of X are weakly compatible, then $g(C(f, g)) \subset C(f, g)$, and hence f and g are P-operators.

Definition 2.32 (Hussain *et al.* [28]) Two self mappings f and g of a metric space (X, d) are said to be JH-operators iff there is a point $w = fx = gx$ in $PC(f, g)$ such that $d(w, x) \leq \delta(PC(f, g))$.

Example 2.32 *[52]* Consider $X = [0, 1)$ with the Euclidean metric d on X. Define $f, g : X \to X$ by

$$fx = x^2 \quad \text{and} \quad gx = 2x \quad \text{for all } x \neq 0,\ f0 = g0 = 1.$$

Then $C(f, g) = \{0, 2\}$ and $PC(f, g) = \{1, 4\}$. Obviously f and g are P-operators and JH-operators but neither occasionally weakly compatible mappings nor weakly compatible mappings. Further note that $Ff = \{1\}$ and $g(1) = 2$ belongs $F(f)$ and shows that (g, f) is not a Banach operator pair.

Definition 2.33 (Bouhadjera and Djoudi [13], Hussain *et al.* [28]) Two self mappings f and g of a metric space (X, d) are said to be occasionally weakly f-biased iff $x \in X$ such that $fx = gx$ and $d(fgx, fx) \leq d(gfx, gx)$.

The definition of occasionally weakly g-biased can be obtained from the definition of occasionally weakly f-biased by interchanging the roles of f and g.

Bouhadjera and Djoudi [13] have shown that if f and g are occasionally weakly compatible mappings, f and g are both occasionally weakly f-biased and g-biased. Hence occasionally weakly compatible mappings constitute a subclass of occasionally weakly biased mappings.

It may be noted that nontrivially ($C(f, g) \neq \phi$) weakly f-biased and g-biased mappings, are occasionally weakly f-biased and g-biased, respectively. However, the reverse implications are not true.

Example 2.33 *[13]* Let $X = [0, \infty)$ and d be the usual metric on X. Define $f, g : X \to X$ as follows

$$f(x) = \begin{cases} 4x^2 & \text{if } 0 \leq x \leq 1, \\ \dfrac{9}{x} & \text{if } 1 < x < \infty, \end{cases} \qquad g(x) = \begin{cases} 1 & \text{if } 0 \leq x \leq 1, \\ 3x & \text{if } 1 < x < \infty. \end{cases}$$

Then we have $fx = gx$ if and only if $x = 1/2$ or $x = 3$ and $d(fg(\sqrt{3}), f(\sqrt{3})) \leq d(gf(\sqrt{3}), g(\sqrt{3}))$, that is, the pair $\{f, g\}$ is occasionally weakly f-biased. But $3 = d(fg(\frac{1}{2}), f(\frac{1}{2})) \nleq 0 = d(gf(\frac{1}{2}), g(\frac{1}{2}))$, i.e., the pair $\{f, g\}$ is not weakly f-biased. Similarly we can show that g-occasionally weakly biased mappings may not be g-weakly biased mappings.

Definition 2.34 (Hussain *et al.* [28]) Let $d : X \times X \to [0, \infty)$ be a mapping such that $d(x, y) = 0$ iff $x = y$. Two self mappings f and g on X are said to be JH-operators iff $w = fx = gx$ in $PC(f, g)$ such that $d(w, x) \leq \delta(PC(f, g))$ and $d(x, w) \leq \delta(PC(f, g))$ where $\delta(A) = \sup\{\max\{d(x, y), d(y, x)\} : x, y \in A\}$ for$A \subset X$.

Definition 2.35 (Sintunavarat and Kumam [79]) Let $d : X \times X \to [0,\infty)$ be a mapping such that $d(x,y) = 0$ iff $x = y$. Two self mappings f and g on X are said to be generalized JH-operators with order n iff $w = fx = gx$ in $PC(f,g)$ such that $d(w,x) \leq (\delta(PC(f,g)))^n$ and $d(x,w) \leq (\delta(PC(f,g)))^n$ for some $n \in \mathbb{N}$.

It is clear from the definition that a JH-operator pair is a generalized JH-operator with order n, but the converse is not true in general.

Definition 2.36 (Pathak and Rai [68]) Let $d : X \times X \to [0,\infty)$ be a mapping such that $d(x,y) = 0$ iff $x = y$. Two self mappings f and g on X are said to be a PD-operator pair if $u \in X$ such that $u \in C(f,g)$ and $d(fgu, gfu) \leq diam(PC(f,g))$ where $\delta(A) = \sup\{\max\{d(x,y), d(y,x)\} : x, y \in A\}$ for $A \subset X$.

Definition 2.37 (Deepmala and Pathak [20]) Let $d : X \times X \to [0,\infty)$ be a mapping such that $d(x,y) = 0$ iff $x = y$. Two self mappings f and g on X are said to form a D-operator pair if there is a point $x \in X$ such that $x \in C(f,g)$ and

$$d(fgx, gfx) \leq R.diam(PC(f,g)), \quad \text{for some} \quad x \in C(f,g) \text{ and } R > 0.$$

Example 2.34 *[68]* Let $X = [0,1]$ and let $d = |x-y|^2$ be the metric. Define $f, g : X \to X$ by $fx = x^2$ and $gx = \frac{x}{2}$, for all $x \neq 0$ and $f0 = g0 = 1$. Since $C(f,g) = \{0, 1/2\}$ and $PC(f,g) = \{1, 1/4\}, diam(C(f,g)) = 1/4$ and $diam(PC(f,g)) = \frac{9}{16}$. Clearly, (f,g) is a PD-operator pair, but not commuting, not weakly compatible and not occasionally weakly compatible.

Definition 2.38 (Pant and Pant [47]) Two self mappings f and g of a metric space (X,d) are called conditionally commuting if they commute on a nonempty subset of the set of coincidence points whenever the set of their coincidences is nonempty.

Example 2.35 *[47]* Let $X = [0,1]$ and d be the usual metric on X. Define self mappings f and g on X as follows:

$$f(x) = \left(\frac{\sqrt{5 - 4(2x-1)^2} - 1}{4}\right), \qquad g(x) = \frac{1-x}{3}.$$

Then $C(f,g) = \{1, \frac{1}{4}\}$ and $fg1 \neq gf1$. Hence f and g are conditionally commuting but not weakly compatible. Further f and g are noncompatible mappings. To see this let us consider a sequence $\{x_n\}$ given by $x_n = 1 - \frac{1}{n}$. Then $fx_n \to 0$, $gx_n \to 0$, $fgx_n \to 0$, $gfx_n \to \frac{1}{3}$ and $\lim_{n\to\infty} d(fgx_n, gfx_n) \neq 0$.

Remark 2.10 It is to be noted that the notions of weak compatibilty, OWC and conditional commutativity are purely set theoretical and do not depend on the metric structure of (X,d). Remaining noncommuting notions are metrical and could vary if the metric of the space is varied (or some other structure of the space is applied) [37].

Remark 2.11 A generalized Lipschitz type pair (f, g) of self mappings of a metric space (X, d) may demonstrate any of the following conditions:

(i) f and g do not possess a coincidence point;

(ii) f and g possess coincidence points and commute at each coincidence point;

(iii) f and g possess multiple coincidence points and commute on a proper subset of the set of their coincidence points.

The notions of commuting, weak commuting, compatible, pointwise R-weak commuting, and weak compatible mappings apply in conditions (i) and (ii) but do not apply in condition (iii). The recently introduced notion of OWC mappings applies in conditions (ii) and (iii) but does not apply in condition (i) as it presupposes the existence of a coincidence point. The notion of conditional commutativity is applicable in each of the three conditions mentioned above.

In 2009, Bouhadjera and Godet-Thobie [11] introduced subcompatible mappings.

Definition 2.39 (Bouhadjera and Godet-Thobie [11]) Two self mappings f and g of a metric space (X, d) are called subcompatible iff there exists a sequence $\{x_n\}$ such that $\lim_{n\to\infty} fx_n = \lim_{n\to\infty} gx_n = t$ and $\lim_{n\to\infty} d(fgx_n, gfx_n) = 0$.

Example 2.36 *[11]* Let $X = [0, \infty)$ and d be the usual metric on X. Define $f, g : X \to X$ as follows:

$$fx = x^2 \text{ for all } x \in X,$$

$$g(x) = \begin{cases} x+2 & \text{if } 0 \le x \le 4 \text{ and } 9 < x < \infty, \\ x+12 & \text{if } 4 < x \le 9. \end{cases}$$

Then f and g are subcompatible but not OWC. To see this, consider the sequence $\{x_n\}$ given by $x_n = 2 + \frac{1}{n}$. Then $fx_n \to 4$, $gx_n \to 4$, $fgx_n \to 16$, $gfx_n \to 16$ and $\lim_{n\to\infty} d(fgx_n, gfx_n) = 0$. On the other hand, we have $fx = gx$ iff $x = 2$ and $fg2 \neq gf2$, hence mappings f and g are not OWC.

Remark 2.12 *[2]* The above example shows that subcompatible mappings need not imply commutativity at the coincidence point. It may be observed that subcompatible mappings are independent from the compatible mappings and in the setting of a unique common fixed point (or unique point of coincidence), subcompatibility does not reduce to the class of compatibility. The following examples illustrate these facts.

Example 2.37 Let $X = [0, \infty)$ and d be the usual metric on X. Define $f, g : X \to X$ by

$$fx = x \text{ for all } x \text{ and } gx = 3x \text{ for all } x.$$

Then f and g are compatible but not subcompatible.

Example 2.38 *[2]* Let $X = [1, 8]$ and d be the usual metric on X. Define $f, g : X \to X$ as follows

$$f(x) = \begin{cases} 2 & \text{if } x \leq 2, \\ 6 & \text{if } x > 2. \end{cases} \qquad g(x) = \begin{cases} 6 - 2x & \text{if } x \leq 2, \\ 8 & \text{if } x > 2. \end{cases}$$

It may be observed that f and g are subcompatible mappings. To see that f and g are subcompatible, let us consider the constant sequence $\{x_n\}$ given by $x_n = 2$. Then $fx_n \to 2$, $gx_n \to 2$, $fgx_n \to 2$, $gfx_n \to 2$, and $\lim\limits_{n\to\infty} d(fgx_n, gfx_n) = 0$. Further, if we consider $\{y_n\}$ given by $y_n = 2 - \frac{1}{n}$: $n > 0$, then $fy_n \to 2$, $gy_n \to 2$, and $\lim\limits_{n\to\infty} d(fgy_n, gfy_n) \neq 0$. Thus f and g are subcompatible but not compatible.

Remark 2.13 The notion of subcompatibility imposes a strong condition on the mappings f and g by requiring the existence of a sequence $\{x_n\}$ such that $fx_n \to t$, $gx_n \to t$. Such a precondition is not required in order that f and g be compatible. In [48] Pant and Bisht introduced conditional compatibility which does not require such a precondition and yet is a proper generalization of both nontrivial compatibility and subcompatibility.

Definition 2.40 (Pant and Bisht [48]) Two self mappings f and g of a metric space (X, d) are called conditionally compatible whenever the set of sequences $\{x_n\}$ satisfying $\lim\limits_{n\to\infty} fx_n = \lim\limits_{n\to\infty} gx_n$ is nonempty, there exists $\{y_n\}$ such that $\lim\limits_{n\to\infty} fy_n = \lim\limits_{n\to\infty} gy_n = t$ (say) and $\lim\limits_{n\to\infty} d(fgy_n, gfy_n) = 0$.

In [50], Pant and Bisht introduced a new notion of pseudo-compatible mappings, which is a stronger version of conditionally compatible mappings.

Let f, g be self mappings of a metric space (X, d). Then for a sequence $\{y_n\}$ in X satisfying $\lim\limits_{n\to\infty} fx_n = \lim\limits_{n\to\infty} gy_n$, a sequence $\{z_n\}$ is called an associated sequence if $fy_n = gz_n$ or $gy_n = fz_n$ and $\lim\limits_{n\to\infty} fz_n = \lim\limits_{n\to\infty} gz_n$.

Definition 2.41 (Pant and Bisht [50]) Two self mappings f and g of a metric space (X, d) are called pseudo-compatible iff whenever the set of sequences $\{x_n\}$ satisfying $\lim\limits_{n\to\infty} fx_n = \lim\limits_{n\to\infty} gx_n$ is nonempty, there exists a sequence $\{y_n\}$ such that $\lim\limits_{n\to\infty} fy_n = \lim\limits_{n\to\infty} gy_n = t$ (say), $\lim\limits_{n\to\infty} d(fgy_n, gfy_n) = 0$ and $\lim\limits_{n\to\infty} d(fgz_n, gfz_n) = 0$ for any associated sequence $\{z_n\}$ of $\{y_n\}$.

Suzuki and Pathak [80] also extended the class of compatible type mappings and several analogous notions to almost compatible mappings and analogous notions, respectively.

Definition 2.42 (Suzuki and Pathak [80]) Two self mappings f and g of a metric space (X,d) are called almost compatible iff f and g are compatible and the following hold for every sequence $\{x_n\}$ in X:

(C1) $fx_{n+1} = gx_n$ for $n \in \mathbb{N}$.

(C1) $\{fx_n\}$ converges.

(C1) $\{gfx_n\}$ is bounded.

Proposition 2.5 *[80] Let f and g be mappings from a metric space (X,d) into itself. Assume that $k \in [0,1)$ such that $d(fx,fy) \leq kd(gx,gy)$ for all $x,y \in X$. Then the following are equivalent:*

(i) f and g are almost g-biased,

(i) f and g are almost compatible,

(i) f and g are almost compatible of type (P),

(ii) f and g are almost f-compatible,

(iii) f and g are almost g-compatible.

Motivated by Pant and Bisht [49, 50], Patel *et al.* [51] introduced conditional sequentially absorbing as well as pseudo-reciprocal continuous mappings which allow us to comparatively study various types of commutative and continuity conditions.

Definition 2.43 *[51]* Two self mappings f and g of a metric space (X,d) are called conditionally sequential absorbing if whenever the set of sequences $\{x_n\}$ satisfying $\lim_{n\to\infty} fx_n = \lim_{n\to\infty} gx_n$ is nonempty, there exists a sequence $\{y_n\}$ satisfying $\lim_{n\to\infty} fy_n = \lim_{n\to\infty} gy_n = t$ (say) such that $\lim_{n\to\infty} d(fy_n, fgy_n) = 0$ and $\lim_{n\to\infty} d(gy_n, gfy_n) = 0$.

Definition 2.44 *[51]* Two self mappings f and g of a metric space (X,d) are called pseudo-reciprocal continuous (PRC) (with respect to conditionally sequential absorbing) if whenever the set of sequences $\{x_n\}$ satisfying $\lim_{n\to\infty} fx_n = \lim_{n\to\infty} gx_n$ is nonempty, there exists a sequence $\{y_n\}$ (satisfying $\lim_{n\to\infty} fy_n = \lim_{n\to\infty} gy_n = t$ (say), $\lim_{n\to\infty} d(fy_n, fgy_n) = 0$ and $\lim_{n\to\infty} d(gy_n, gfy_n) = 0$) such that $\lim_{n\to\infty} fgy_n = ft$ and $\lim_{n\to\infty} gfy_n = gt$.

To compare the applicability of the above notion over other known commuting and continuity conditions, we state the following theorem:

Theorem 2.2 *[70] Let (X,d) be a complete metric space, f and g two noncompatible self mappings on X satisfying*

$$d(fx,fy) \leq \varphi(d(gx,gy)) \quad \textit{for all} \quad x,y \in X,$$

where $\varphi : [0,\infty) \to [0,\infty)$ *is a continuous from right and nondecreasing function such that* $\varphi(t) < t$ *for all* $t > 0$. *Assume that*

(i) $\overline{f(X)} \subseteq g(X)$,

(ii) $\max\{d(ggx, fgx),\ d(ffx, gfx)\} \leq \varphi(d(fx, gx))$ *for all* $x \in X$,

(iii) $$\varphi(d(fx, f^2x)) \neq \varphi(\max\{d(gx, gfx), d(g^2x, gfx), d(fx, gx), d(f^2x, gfx), d(fx, gfx), d(gx, f^2x)\}),$$ *whenever* $fx \neq f^2x$.

Then, f *and* g *have a unique common fixed point. Also,* f *and* g *are discontinuous at the common fixed point.*

Theorem 2.3 *[51] Let* f *and* g *be two pseudo-reciprocal continuous (w.r.t. conditionally sequential absorbing) self mappings of a complete metric space* (X, d) *such that* $fX \subseteq gX$ *and let* $\phi, \psi : [0,\infty) \to [0,\infty)$ *be two functions satisfying:*

(a) ϕ *is nondecreasing, continuous and* $\phi(0) = 0 < \phi(t)$ *for every* $t > 0$;

(b) ψ *is nondecreasing, right-continuous, and* $\psi(t) < t$ *for every* $t > 0$.

If for all $x, y \in X$,

$$\phi(d(fx, fy)) \leq \psi(\phi(M(x, y))),$$

where

$$M(x, y) = \max\left\{d(gx, gy), d(fx, gx), d(fy, gy), \frac{d(gx, fy) + d(fx, gy)}{2}\right\},$$

then f *and* g *have a unique common fixed point provided* (f, g) *is conditionally sequential absorbing.*

Example 2.39 *[51]* Let $X = [0, 1]$ with the usual metric d. Define self mappings f and g as follows:

$$fx = \begin{cases} \dfrac{1+x}{2} & \text{if } x \in [0, \frac{1}{2}), \\ \dfrac{1}{2} & \text{if } x \in [\frac{1}{2}, 1], \end{cases} \qquad gx = \begin{cases} \dfrac{1}{2} + x & \text{if } x \in [0, \frac{1}{2}), \\ \dfrac{1}{2} & \text{if } x = \frac{1}{2}, \\ \dfrac{4}{5} & \text{if } x \in (\frac{1}{2}, 1]. \end{cases}$$

Then f and g satisfy all the conditions of Theorem 2.3 with $f(X) = [\frac{1}{2}, \frac{3}{4}) \subseteq [\frac{1}{2}, 1) = g(X)$. Here f and g are conditionally sequential absorbing and pseudo-reciprocal continuous (w.r.t. conditionally sequential absorbing) in respect of the constant sequence $x_n = \frac{1}{2}$. Let us consider the sequence $x_n = \frac{1}{n+2}$, then

$$\lim_{n\to\infty} fx_n = \lim_{n\to\infty}\left(\frac{1}{2} + \frac{1}{2n+4}\right) = \frac{1}{2} = \lim_{n\to\infty}\left(\frac{1}{2} + \frac{1}{n+2}\right) = \lim_{n\to\infty} gx_n,$$

$$\lim_{n\to\infty} fgx_n = \lim_{n\to\infty} f\left(\frac{1}{2}+\frac{1}{n+2}\right) = \frac{1}{2} = f\left(\frac{1}{2}\right),$$

$$\lim_{n\to\infty} gfx_n = \lim_{n\to\infty} g\left(\frac{1}{2}+\frac{1}{2n+4}\right) = \frac{4}{5} \neq g\left(\frac{1}{2}\right),$$

$$\lim_{n\to\infty} ffx_n = \lim_{n\to\infty} f\left(\frac{1}{2}+\frac{1}{2n+4}\right) = \frac{1}{2} = f\left(\frac{1}{2}\right),$$

$$\lim_{n\to\infty} ggx_n = \lim_{n\to\infty} g\left(\frac{1}{2}+\frac{1}{n+2}\right) = \frac{4}{5} \neq g\left(\frac{1}{2}\right).$$

Thus, (f, g) is neither a reciprocal continuous nor g-reciprocal continuous pair. Also the pair (f, g) is neither compatible, f compatible nor g-compatible.

If $\phi(t) = t$ and $\psi(t) = kt$, $k \in [0, 1)$, it can be verified that f and g satisfy the contraction condition with $k = \frac{5}{6}$. Here $x = \frac{1}{2}$ is the unique common fixed point of f and g and which is also a point of discontinuity.

On the other hand, notice that at $x = 1$, f and g do not satisfy the condition

$$\max\{d(ggx, fgx), d(ffx, gfx)\} \leq \varphi(d(fx, gx))$$

used in Theorem 2.2.

In 2013, Bisht and Shahzad [10] redefined conditionally compatible mappings as faintly compatible mappings.

Definition 2.45 (Bisht and Shahzad [10]) Two self mappings f and g of a metric space (X, d) are called faintly compatible iff f and g are conditionally compatible and f and g commute at the subset of coincidence points whenever the set of coincidence points is nonempty.

Example 2.40 Let $X = [0, 1]$ and d be the usual metric on X. Define self mappings f and g on X as follows:

$$f(x) = \tfrac{1}{2} - \left|\tfrac{1}{2} - x\right|,$$
$$g(x) = \tfrac{2}{3}(1 - x).$$

Then f and g are faintly compatible but neither compatible nor weakly compatible.

Definition 2.46 (Moeini and Razani [41]) Two self mappings f and g of a metric space (X, d) are called JH-operator pairs of type (R) if $PC(f, g) \neq \phi$ and

$$\lim_{n\to\infty} d(x_n, t) \leq diam(PC(f, g))$$

for some sequence $\{x_n\}$ in X with $\lim_{n\to\infty} fx_n = \lim_{n\to\infty} gx_n = t \in X$.

In view of the following example we can observe that the class of JH-operator pairs is a proper subclass of JH-operator pairs of type (R).

Example 2.41 *[41]* Let $X = [0,1]$ and d be the usual metric on X. Define self mappings f and g on X as follows:

$$f(x) = \begin{cases} \frac{x}{2} & \text{if} 0 < x \leq 1, \\ 1 & \text{if} x = 0, \end{cases} \qquad g(x) = x^2 \text{ for all } x.$$

Then $C(f,g) = \{\frac{1}{2}\}$ and $PC(f,g) = \{\frac{1}{4}\}$ and $d(\frac{1}{4}, \frac{1}{2}) = \frac{1}{4} > diam(PC(f,g)) = 0$.

Thus, (f,g) is not a JH-operator pair. Now, we define a sequence $x_n = \frac{1}{n+1}$. We have

$$\lim_{n\to\infty} fx_n = \lim_{n\to\infty} gx_n = 0 \in X$$

and

$$\lim_{n\to\infty} d(x_n, t) = 0 \leq diam(PC(f,g)) = 0.$$

Therefore, the pair (f,g) is a JH-operator pair of type (R).

In 2013, Pant and Pant [46] introduced a new class of noncommuting mappings which are stronger than weak compatibility but unify R-weakly commuting and its analogous notions.

Definition 2.47 (Pant and Pant [46]) Two self mappings f and g of a metric space (X,d) are called:

(i) R-commuting provided there exists a positive real number R such that $d(fgx, ggx) \leq Rd(fx, gx), d(ffx, gfx) \leq Rd(fx, gx), d(fgx, gfx) \leq Rd(fx, gx)$ and $d(ffx, ggx) \leq Rd(fx, gx)$ for all $x \in X$.

(ii) Quasi-R-commuting provided there exists a positive real number R such that $d(fgx, ggx) \leq Rd(fx, gx)$ or $d(ffx, gfx) \leq Rd(fx, gx)$ or $d(fgx, gfx) \leq Rd(fx, gx)$ or $d(ffx, ggx) \leq Rd(fx, gx)$ for all $x \in X$.

They [46] also unified the notions of compatible mappings and analogous notions of compatible mappings by defining the notion of quasi-α-compatible mappings. Under contractive conditions, the quasi-α-compatible mappings are stronger than weakly compatible mappings, i.e., α-compatible mappings imply weak compatibility but the converse need not be true.

Definition 2.48 *[46]* Two self mappings f and g of a metric space (X,d) are called:

(i) α-compatible iff $\lim_{n\to\infty} d(fgx_n, ggx_n) = 0$, $\lim_{n\to\infty} d(ffx_n, gfx_n) = 0$, $\lim_{n\to\infty} d(fgx_n, gfx_n) = 0$, $\lim_{n\to\infty} d(ffx_n, ggx_n) = 0$, whenever $\{x_n\}$ is a sequence in X such that $\lim_{n\to\infty} fx_n = \lim_{n\to\infty} gx_n = t$ for some t in X.

(ii) quasi-α-compatible provided every sequence $\{x_n\}$ in X satisfying $\lim_{n\to\infty} fx_n = \lim_{n\to\infty} gx_n = t$ for some t in X splits into at most four subsequences such that any of these subsequences, say $\{x_{n_i}\}$, satisfies at least one of the four conditions $\lim_{n_i\to\infty} d(fgx_{n_i}, ggx_{n_i}) = 0$, $\lim_{n_i\to\infty} d(ffx_{n_i}, gfx_{n_i}) = 0$, $\lim_{n_i\to\infty} d(fgx_{n_i}, gfx_{n_i}) = 0$, $\lim_{n_i\to\infty} d(ffx_{n_i}, ggx_{n_i}) = 0$.

Pant and Pant [46] gave the following example to justify their definition of R-commuting mappings (α-compatible mappings). It is interesting that quasi-R-commuting (quasi-α-compatible) implies R-weakly commuting (α-compatible) but not the converse.

Example 2.42 *[49, 46]* Let $X = [0, \infty)$ be endowed with the usual metric. Define $f, g : X \to X$ by

$$fx = 2x - 1 \text{ and } gx = 3x - 2 \text{ for all } x \in X.$$

In this example f and g are R-commuting with $R = 5$. Also in this example $\lim_{n\to\infty} fx_n = \lim_{n\to\infty} gx_n = t$ for some sequence $\{x_n\}$ in X implies $t = 1, \{x_n\} \to 1$, $\lim_{n\to\infty} d(fgx_n, ggx_n) = 0$, $\lim_{n\to\infty} d(ffx_n, gfx_n) = 0$, $\lim_{n\to\infty} d(fgx_n, gfx_n) = 0$ and $\lim_{n\to\infty} d(ffx_n, ggx_n) = 0$. Hence f and g are α-compatible mappings.

Example 2.43 *[46]* Let $X = [0, 20]$ and d be the Euclidean metric. Define $f, g : X \to X$ by

$$f(x) = \begin{cases} 2 + \frac{(2-x)}{2} & \text{if } x \leq 2, \\ 6 & \text{if } 2 < x \leq 5, \\ 2 & \text{if } x > 5 \end{cases} \qquad g(x) = \begin{cases} 2 & \text{if } x \leq 2, \\ 12 & \text{if } 2 < x \leq 5, \\ 2 + \frac{(x-5)}{2} & \text{if } x > 5. \end{cases}$$

Then $d(fgx, ggx) \leq d(fx, gx)$ for each $x \leq 2$ and $d(ffx, gfx) \leq d(fx, gx)$ for each $x > 2$. We thus see that the mappings f and g are quasi-R-commuting with $R = 1$ but the mappings are neither R-weakly commuting nor analogous notions of R-weakly commuting. It may also be noted that the mappings f and g are quasi-α-compatible but are neither compatible nor f-compatible, g-compatible nor compatible of type (P). Also in this example f and g are weakly compatible mappings as they commute on $x = 2$.

Remark 2.14 Several authors claimed to introduce weaker noncommuting notions and described weak compatibility as a proper subclass of their weaker notions. This is, however, not true. In view of the results of Dorić *et al.* [23] (see also Alghamdi *et al.* [5], Kadelburg *et al.* [37]), Bisht and Rakocević [8]), most of the generalized commutativity notions fall in the subclass of weak compatibility in the setting of a unique common fixed point (or unique point of coincidence). Also, Pant and Bisht [48] (see also [9, 8]) have shown that under contractive conditions proving existence of common fixed points by assuming several weaker noncommuting notions is equivalent to proving the existence of common fixed points by assuming the existence of common fixed points.

The following lemma was proved by Abbas and Jungck [3].

Lemma 2.2 *[3] If a weakly compatible (WC) pair (f, g) of self mappings on X has a unique POC, then it has a unique common fixed point.*

Jungck and Rhoades [36] established the following lemma.

Lemma 2.3 *[36] If an OWC pair (f, g) of self mappings on X has a unique POC, then it has a unique common fixed point.*

The following result is due to Dorić *et al.* [23].

Proposition 2.6 *[23] Let a pair of mappings (f, g) have a unique POC. Then the pair is WC if and only if it is OWC.*

Proposition 2.7 *[5] Let $d : X \times X \to [0, \infty)$ be a mapping such that $d(x, y) = 0$ if and only if $x = y$. Let a pair of mappings (f, g) have a unique POC. If it is a pair of JH-operators, it is WC.*

Proposition 2.8 *[5] Let d be symmetric on X. Let a pair of mappings (f, g) have a unique POC which belongs to $F(f)$. If it is a pair of occasionally weakly g-biased mappings, it is WC.*

Proposition 2.9 *[5] Let d be symmetric on X, and let a pair of mappings (f, g) have a unique CP, that is, $C(f, g)$ is a singleton. If (f, g) is a P-operator pair, it is WC.*

Proposition 2.10 *[5] Let d be symmetric on X, and let a pair of mappings (f, g) have a unique POC. Then it is weakly g-biased if and only if it is occasionally weakly g-biased.*

Proposition 2.11 *[8] Let $d : X \times X \to [0, \infty)$ be a mapping such that $d(x, y) = 0$ if and only if $x = y$. Let a pair of mappings (f, g) have a unique POC. If it is a pair of PD-operators or D-operators, it is WC.*

Let $\phi : \mathbb{R}^+ \to \mathbb{R}^+$ be a nondecreasing function satisfying the condition $\phi(t) < t$, for each $t > 0$.

Proposition 2.12 *[8] Let $d : X \times X \to [0, \infty)$ be a mapping such that $d(x, y) = 0$ if and only if $x = y$. Suppose (f, g) is a PD- or D-operator pair and satisfy the condition:*

$$d(fx, fy) \leq \phi(\max\{d(gx, gy), d(gx, fy), d(fx, gy), d(gy, fy)\}),$$

for each $x, y \in X$. Then f and g are WC.

Proposition 2.13 *[4] Let (X, d) be a metric space. Suppose f and g satisfy the following condition:*

$$\phi(d(fx, fy), d(gx, gy), d(fx, gx), d(fy, gy), d(gx, fy), d(fx, gy)) \leq 0,$$

for all $x, y \in X$, where $\phi \in \Phi$. If the pair (f, g) is subcompatible and reciprocally continuous, the pair (f, g) is WC.

Theorem 2.4 *[8] Under the contractive condition assumed in Proposition 2.12, the assumption of PD-operator or D-operator and the existence of a unique common fixed point are equivalent conditions.*

More recently, Kadelburg *et al.* [37] have shown that some noncommuting conditions which are formally distinct coincide when the given mappings have a unique point of coincidence. Hence, many new common fixed point results cannot be incurred in this way. The following proposition was proved in [37].

Proposition 2.14 *Let (X, d) be a metric space, and let $f, g : X \to X$. Let the pair (f, g) have exactly one point of coincidence. Then noncommuting conditions that are weakly compatible, occasionally weakly compatible, conditionally commuting and PD-operators are equivalent to each other and with the condition that the pair (f, g) has a unique common fixed point.*

Remark 2.15 It is important to note that most noncommuting conditions before weak compatibility do not presume the existence of a coincidence point but establish a coincidence point by using these conditions. In fact, some of these conditions, e.g., weakly commuting, imply commutativity at coincidence points and in combination with other sufficient conditions imply the existence of the coincidence point. Thus we can say that with the inception of occasionally weakly compatible mappings more focus would be given to noncommuting conditions which directly assume the existence of coincidence point, which is relatively strong in comparison to weak compatible mappings.

Remark 2.16 In [37] the authors showed that few noncommuting conditions which are formally distinct coincide when the given mappings have a unique point of coincidence. Hence, they declared that those new common fixed point results are not real generalizations. This assertion is not true in any respect. In view of new noncommuting conditions, fixed point theorems satisfying contractive mappings are studied under various independent sufficient conditions that ensure a unique point of coincidence. These noncommuting conditions are useful in establishing the existence of point of coincidence and imply commutativity at coincidence points. However, the minimal noncommuting notion, i.e., weak compatibility, may not ensure coincidence points.

2.3 Motivation and further scope

The existence of common fixed points of commuting continuous self mappings of a compact convex set has been a question of interest for a long time. In 1969, Boyce [14] and Huneke [27] independently proved two continuous commuting self mappings of the unit interval $[0, 1]$ without a common fixed point. Thus, in the search of common fixed points of a pair of self mappings on the unit interval $[0, 1]$, what additional conditions guarantee that f and g have a common fixed point? The earliest result addressing this problem is due to DeMarr [21] who in 1963 proved the following result:

For any $\alpha \geq 1$ and any $0 \leq \beta < \frac{\alpha+1}{\alpha-1}$, f and g have a common fixed point provided that $|f(x) - f(y)| \leq \alpha|x - y|$ and $|g(x) - g(y)| \leq \beta$ for all $x, y \in [0, 1]$.

The following result is proved by Jungck [30]:

If $|f(x) - f(y)| \leq |g(x) - g(y) + |x - y|$ for all $x, y \in [0, 1]$ such that $f(x) = g(x)$ or $f(y) = g(y)$, then f and g have a common fixed point.

In 1976, Jungck's remarkable result [31] promoted study of common fixed point theorems for commuting mappings. In recent years several authors (see Table 2.1) tried to weaken the notion of commuting mappings by several noncommuting mappings satisfying contractive or noncontractive conditions. Therefore, the study of common fixed point for a pair of self mappings satisfying contractive or noncontractive conditions becomes interesting and practical in view of the fact that even commuting continuous mappings on such well behaved entities as compact convex sets may fail to have a common fixed point.

TABLE 2.1: Weaker forms of commuting mappings

S. No.	Forms	Year	References
1	Weakly commuting mappings	1982	Sessa [73]
2	Asymptotically commutings mappings	1986	Tiwari and Singh [81]
3	Compatible mappings	1986	Jungck [32]
4	Weakly* commuting mappings	1986	Pathak [55]
5	Weakly uniformly contraction mappings	1990	Pathak [60]
6	Weakly** commuting mappings	1992	Pathak [57]
7	Compatible mappings of type (A)	1993	Jungck *et al.* [34]
8	Preorbitally commuting mappings	1994	Singh and Mishra [75]
9	R-weakly commuting mappings	1994	Pant [44]
10	Pointwise R-weakly commuting mappings	1994	Pant [44]
11	Biased and weakly g-biased mappings	1995	Jungck and Pathak [35]
12	Semicompatible mappings	1995	Cho *et al.* [18]
13	Compatible mappings of type (B)	1995	Pathak and Khan [53]
14	Compatible mappings of type (P)	1995	Pathak *et al.* [62]

TABLE 2.1: (*Continued*)

15.	Compatible mappings of type (f) or type (g)	1995	Pathak *et al.* [58]
16	Weakly compatible mappings	1996	Jungck [33]
17	f-compatible or g-compatible mappings	1997	Pathak and Khan [56]
18	R-weakly commuting mappings of type (A_f) or (A_g)	1997	Pathak *et al.* [61]
19	g-biased mappings of type (A)	1998	Pathak *et al.* [65]
20	Compatible mappings of type (C)	1998	Pathak *et al.* [54]
21	Biased mappings of type (A_f) or (A_g)	1999	Fisher and Murthy [24]
22	Coincidentally commuting mappings	1999	Dhage [22]
23	Partially commuting mappings	2000	Sastry *et al.* [72]
24	Compatible mappings of type (N)	2000	Srivastava *et al.* [74]
25	Intimate mappings	2001	Sahu *et al.* [71]
26	Weakly compatible mappings of type (f) or (g)	2007	Pathak and Ume [63]
27	Banach operator pairs	2007	Chen and Li [16]
28	Absorbing mappings	2008	Gopal *et al.* [25]
29	Occasionally weakly compatible mappings	2008	Al-Thagafi and Shahzad [6]
30	Conditionally commuting mappings	2009	Pant and Pant [47]
31	Subcompatible mappings	2009	Bouhadjera and Godet-Thobie [11]
32	P-operators	2010	Pathak and Hussain [52]
33	Almost compatible and almost biased mappings	2010	Suzuki and Pathak [17]
34	ϕ-weakly compatible of type (f, g)	2011	Pathak and Tiwari [67]
35	JH-operators	2011	Hussain *et al.* [28]
36	Generalized JH-operators with order n	2011	Sintunavarat and Kumam [79]
37	Conditionally compatible mappings	2012	Pant and Bisht [48]
38	Occasionally weakly biased pairs	2012	Bouhadjera and Djoudi [13]
39	Pseudo compatible mappings	2013	Pant and Bisht [50]
40	Sequentially absorbing map	2013	Patel *et al.* [51]
41	PD-operators	2013	Pathak and Rai [68]
42	D-operators	2013	Deepmala and Pathak [20]
43	Quasi-α compatible	2013	Pant and Pant [46]
44	Faintly compatible	2013	Bisht and Shahzad [10]
45	JH-operator pairs of type (R)	2015	Moein and Razani [41]

References

[1] Aamri M., Moutawakil D.E., Some new common fixed point theorems under strict contractive conditions. J. Math. Anal. Appl. 2002; 270:181-188.

[2] Agrawal R.P., Bisht R.K., Shahzad N., A comparison of various noncommuting conditions in metric fixed point theory and their applications. Fixed Point Theory Appl. 2014; Art. ID 38.

[3] Abbas M., and Jungck G., Common fixed point results for noncommuting mappings without continuity in cone metric spaces. J. Math. Anal. Appl. 2008; 341(1):416-420.

[4] Abbas M., Gopal D., Radenović S., A note on recently introduced commutative conditions, Indian *J. Math.* 2013; 55:195-202.

[5] Alghamdi M.A., Radenović S., Shahzad N., On some generalizations of commuting mappings. Abst. Appl. Anal. 2012; Art. ID 952052.

[6] Al-Thagafi M.A., Shahzad N., Generalized I-nonexpansive self maps and invariant approximations. Acta Math. Sin. 2008; 24:867-876.

[7] Al-Thagafi M.A., Shahzad N., A note on occasionally weakly compatible maps. Int. *J. Math.* Anal. 2009; 2, 55-58.

[8] Bisht R.K., Rakocević V., Some notes on PD-operator pairs. Math. Commun. 2013; 18, 441-445.

[9] Bisht R.K., Pant R.P., A critical remark on "Fixed point theorems for occasionally weakly compatible mappings". J. Egypt. Math. Soc. 2013; 21, 273-275.

[10] Bisht R.K., Shahzad N. Faintly compatible mappings and common fixed points. Fixed Point Theory Appl. 2013; 156, doi:10.1186/1687-1812-2013-156.

[11] Bouhadjera H., Godet-Thobie C., Common fixed point theorems for pair of subcompatible maps. June 2009, arXiv: 0906.3159v1 [math. FA].

[12] Bouhadjera H., Common fixed point theorems for compatible mappings of type (C). Sarajevo *J. Math.* 2005; 1(14):261-270.

[13] Bouhadjera H., Djoudi A, Fixed point for occasionally weakly biased maps. SE Asian Bull. 2012; 36:489-500.

[14] Boyce W.B., Commuting functions with common fixed point. Trans. Amer. Math. Soc. 1969; 137:77-92.

[15] Boyd D.W., Wong J. S., On nonlinear contractions. Proc. Amer. Math. Soc. 1969; 20:458-464.

[16] Chen J., Li Z., Common fixed points for Banach operator pairs in best approximation. J. Math. Anal. Appl. 2007; 336:1466-1475.

[17] Cho Y.J., Pathak H.K., Kang S.M., Jung J.S., Common fixed points of compatible maps of type (P) on fuzzy metric spaces. Fuzzy Sets Syst. 1998; 93(1):99-111.

[18] Cho Y.J., Sharma B.K., Sahu D.R., Semi-compatibility and fixed points. Math. Japon. 1995; 42:91-98.

[19] Chugh R., Kumar S., Common fixed points for weakly compatible maps. Proc. Indian Acad. Sci. (Math. Sci.). 2001; 111(2):241-247.

[20] Deepmala, Pathak H.K., Some common fixed point theorems for D-operator pair with applications to nonlinear integral equations. Nonlinear Funct. Anal. Appl. 2013; 18(2):205-218.

[21] DeMarr R., A common fixed point theorem for commuting mappings. Amer. Math. Monthly. 1963; 70:535-537.

[22] Dhage B.C., On common fixed point of coincidentally commuting mappings in D-metric space. Indian J. Pure Appl. Math. 1999; 30(4):395-406.

[23] Dorić D., Kadelburg Z., Radenović S., A note on occasionally weakly compatible and common fixed points. Fixed Point Theory. 2012; 13:475-480.

[24] Fisher B., Murthy P.P., Biased maps of type (A_f) and fixed points. J. Nat. Sci. Math. 1999; 39:33-42.

[25] Gopal D., Ranadive A.S., Common fixed points of absorbing maps. Bull. Marathwada Math. Soc. 2008; 9:43-48.

[26] Haghi R.H., Rezapour S., Shahzad N., Some fixed point generalizations are not real generalizations. Nonlinear Anal. 2011; 74:1799-1803.

[27] Huneke J.P., On common fixed points of commuting continuous functions on an interval. Trans. Amer. Math. Soc. 1969; 139:371-381.

[28] Hussain N., Khamsi M.A., Latif A., Common fixed points for JH-operators and occasionally weakly biased pairs under relaxed conditions. Nonlinear Anal. 2011; 74:2133-2140.

[29] Joshi M.C., Bose R.K., Some Topics in Nonlinear Functional Analysis. Wiley, 1985.

[30] Jungck G., Commuting mappings and common fixed points. Amer. Math. Monthly. 1966; 73:735-738.

[31] Jungck G., Commuting mappings and fixed points. Amer. Math. Monthly. 1976; 83(4):261-263.

[32] Jungck G., Compatible mappings and common fixed points. Intl. J. Math. Sci. 1986; 9(4):771-779.

[33] Jungck G., Common fixed points for noncontinuous nonself maps on non-metric spaces. Far East J. Math. Sci. 1996; 4:199-215.

[34] Jungck G., Murthy P.P., Cho Y.J., Compatible mappings of type (A) and common fixed points. Math. Japon. 1995; 38(2):381-390.

[35] Jungck G., Pathak H.K., Fixed points via biased maps. Proc. Amer. Math. Soc. 1995; 123(7):2049-2060.

[36] Jungck G., Rhoades B.E., Fixed point theorems for occasionally weakly compatible mappings. Fixed Point Theory. 2006; 7(2):287-296.

[37] Kadelburg Z., Radenović S., Shahzad N., A note on various classes of compatible-type pairs of mappings and common fixed point theorems. Abstract Appl. Anal. 2013; Art. 697151.

[38] Raul M., Mathematical Notes: A coincidence theorem. Amer. Math. Monthly. 1967; 74(5):469.

[39] Matkowski J., Integrable solutions of functional equations. Diss. Math. 1975; 127:1-68.

[40] Mcdowell E.L., Coincidence values of commuting functions. Topo. Proc. 2009; 34:365-384.

[41] Bahman M., Abdolrahman R., JH-operator pairs of type (R) with application to nonlinear integral equations. Vietnam J. Math. 2015; DOI 10.1007/s10013-015-0156-9.

[42] Pant R.P., Discontinuity and fixed points. J. Math. Anal. Appl. 1999; 240:284-289.

[43] Pant R.P., Common fixed points of Lipschitz type mapping pairs. J. Math. Anal. Appl. 1999; 240:280-283.

[44] Pant R.P., Common fixed points of noncommuting mappings. J. Math. Anal. Appl. 1994; 188:436-440.

[45] Pant R.P., A Study on Fixed Points of Noncommuting Mappings. D.Sc. thesis, Kumaun University, Nainital, India, 2000.

[46] Pant R.P., Pant A., Fixed point theorems under new commuting conditions. J. Intl. Acad. Phys. Sci. 2013; 17(2):107-112.

[47] Pant V., Pant R.P., Common fixed points of conditionally commuting maps. Fixed Point Theory. 2010; 1:113-118.

[48] Pant R.P., Bisht R. K., Occasionally weakly compatible mappings and fixed points. Bull. Belg. Math. Soc. 2012; 19:655-661.

[49] Pant R.P. and Bisht R.K., Common fixed point theorems under a new continuity condition. Ann. Univ. Ferrara. 2012; 58:127-141.

[50] Pant R.P.and Bisht R.K., Common fixed points of pseudo-compatible mappings. RACSAM. 2013; DOI 10.1007/s13398-013-0119-5.

[51] Patel D.K., Kumam P., Gopal D., Some discussion on the existence of common fixed points for a pair of maps. Fixed Point Theory Appl. 2013; 187: doi:10.1186/1687-1812-2013-187.

[52] Pathak H.K., Hussain N., Common fixed points for P-operator pair with applications. Appl. Math. Comput. 2010; 217:3137-3143.

[53] Pathak H.K., Khan M.S., Compatible mappings of type (B) and common fixed point theorems of Greguś type. Czech. Math. J. 1995; 45(4):685-698.

[54] Pathak H.K., Cho Y.J., Kang S.M., Madharia B., Compatible mappings of type (C) and common fixed point theorems of Greguś type. Demonstr. Math. 1998; 31(3):499-518.

[55] Pathak H.K., Weak* commuting mappings and fixed points. Indian J. Pure Appl. Math. 1986; 17(2):201-211.

[56] Pathak H.K., Khan M., A comparison of various types of compatible maps and common fixed points. Indian J. Pure Appl. Math. 1997; 28(4):477-485.

[57] Pathak H.K., Weak** commuting mappings and fixed points. J. Indian Acad. Math. 1992; 14(2):94-98.

[58] Pathak H.K., Kang S.M., Cho Y.J., Jung J.S., Gregus type common fixed point theorems for compatible mappings of type (T) and variational inequalities. Publ. Math. Debrecen. 1995; 46:285-299.

[59] Pathak H.K., Khan M.S., Tiwari R., A common fixed point theorem and its application to nonlinear integral equations. Comput. Math. Appl. 2007; 53:961-971.

[60] Pathak H.K., A Meir-Keeler type fixed point theorem for weakly uniformly contraction maps. Bull. Malaysian Math. Soc. 1990; 13(1):21-29.

[61] Pathak H.K., Cho Y.J., Kang S.M., Remarks on R-weakly commuting mappings and common fixed point theorems. Bull. Korean Math. Soc. 1997; 34(2):247-257.

[62] Pathak H.K., Cho Y.J., Kang S.M., Lee B.S., Fixed point theorems for compatible mappings of type (P) and applications to dynamic programming. Matematiche (Catania). 1995; 50(1):15-33.

[63] Pathak H.K., Ume J.S., Weakly compatible mappings and common fixed point theorems with applications to variational inequalities. Adv. Nonlinear Variational Inequalities. 2007; 10:55-68.

[64] Pathak H.K., Cho Y.J., Chang S.S., Kang S.M., Compatible mappings of type (P) and fixed point theorems in metric spaces and probabilistic metric spaces. Novi Sad J. Math. 1996; 26:87-109.

[65] Pathak H.K., Cho Y.J., Kang S.M., Common fixed points of biased maps of type (A) and applications. Int. J. Math. Sci. 1998; 21:681-693.

[66] Pathak H.K., Verma R.K., Weakly compatible mappings and Altman type contraction. Filomat. 2008; 22(1):33-46.

[67] Pathak H.K., Tiwari R., A Gregus type common fixed point theorem in normed spaces with application. Banach J. Math. Anal. 2011; 5(1):136-147.

[68] Pathak H.K., Rai D., Common fixed point theorems for PD-operator pairs under relaxed conditions with applications. J. Comput. Applied Math. 2013; 239:103-113.

[69] Pfeffer W., More on involution of a circle. Amer. Math. Monthly 1974; 81:613-616.

[70] Rezapour S.H., Shahzad N., Common fixed point of (ψ, ϕ)-type contractive maps. Appl. Math. Lett. 2011; doi:10.1016/j.aml.2011.11.004.

[71] Sahu D.R., Dhagat V., Srivastava M., Fixed points with intimate mappings (I). Bull. Calcutta Math. Soc. 2001; 93(2):107-114.

[72] Sastry K.P.R., Krishna Murthy I.S.R., Common fixed points of two partially commuting tangential self maps on a metric space. J. Math. Anal. and Appl. 2000; 250(2):731-734.

[73] Sessa S., On a weak commutativity condition of mappings in fixed point consideration. Publ. Inst. Math. 1982; 32:149-153.

[74] Shrivastava P.K., Bawa N.P.S., Singh P., Coincidence theorems for hybrid contraction II. Soochow J. Math. 2000; 26(4):411-421.

[75] Singh S.L., Mishra S.N., Coincidence points, hybrid fixed and stationary points of orbitally weakly dissipative maps. Math. Japon. 1994; 39(3):451-459.

[76] Singh S.L., Tomar A., Weaker forms of commuting maps and existence of fixed points. J. Korea. Soc. Math. Educ. Ser. B. 2003; 10(3):145-160.

[77] Singh Y.R., Studies on fixed points, common fixed points and coincidences. Ph.D. Thesis, Manipur University, Kanchipur, India; 2002.

[78] Sintunavarat W., Kumam P., Common fixed point theorems for a pair of weakly compatible mappings in fuzzy metric spaces. J. Appl. Math. 2011; Art. 637958.

[79] Sintunavarat W., Kumam P., Common fixed point theorems for generalized JH-operator classes and invariant approximations. J. Inequal. Appl. 2011; Art. 67.

[80] Suzuki T., Pathak H.K., Almost biased mappings and almost compatible mappings are equivalent under some condition. J. Math. Anal. Appl. 2010; 368:211-217.

[81] Tivari B.M.L., Singh S.L., A note on recent generalizations of Jungck contraction principle. J. Uttar Pradesh Gov. Coll. Acad. Soc. 1986; 3(1):13-18.

Chapter 3

α-Admissibility and Fixed Points

Deepesh Kumar Patel and Wutiphol Sintunavarat

3.1 Introduction

Let (X, d) be a complete metric space and $f : X \to X$ be a mapping. If there exists a real number $k \in [0, 1)$ such that for all $x, y \in X$, the following inequality holds:

$$d(fx, fy) \leq kd(x, y), \tag{3.1}$$

then f has a unique fixed point in X.

This famous result is called the Banach contraction mapping principle named for Polish mathematician Stefan Banach in 1922 (see [10]), and is one of the most important theorems in the field of fixed point theory. Due to its strong applications in proving the existence and uniqueness of solutions of integral equations, people have looked more closely at it for many years.

Notice that the contractive condition (3.1) is satisfied for all $x, y \in X$, which forces the mapping f to be continuous, and so the principle is not applicable where f is discontinuous. By virtue of this drawback, it is sometimes inconvenient to use this principle. Recently, several authors attempted to overcome this difficulty, see, e.g., Ran and Reurings [50], Kirk *et al.* [35], Jachymski [27] and Samet *et al.* [58]. The notion of α-admissible mapping is an interesting improvement of Banach contraction mapping that makes it more general whether a case is continuous or discontinuous. An extensive literature deals with fixed point problems via α-admissible mappings (see [14, 36, 38, 44, 53, 63, 64, 67, 68]) and issued often to generalize the results on various contractions.

3.2 Fixed point of α-ψ-contractive type mappings

Let Ψ be the family of functions $\psi : [0, \infty) \to [0, \infty)$ satisfying the following conditions:

(Ψ_1) ψ is nondecreasing,

(Ψ_2) $\sum_{n=1}^{+\infty} \psi^n(t) < \infty$ for each $t > 0$, where ψ^n is the n-th iterate of ψ.

These functions are known in the literature as (c)-comparison functions.

Lemma 3.1 *If ψ is a (c)-comparison function, for each $t > 0$, $\lim_{n\to+\infty} \psi^n(t) = 0$ implies $\psi(t) < t$.*

Definition 3.1 Let (X, d) be a metric space and $f : X \to X$ be a given mapping. We say that f is an α-ψ-contractive mapping if there exist two functions $\alpha : X \times X \to [0, +\infty)$ and $\psi \in \Psi$ such that

$$\alpha(x, y)d(fx, fy) \leq \psi(d(x, y)), \tag{3.2}$$

for all $x, y \in X$.

Remark 3.1 If $f : X \to X$ satisfies the Banach contraction principle, f is an α-ψ-contractive mapping, where $\alpha(x, y) = 1$ for all $x, y \in X$ and $\psi(t) = kt$ for all $t \geq 0$ and some $k \in [0, 1)$.

Definition 3.2 Let $f : X \to X$ and $\alpha : X \times X \to [0, +\infty)$. We say that f is α-admissible if $x, y \in X$, $\alpha(x, y) \geq 1 \Longrightarrow \alpha(fx, fy) \geq 1$.

Example 3.1 Let $X = (0, +\infty)$. Define $f : X \to X$ and $\alpha : X \times X \to [0, +\infty)$ by $fx = \ln x$ for all $x \in X$ and

$$\alpha(x, y) = \begin{cases} 2 & \text{if } x \geq y, \\ 0 & \text{if } x < y. \end{cases}$$

Then, f is α-admissible.

Example 3.2 Let $X = (0, +\infty)$. Define $f : X \to X$ and $\alpha : X \times X \to [0, +\infty)$ by $fx = \sqrt{x}$ for all $x \in X$ and

$$\alpha(x, y) = \begin{cases} e^{x-y} & \text{if } x \geq y, \\ 0 & \text{if } x < y. \end{cases}$$

Then, f is α-admissible.

Remark 3.2 In the setting of Examples 3.1 and 3.2 every nondecreasing self mapping f is α-admissible.

Theorem 3.1 *Let (X, d) be a complete metric space and $f : X \to X$ be an α-ψ-contractive mapping satisfying the following conditions:*

(i) f is α-admissible;

(ii) $x_0 \in X$ such that $\alpha(x_0, fx_0) \geq 1$;

(iii) f is continuous.

Then, f has a fixed point, that is, $x^* \in X$ such that $fx^* = x^*$.

Proof. Let $x_0 \in X$ such that $\alpha(x_0, fx_0) \geq 1$. Define the sequence $\{x_n\}$ in X by $x_{n+1} = fx_n$, for all $n \in \mathbb{N}$. If $x_n = x_{n+1}$ for some $n \in \mathbb{N}$, then $x^* = x_n$ is a fixed point for f. Assume that $x_n \neq x_{n+1}$ for all $n \in \mathbb{N}$. Since f is α-admissible, we have

$$\alpha(x_0, x_1) = \alpha(x_0, fx_0) \geq 1 \Rightarrow \alpha(fx_0, fx_1) = \alpha(x_1, x_2) \geq 1.$$

By induction, we get

$$\alpha(x_n, x_{n+1}) \geq 1, \text{for all } n \in \mathbb{N}. \tag{3.3}$$

Applying the inequality (3.2) with $x = x_{n-1}$ and $y = x_n$, and using (3.3), we obtain

$$d(x_n, x_{n+1}) = d(fx_{n-1}, fx_n) \leq \alpha(x_{n-1}, x_n) d(fx_{n-1}, fx_n) \leq \alpha(d(x_{n-1}, x_n)).$$

By induction, we get $d(x_n, x_{n+1}) \leq \psi^n(d(x_0, x_1))$, for all $n \in N$.

Fix $\epsilon > 0$ and let $n(\epsilon) \in \mathbb{N}$ such that $\sum_{n \geq n(\epsilon)} \psi^n(d(x_0, x_1)) < \epsilon$. Let $n, m \in \mathbb{N}$ with $m > n > n(\epsilon)$, using the triangular inequality, we obtain

$$\begin{aligned} d(x_n, x_m) &\leq \sum_{k=n}^{m-1} d(x_k, x_{k+1}) \\ &\leq \sum_{k=n}^{m-1} \psi^k(d(x_0, x_1)) \\ &\leq \sum_{n \geq n(\epsilon)} \psi^n(d(x_0, x_1)) < \epsilon. \end{aligned}$$

Thus we proved that $\{x_n\}$ is a Cauchy sequence in the metric space (X,d). Since (X,d) is complete, there exists $x^* \in X$ such that $x_n \to x^*$ as $n \to +\infty$. From the continuity of f, it follows that $x_{n+1} = fx_n \to fx^*$ as $n \to +\infty$. By the uniqueness of the limit, we get $x^* = fx^*$, that is, x^* is a fixed point of f.

In the next theorem, we omit the continuity hypothesis of f.

Theorem 3.2 *Let (X,d) be a complete metric space and $f : X \to X$ be an α-ψ-contractive mapping satisfying the following conditions:*

(i) f is α-admissible;

(ii) $x_0 \in X$ such that $\alpha(x_0, fx_0) \geq 1$;

(iii) if $\{x_n\}$ is a sequence in X such that $\alpha(x_n, x_{n+1}) \geq 1$ for all n and $x_n \to x \in X$ as $n \to +\infty$, then $\alpha(x_n, x) \geq 1$ for all n.

Then f has a fixed point.

Proof. Following the proof of Theorem 3.1, we know that $\{x_n\}$ is a Cauchy sequence in the complete metric space (X,d). Then $x^* \in X$ such that $x_n \to x^*$ as $n \to +\infty$. On the other hand, from (3.3) and the hypothesis (iii), we have

$$\alpha(x_n, x^*) \geq 1, \text{for all } \ n \in \mathbb{N}. \tag{3.4}$$

Now, using the triangular inequality, (3.2) and (3.4), we get

$$\begin{aligned} d(fx^*, x^*) &\leq d(fx^*, fx_n) + d(x_{n+1}, x^*) \\ &\leq \alpha(x_n, x^*) d(fx_n, fx^*) + d(x_{n+1}, x^*) \\ &\leq \psi(d(x_n, x^*)) + d(x_{n+1}, x^*). \end{aligned}$$

Letting $n \to +\infty$, since ψ is continuous at $t = 0$, we obtain $d(fx^*, x^*) = 0$, that is, $fx^* = x^*$.

Example 3.3 Let $X = \mathbb{R}$ be endowed with the standard metric $d(x,y) = |x - y|$ for all $x, y \in \mathbb{R}$. Define the mapping $f : X \to X$ by

$$fx = \begin{cases} 2x - \frac{3}{2} & \text{if } x > 1, \\ \frac{x}{2} & \text{if } 0 \leq x \leq 1, \\ 0 & \text{if } x < 0. \end{cases}$$

At first, we observe that the Banach contraction principle cannot be applied since we have $d(f1, f2) = 2 > 1 = d(2,1)$. Now we define the mapping $\alpha : X \times X \to [0, +\infty)$ by

$$\alpha(x,y) = \begin{cases} 1 & \text{if } x, y \in [0,1], \\ 0 & \text{otherwise.} \end{cases}$$

Clearly f is an α-ψ-contractive mapping with $\psi(t) = t/2$ for all $t \geq 0$. In fact, for all $x, y \in X$, we have

$$\alpha(x, y)d(fx, fy) \leq \frac{1}{2}d(x, y).$$

Moreover, there exists $x_0 \in X$ such that $\alpha(x_0, fx_0) \geq 1$. In fact, for $x_0 = 1$, we have

$$\alpha(1, f1) = \alpha(1, \frac{1}{2}) = 1.$$

Obviously f is continuous and so it remains to show that f is α-admissible. In so doing, let $x, y \in X$ such that $\alpha(x, y) \geq 1$. This implies that $x, y \in [0, 1]$ and by the definitions of f and α we have $fx = x/2 \in [0, 1]$, $fy = y/2 \in [0, 1]$, and $\alpha(fx, fy) = 1$. and f is α-admissible. Now, all the hypotheses of Theorem 3.1 are satisfied. Consequently, f has a fixed point.

Note that Theorem 3.1 and Theorem 3.2 guarantee only the existence of a fixed point but not the uniqueness. In this example, 0 and 3/2 are two fixed points of f.

Our next example involves a function f that is not continuous.

Example 3.4 Let $X = \mathbb{R}$ be endowed with the standard metric $d(x, y) = |x - y|$ for all $x, y \in \mathbb{R}$. Define the mapping $f : X \to X$ by

$$fx = \begin{cases} 2x - \frac{3}{2} & \text{if } x > 1, \\ \frac{x}{4} & \text{if } 0 \leq x \leq 1, \\ 0 & \text{if } x < 0. \end{cases}$$

It is clear that f is not continuous at 1. The Banach contraction principle and also Theorem 3.1 are not applicable. Define the mapping $\alpha : X \times X \to [0, +\infty)$ by

$$\alpha(x, y) = \begin{cases} 1 & \text{if } x, y \in [0, 1], \\ 0 & \text{otherwise.} \end{cases}$$

Clearly f is an α-ψ-contractive mapping with $\psi(t) = t/4$ for all $t \geq 0$. In fact, for all $x, y \in X$, we have

$$\alpha(x, y)d(fx, fy) \leq \frac{1}{4}d(x, y).$$

Moreover, $x_0 \in X$ such that $\alpha(x_0, fx_0) \geq 1$. In fact, for $x_0 = 1$, we have

$$\alpha(1, f1) = \alpha(1, \frac{1}{4}) = 1.$$

Now, let $x, y \in X$ such that $\alpha(x, y) \geq 1$.
This implies that $x, y \in [0, 1]$ and by the definitions of f and α we have

$$fx = x/4 \in [0, 1], \quad fy = y/4 \in [0, 1], \quad \text{and} \quad \alpha(fx, fy) = 1,$$

and f is α-admissible.

Finally, let $\{x_n\}$ be a sequence in X such that $\alpha(x_n, x_{n+1}) \geq 1$ for all n and $x_n \to x \in X$ as $n \to +\infty$. Since $\alpha(x_n, x_{n+1}) \geq 1$ for all n, by the definition of α, we have $x_n \in [0, 1]$ for all n and $x \in [0, 1]$. Then, $\alpha(x_n, x) = 1$.

Therefore, all the required hypotheses of Theorem 3.2 are satisfied, and so f has a fixed point. Here, 0 and 3/2 are two fixed points of f.

To assure the uniqueness of the fixed point, we will consider the following hypothesis. (H): For all $x, y \in X$, there exists $z \in X$ such that $\alpha(x, z) \geq 1$ and $\alpha(y, z) \geq 1$.

Theorem 3.3 *Adding condition (H) to the hypotheses of Theorem 3.1 (resp. Theorem 3.2) we obtain uniqueness of the fixed point of f.*

Proof. Suppose that x^* and y^* are two fixed points of f. From (H), there exists $z \in X$ such that

$$\alpha(x^*, z) \geq 1 \quad \text{and} \quad \alpha(y^*, z) \geq 1. \tag{3.5}$$

Since f is α-admissible, from (3.5), we get

$$\alpha(x^*, f^n z) \geq 1 \quad \text{and} \quad \alpha(y^*, f^n z) \geq 1, \quad \text{for all} \quad n \in \mathbb{N}. \tag{3.6}$$

Using (3.6) and (3.2), we have

$$\begin{aligned} d(x^*, f^n z) &= d(fx^*, f(f^{n-1}z)) \\ &\leq \alpha(x^*, f^{n-1}z) d(fx^*, f(f^{n-1}z)) \\ &\leq \psi(d(x^*, f^{n-1}z)). \end{aligned}$$

This implies that

$$d(x^*, f^n z) \leq \psi_n(d(x^*, z)) \quad \text{for all} \quad n \in \mathbb{N}.$$

Then, letting $n \to +\infty$, we have

$$f^n z \to x^*. \tag{3.7}$$

Similarly, using (3.6) and (3.2), we get

$$f^n z \to y^* \quad \text{as} \quad n \to +\infty. \tag{3.8}$$

Using (3.7) and (3.8), the uniqueness of the limit gives us $x^* = y^*$. This finishes the proof.

Now, we will show that many existing results in the literature can be easily deduced from the above theorem of Samet *et al.* [58].

Theorem 3.4 (Banach [10]) *Let (X, d) be a complete metric space and $f : X \to X$ be a given mapping satisfying $d(fx, fy) \leq kd(x, y)$ for all $x, y \in X$, where $k \in [0, 1)$. Then f has a unique fixed point.*

Proof. Let $\alpha : X \times X \to [0, +\infty)$ be the mapping defined by $\alpha(x, y) = 1$ for all $x, y \in X$ and $\psi : [0, +\infty) \to [0, +\infty)$ defined by $\psi(t) = kt$. Then f is an α-ψ-contractive mapping. It is easy to show that all the hypotheses of Theorems 3.1 and 3.3 are satisfied. Consequently, f has a unique fixed point.

Remark 3.3 As noted in Example 3.3, the Banach contraction principle cannot be applied since $d(f1, f2) > d(2, 1)$. However, using our Theorem 3.1, we obtain the fixed point of f.

Many fixed point results in ordered metric spaces can be deduced easily from Theorems 3.1 through 3.3.

Theorem 3.5 (Ran and Reurings [50]) *Let $(X, \preceq)$ be a partially ordered set and assume a metric d in X such that the metric space (X, d) is complete. Let $f : X \to X$ be a continuous and nondecreasing mapping w.r.t. $\preceq$. Suppose that the following two assertions hold:*

(i) There exists $k \in [0, 1)$ such that $d(fx, fy) \leq k(d(x, y))$ for all $x, y \in X$ with $x \preceq y$;

(ii) $x_0 \in X$ such that $x_0 \preceq fx_0$;

(iii) f is continuous.

Then f has a fixed point.

Proof. Define the mapping $\alpha : X \times X \to [0, +\infty)$ by

$$\alpha(x, y) = \begin{cases} 1 & \text{if } x \preceq y, \\ 0 & \text{otherwise.} \end{cases}$$

From (i), we have $\alpha(x, y)d(fx, fy) \leq kd(x, y)$, for all $x, y \in X$. Then f is an α-ψ-contractive mapping with $\psi(t) = kt$ for all $t \geq 0$. Now, let $x, y \in X$ such that $\alpha(x, y) \geq 1$. By the definition of α, this implies that $x \preceq y$. Since f is a nondecreasing mapping w.r.t. $\preceq$, we have $fx \preceq fy$, which gives us that $\alpha(fx, fy) = 1$. Then f is α-admissible. From (ii), there exists $x_0 \in X$ such that $x_0 \preceq fx_0$. This implies that $\alpha(x_0, fx_0) = 1$. Therefore, all the hypotheses of Theorem 3.1 are satisfied, and f has a fixed point.

Remark 3.4 A generalization of Theorem 3.5 can be given by replacing condition (i) with the following:

(i′) There exists $\psi \in \Psi$ such that $d(fx, fy) \leq \psi(d(x, y))$ for all $x, y \in X$ with $x \preceq y$.

Example 3.5 Let $X = [0, +\infty)$ be endowed with the standard metric $d(x, y) = |x - y|$ for all $x, y \in X$, and $f : X \to X$ be the mapping defined by

$$fx = 2x \quad \text{for all } x \in X.$$

Clearly f is a continuous mapping. We endow X with the usual order $\leq$ in $\mathbb{R}$. Condition (i) of Theorem 3.5 is not satisfied for $x = 1 \leq 3 = y$. Then, we cannot apply Theorem 3.5 to prove the existence of a fixed point of f. Define $\alpha : X \times X \to [0, +\infty)$ by

$$\alpha(x,y) = \begin{cases} \frac{1}{4} & \text{if } (x,y) \neq (0,0), \\ 1 & \text{if } (x,y) = (0,0). \end{cases}$$

It is clear that

$$\alpha(x,y)d(fx,fy) \leq \frac{1}{2}d(x,y) \quad \text{for all } x,y \in X.$$

Then f is an α-ψ-contractive mapping with $\psi(t) = t/2$ for all $t \geq 0$. Now let $x, y \in X$ such that $\alpha(x,y) \geq 1$. The definition of f implies that $x = y = 0$. Then we have $\alpha(fx, fy) = \alpha(0,0) = 1$, and so f is α-admissible. Also, for $x_0 = 0$, we have $\alpha(x_0, fx_0) = 1$. Consequently, all the hypotheses of Theorem 3.1 are satisfied, and we deduce the existence of a fixed point of f. Here 0 is a fixed point of f.

Theorem 3.6 (Nieto and Rodriguez-Lopez [43]) *Let $(X, \preceq)$ be a partially ordered set and suppose that there exists a metric d in X such that the metric space (X,d) is complete. Let $f : X \to X$ be a nondecreasing mapping w.r.t. $\preceq$. Suppose that the following assertions hold:*

(i) There exists $k \in [0,1)$ such that $d(fx,fy) \leq kd(x,y)$ for all $x, y \in X$ with $x \preceq y$;

(ii) $x_0 \in X$ such that $x_0 \preceq fx_0$;

(iii) if $\{x_n\}$ is a nondecreasing sequence in X such that $x_n \to x \in X$ as $n \to +\infty$, then $x_n \preceq x$ for all n.

Then f has a fixed point.

Proof. Define the mapping $\alpha : X \times X \to [0, +\infty)$ by

$$\alpha(x,y) = \begin{cases} 1 & \text{if } x \preceq y, \\ 0 & \text{otherwise.} \end{cases}$$

We can show easily that f is α-ψ-contractive and α-admissible. Now, let $\{x_n\}$ be a sequence in X such that $\alpha(x_n, x_{n+1}) \geq 1$ for all n and $x_n \to x \in X$ as $n \to +\infty$. By the definition of α, we have $x_n \preceq x_{n+1}$ for all n. Assertion (iii) implies that $x_n \preceq x$ for all n, which gives us $\alpha(x_n, x) = 1$ for all n. Thus all the hypotheses of Theorem 3.2 are satisfied, and f has a fixed point.

Remark 3.5 Again, a generalization of Theorem 3.6 can be also given by replacing condition (i) with the following:

(i′) There exists $\psi \in \Psi$ such that $d(fx, fy) \leq \psi(d(x, y))$ for all $x, y \in X$ with $x \preceq y$.

Remark 3.6 In Example 3.5, Theorem 3.6 cannot be applied since condition (i) is not satisfied.

Remark 3.7 To establish the uniqueness of the fixed point, Ran and Reurings and Nieto and Rodriguez-Lopez [50, 43] considered the following hypothesis: for all $x, y \in X$, there exists $z \in X$ such that $x \preceq z$ and $y \preceq z$.

This result is a specific case of Theorem 3.3. To be precise, if $x, y \in X$, since $z \in X$ such that $x \preceq z$ and $y \preceq z$, we have $\alpha(x, z) \geq 1$ and $\alpha(y, z) \geq 1$. Here, we consider the same function α used in the previous proof. Then the hypothesis (H) of Theorem 3.3 is satisfied, and we deduce the uniqueness of the fixed point.

Remark 3.8 Theorems 3.5 and 3.6 are only extensions of the Banach contraction principle. However, Theorem 3.3 is an extension and generalization of the Banach contraction principle.

Further, we show that the coupled fixed point results in metric and/or ordered metric spaces can also be deduced easily from Theorems 3.1 through 3.3. At first we recall the following definitions and results on coupled fixed points. For more details about coupled fixed point theory, we refer to [2, 17, 19, 21, 32, 55, 56, 57].

Definition 3.3 (Bhaskar and Lakshmikantham [12]) Let $F : X \times X \to X$ be a given mapping. We say that $(x, y) \in X \times X$ is a coupled fixed point of F if

$$F(x, y) = x \quad \text{and} \quad F(y, x) = y.$$

Lemma 3.2 (A coupled fixed point is a fixed point) *Let $F : X \times X \to X$ be a given mapping. Define the mapping $T : X \times X \to X \times X$ by*

$$T(x, y) = (F(x, y), F(y, x)), \quad \text{for all } (x, y) \in X \times X. \tag{3.9}$$

Then (x, y) is a coupled fixed point of F if and only if (x, y) is a fixed point of T.

Definition 3.4 (Bhaskar and Lakshmikantham [12]) Let $(X, \preceq)$ be a partially ordered set and $F : X \times X \to X$ be a given mapping. We say that F has the mixed monotone property if for all $x, y \in X$, we have

$$x_1, x_2 \in X, \quad x_1 \preceq x_2 \Rightarrow F(x_1, y) \preceq F(x_2, y)$$

and

$$y_1, y_2 \in X, \quad y_1 \preceq y_2 \Rightarrow F(x, y_2) \preceq F(x, y_1).$$

Theorem 3.7 *Let (X, d) be a complete metric space and $F : X \times X \to X$ be a given mapping. Suppose that there exist $\psi \in \Psi$ and a function $\alpha : X^2 \times X^2 \to [0, +\infty)$ such that*

$$\alpha((x, y), (u, v))d(F(x, y), F(u, v)) \leq \frac{1}{2}\psi(d(x, u) + d(y, v)), \tag{3.10}$$

for all $(x, y), (u, v) \in X \times X$. Suppose also that

(i) *for all $(x, y), (u, v) \in X \times X$, we have*

$$\alpha((x, y), (u, v)) \geq 1 \Rightarrow \alpha((F(x, y), F(y, x)), (F(u, v), F(v, u))) \geq 1,$$

(ii) *there exists $(x_0, y_0) \in X \times X$ such that*

$$\alpha((x_0, y_0), (F(x_0, y_0), F(y_0, x_0))) \geq 1 \quad \textit{and}$$
$$\alpha((F(y_0, x_0), F(x_0, y_0)), (y_0, x_0)) \geq 1,$$

(iii) *F is continuous.*

Then F has a coupled fixed point, that is, $(x^*, y^*) \in X \times X$ such that $x^* = F(x^*, y^*)$ and $y^* = F(y^*, x^*)$.

Proof. The concept involves transporting the problem to the complete metric space (Y, δ), where $Y = X \times X$ and $\delta((x, y), (u, v)) = d(x, u) + d(y, v)$ for all $(x, y), (u, v) \in X \times X$. From (3.10), we have

$$\alpha((x, y), (u, v))d(F(x, y), F(u, v)) \leq \frac{1}{2}\psi(\delta((x, y), (u, v))) \tag{3.11}$$

and

$$\alpha((v, u), (y, x))d(F(v, u), F(y, x)) \leq \frac{1}{2}\psi(\delta((x, y), (u, v))). \tag{3.12}$$

Adding (3.11) to (3.12), we get

$$\beta(\xi, \eta)\delta(T\xi, T\eta) \leq \psi(\delta(\xi, \eta)), \tag{3.13}$$

for all $\xi = (\xi_1, \xi_2), \eta = (\eta_1, \eta_2) \in Y$, where $\beta : Y \times Y \to [0, +\infty)$ is the function defined by

$$\beta((\xi_1, \xi_2), (\eta_1, \eta_2)) = \min\{\alpha((\xi_1, \xi_2), (\eta_1, \eta_2)), \alpha((\eta_1, \eta_2), (\xi_1, \xi_2))\}$$

and $T : Y \to Y$ is given by (3.9). Then T is a continuous and β-ψ-contractive mapping.

Let $\xi = (\xi_1, \xi_2), \eta = (\eta_1, \eta_2) \in Y$, such that $\beta(\xi, \eta) \leq 1$. Using condition (i), we obtain immediately that $\beta(T\xi, T\eta) \geq 1$. Then T is β-admissible. Moreover, from condition (ii), we know that $(x_0, y_0) \in Y$ such that $\beta((x_0, y_0), T(x_0, y_0)) \geq 1$. All the hypotheses of Theorem 3.1 are satisfied, and so we deduce the existence of a fixed point of T that gives us from Lemma 3.2 the existence of a coupled fixed point of F.

Now, we omit the continuity hypothesis of F.

Theorem 3.8 *Let (X,d) be a complete metric space and $F : X\times X \to X$ be a given mapping. Suppose that there exist $\psi \in \Psi$ and a function $\alpha : X^2 \times X^2 \to [0,+\infty)$ such that*

$$\alpha((x,y),(u,v))d(F(x,y),F(u,v)) \leq \frac{1}{2}\psi(d(x,u)+d(y,v)), \tag{3.14}$$

for all $(x,y),(u,v) \in X\times X$. Suppose also that

(i) for all $(x,y),(u,v) \in X\times X$, we have

$$\alpha((x,y),(u,v)) \geq 1 \Rightarrow \alpha((F(x,y),F(y,x)),(F(u,v),F(v,u))) \geq 1,$$

(ii) $(x_0,y_0) \in X\times X$ such that

$$\alpha((x_0,y_0),(F(x_0,y_0),F(y_0,x_0))) \geq 1 \quad \text{and}$$

$$\alpha((F(y_0,x_0),F(x_0,y_0)),(y_0,x_0)) \geq 1,$$

(iii) if $\{x_n\}$ and $\{y_n\}$ are sequences in X such that $\alpha((x_n,y_n),(x_{n+1},y_{n+1})) \geq 1$ and $\alpha((y_{n+1},x_{n+1}),(y_n,x_n)) \geq 1$, $x_n \to x \in X$ and $y_n \to y \in X$ as $n \to +\infty$, then $\alpha((x_n,y_n),(x,y)) \geq 1$ and $\alpha((y,x),(y_n,x_n)) \geq 1$ for all $n \in \mathbb{N}$.

F has a coupled fixed point.

Proof. We use the same notations from the proof of Theorem 3.7. Let $\{(x_n,y_n)\}$ be a sequence in Y such that $\beta((x_n,y_n),(x_{n+1},y_{n+1})) \geq 1$ and $(x_n,y_n) \to (x,y)$ as $n \to +\infty$. Using condition (iii), we obtain easily that $\beta((x_n,y_n),(x,y)) \geq 1$. Then all the hypotheses of Theorem 3.2 are satisfied. We deduce the existence of a fixed point of T that gives us from Lemma 3.2 the existence of a coupled fixed point of F.

For the uniqueness of the coupled fixed point, we consider the following hypothesis. (H′): For all $(x,y),(u,v) \in X\times X$, there exists $(z_1,z_2) \in X\times X$ such that

$$\alpha((x,y),(z_1,z_2)) \geq 1, \alpha((z_2,z_1),(y,x)) \geq 1$$

and

$$\alpha((u,v),(z_1,z_2)) \geq 1, \alpha((z_2,z_1),(v,u)) \geq 1.$$

Theorem 3.9 *Adding condition (H′) to the hypotheses of Theorem 3.7 (resp. Theorem 3.8) we obtain the uniqueness of the coupled fixed point of F.*

Proof. It is easy to show that under hypothesis (H′), T and β satisfy the hypothesis (H). Then, the result follows immediately from Theorem 3.3 and Lemma 3.2.

Example 3.6 Let $X = [0, +\infty)$ be endowed with the standard metric $d(x, y) = |x - y|$ for all $x, y \in X$. Then (X, d) is a complete metric space. Define the mapping $F : X \times X \to X$ by

$$F(x, y) = \begin{cases} \frac{x-y}{4} & \text{if } x \geq y, \\ 0 & \text{if } x < y. \end{cases}$$

Clearly F is a continuous mapping. Define $\alpha : X^2 \times X^2 \to [0, +\infty)$ by

$$\alpha((x, y), (u, v)) = \begin{cases} 1 & \text{if } x \geq y, u \geq v, \\ 0 & \text{otherwise.} \end{cases}$$

It is easy to show that for all $(x, y), (u, v) \in X \times X$, we have

$$\alpha((x, y), (u, v))d(F(x, y), F(u, v)) \leq \frac{1}{4}[d(x, u) + d(y, v)].$$

Then (3.10) is satisfied with $\psi(t) = t/2$ for all $t \geq 0$. On the other hand, condition (i) of Theorem 3.7 holds and condition (ii) of the same theorem is also satisfied with $(x_0, y_0) = (1, 1)$. All the required hypotheses of Theorem 3.7 are true and so we deduce the existence of a coupled fixed point of F. Here $(0, 0)$ is a coupled fixed point of F.

Theorem 3.10 (Bhaskar and Lakshmikantham [12]) *Let $(X, \preceq)$ be a partially ordered set and suppose there is a metric d on X such that (X, d) is complete metric space. Let $F : X \times X \to X$ be a continuous mapping having the mixed monotone property on X. Assume that there exists $k \in [0, 1)$ with*

$$d(F(x, y), F(u, v)) \leq \frac{k}{2}[d(x, u) + d(y, v)], \tag{3.15}$$

for all $x, y, u, v \in X$ such that $x \succeq u$ and $y \preceq v$. If there exists $(x_0, y_0) \in X \times X$ such that $x_0 \preceq F(x_0, y_0)$ and $y_0 \succeq F(y_0, x_0)$, then F has a coupled fixed point.

Proof. Define the mapping $\alpha : X^2 \times X^2 \to [0, +\infty)$ by

$$\alpha((x, y), (u, v)) = \begin{cases} 1 & \text{if } x \succeq u, \ y \preceq v, \\ 0 & \text{otherwise.} \end{cases} \tag{3.16}$$

It follows from (3.15) that

$$\alpha((x, y), (u, v))d(F(x, y), F(u, v)) \leq \frac{k}{2}[d(x, u) + d(y, v)],$$

for all $x, y, u, v \in X$. It implies that F is α-ψ-contractive mapping with $\psi(t) = kt$ for all $t \geq 0$. Let $(x, y), (u, v) \in X \times X$ such that $\alpha((x, y), (u, v)) \geq 1$. By the definition of α, this implies that $x \succeq u$ and $y \preceq v$. Since F satisfies the mixed monotone property, we have $F(x,y) \succeq F(u,v)$ and $F(y,x) \preceq F(v,y)$. Then we get $\alpha((F(x,y),F(y,x)),(F(u,v),F(v,u))) \geq 1$.

By hypothesis, we know that $(x_0, y_0) \in X \times X$ such that $x_0 \preceq F(x_0, y_0)$ and $y_0 \succeq F(y_0, x_0)$. Now, by the definition of α, this implies that

$$\alpha((a,b),(F(a,b),F(b,a))) \geq 1 \text{ and } \alpha((F(b,a),F(a,b)),(b,a)) \geq 1,$$

where $a = y_0$ and $b = x_0$. All the hypotheses of Theorem 3.7 are satisfied, then F has a coupled fixed point.

Theorem 3.11 (Bhaskar and Lakshmikantham [12]) *Let $(X, \preceq)$ be a partially ordered set and suppose there is a metric d on X such that (X,d) is a complete metric space. Let $F : X \times X \to X$ be a mapping having the mixed monotone property on X. Assume that $k \in [0,1)$ with*

$$d(F(x,y),F(u,v)) \leq \frac{k}{2}[d(x,u)+d(y,v)],$$

for all $x,y,u,v \in X$ such that $x \succeq u$ and $y \preceq v$. Assume that X has the following properties:

(i) if a nondecreasing sequence $\{x_n\} \to x \in X$, then $x_n \preceq x$ for all n;

(ii) if a nonincreasing sequence $\{y_n\} \to y \in X$, then $y_n \succeq y$ for all n.

If there exists $(x_0,y_0) \in X \times X$ such that $x_0 \preceq F(x_0,y_0)$ and $y_0 \succeq F(y_0,x_0)$, then F has a coupled fixed point.

Proof. We have only to check that condition (iii) of Theorem 3.8 holds with α given by (3.16). Let $\{a_n\}$ and $\{b_n\}$ two sequences in X such that $a_n \to a \in X$ and $b_n \to b \in X$ as $n \to +\infty$. Suppose that

$$\alpha((a_n,b_n),(a_{n+1},b_{n+1})) \geq 1 \text{ and } \alpha((b_{n+1},a_{n+1}),(b_n,a_n)) \geq 1.$$

By the definition of α, this implies that $\{a_n\}$ is a nonincreasing sequence and $\{b_n\}$ is a nondecreasing sequence. From hypotheses (i) and (ii), $a_n \succeq a$ and $b_n \preceq b$ for all n. This implies that

$$\alpha((a_n,b_n),(a,b)) = 1 \text{ and } \alpha((b,a),(b_n,a_n)) = 1.$$

Then condition (ii) of Theorem 3.8 holds and F has a coupled fixed point.

Remark 3.9 Clearly, a generalization of Theorem 3.10 (resp. Theorem 3.11) can also be given, assuming that there exist $\psi \in \Psi$ with $d(F(x,y),F(u,v)) \leq \frac{1}{2}\psi(d(x,u)+d(y,v))$ for all $x,y,u,v \in X$ such that $x \succeq u$ and $y \preceq v$.

Theorem 3.12 (Bhaskar and Lakshmikantham [12]) *In addition to the hypotheses of Theorem 3.10 (resp. Theorem 3.11) suppose that for all $(x,y),(u,v) \in X \times X$, there exists $(z_1,z_2) \in X \times X$ such that $x \succeq z_1$, $y \preceq z_2$ and $u \succeq z_1$, $v \preceq z_2$. Then F has a unique coupled fixed point.*

Proof. We have only to check hypothesis (H′) of Theorem 3.9. Let $(x,y),(u,v) \in X \times X$. By hypothesis, there exists $(z_1,z_2) \in X \times X$ such that $x \succeq z_1$, $y \preceq z_2$ and $u \succeq z_1$, $v \preceq z_2$. By the definition of α, we have

$$\alpha((x,y),(z_1,z_2)) = 1, \quad \alpha((z_2,z_1),(y,x)) = 1$$

and

$$\alpha((u,v),(z_1,z_2)) = 1, \quad \alpha((z_2,z_1),(v,u)) = 1.$$

Then condition (H′) of Theorem 3.9 is satisfied, and F has a unique coupled fixed point.

Example 3.7 Let $X = [0,+\infty)$ endowed with the usual order $\leq$ in $\mathbb{R}$ and the standard metric $d(x,y) = |x-y|$ for all $x,y \in X$. Define the continuous mapping $F : X \times X \to X$ by

$$F(x,y) = 2|x-y| \quad \text{for all } x,y \in X.$$

Suppose that the inequality (3.15) in Theorem 3.10 of Bhaskar and Lakshmikantham is satisfied for all $x \geq u$ and $y \leq v$. Taking $(x,y) = (6,1)$ and $(u,v) = (5,2)$, we get

$$\begin{aligned} d(F(x,y),F(u,v)) = d(10,6) = 4 &\leq \frac{k}{2}[d(x,u) + d(y,v)] \\ &= \frac{k}{2}[d(6,5) + d(1,2)] = k < 1, \end{aligned}$$

which is a contradiction. Theorem 3.10 and Theorem 3.11 are not applicable in this case. Define the mapping $\alpha : X^2 \times X^2 \to [0,+\infty)$ by

$$\alpha((x,y),(u,v)) = \begin{cases} \frac{1}{8} & \text{if } (x,y,u,v) \neq (0,0,0,0), \\ 1 & \text{if } x = y = u = v = 0. \end{cases} \tag{3.17}$$

We can check easily that for all $x,y,u,v \in X$, we have

$$\alpha((x,y),(u,v))d(F(x,y),F(u,v)) \leq \frac{1}{2}\frac{d(x,u) + d(y,v)}{2}.$$

Then condition (3.10) of our Theorem 3.7 is satisfied with $\psi(t) = t/2$ for all $t \geq 0$. Let $(x,y),(u,v) \in X \times X$ such that $\alpha((x,y),(u,v)) \geq 1$. This implies that $x = y = u = v = 0$ and $F(x,y) = F(y,x) = F(u,v) = F(v,u) = 0$. Moreover, $\alpha((F(x,y),F(y,x)),(F(u,v),F(v,u))) = 1$. Then condition (i) of Theorem 3.7 holds, and condition (ii) of the same theorem is also satisfied with $x_0 = y_0 = 0$. Now, all the hypotheses of Theorem 3.7 are true, and we deduce the existence of a coupled fixed point of F. Here $(0,0)$ is a coupled fixed point.

In 2012, Karapinar and Samet [34] generalized the results of Samet *et al.* [58] by introducing generalized α-ψ contractive type mappings as follows.

Definition 3.5 Let (X, d) be a metric space and $f : X \to X$ be a given mapping. We say that f is a generalized α-ψ contractive mapping if there exist two functions $\alpha : X \times X \to [0, \infty)$ and $\psi \in \Psi$ such that for all $x, y \in X$, and we have

$$\alpha(x, y)d(fx, fy) \leq \psi(M(x, y)), \tag{3.18}$$

where $M(x, y) = max\{d(x, y), (d(x, fx)+d(y, fy))/2, (d(x, fy)+d(y, fx))/2\}$.

Remark 3.10 Clearly, since ψ is nondecreasing, every α-ψ contractive mapping is a generalized α-ψ contractive mapping.

Theorem 3.13 *Let (X, d) be a complete metric space. Suppose that $f : X \to X$ is a generalized α-ψ contractive mapping and satisfies the following conditions:*

(i) f is α-admissible;

(ii) $x_0 \in X$ such that $\alpha(x_0, fx_0) \geq 1$;

(iii) f is continuous.

Then there exists $u \in X$ such that $fu = u$.

The following theorem does not require the continuity of f.

Theorem 3.14 *Let (X, d) be a complete metric space. Suppose that $f : X \to X$ is a generalized α-ψ contractive mapping and the following conditions hold:*

(i) f is α-admissible;

(ii) $x_0 \in X$ such that $\alpha(x_0, fx_0) \geq 1$;

(iii) if $\{x_n\}$ is a sequence in X such that $\alpha(x_n, x_{n+1}) \geq 1$ for all n and $x_n \to x \in X$ as $n \to \infty$, then there exists a subsequence $\{x_{n(k)}\}$ of $\{x_n\}$ such that $\alpha(x_{n(k)}, x) \geq 1$ for all k.

Then there exists $u \in X$ such that $fu = u$.

The following example shows that hypotheses in Theorems 3.13 and 3.14 do not guarantee uniqueness of the fixed point.

Example 3.8 Let $X = \{(1, 0), (0, 1)\} \subset \mathbb{R}^2$ be endowed with the Euclidean distance $d((x, y), (u, v)) = |x - u| + |y - v|$ for all $(x, y), (u, v) \in X$. Obviously, (X, d) is a complete metric space. The mapping $f(x, y) = (x, y)$ is trivially continuous and satisfies for any $\psi \in \Psi$

$$\alpha((x, y), (u, v))d(f(x, y), f(u, v)) \leq \psi(M((x, y), (u, v)))), \tag{3.19}$$

for all $(x, y), (u, v) \in X$, where

$$\alpha((x, y), (u, v)) = \begin{cases} 1, & \text{if } (x, y) = (u, v), \\ 0, & \text{if} (x, y) \neq (u, v). \end{cases} \tag{3.20}$$

Thus f is a generalized α-ψ contractive mapping. On the other hand, for all $(x,y),(u,v)\in X$, we have

$$\alpha((x,y),(u,v))\geq 1\rightarrow (x,y)=(u,v)\rightarrow f(x,y)=f(u,v)\rightarrow \alpha(f(x,y),f(u,v))\geq 1. \tag{3.21}$$

Thus f is α-admissible. Moreover, for all $(x,y)\in X$, we have $\alpha((x,y),f(x,y))\geq 1$. Then the assumptions of Theorem 3.13 are satisfied. Note that the assumptions of Theorem 3.14 are also satisfied; indeed if $\{(x_n,y_n)\}$ is a sequence in X that converges to some point $(x,y)\in X$ with $\alpha((x_n,y_n),(x_{n+1},y_{n+1}))\geq 1$ for all n, from the definition of α, $(x_n,y_n)=(x,y)$ for all n, which implies that $\alpha((x_n,y_n),(x,y))=1$ for all n. However, in this case, f has two fixed points in X.

For the uniqueness of a fixed point of a generalized α-ψ contractive mapping, we will consider the following hypothesis (H): For all $x,y\in Fix(f)$, there exists $z\in X$ such that $\alpha(x,z)\geq 1$ and $\alpha(y,z)\geq 1$.

Theorem 3.15 *Adding condition (H) to the hypotheses of 3.13 (resp., 3.14), u is the unique fixed point of f.*

Example 3.9 Let $X=[0,1]$ be endowed with the standard metric $d(x,y)=|x-y|$ for all $x,y\in X$. Obviously, (X,d) is a complete metric space. Define the mapping $f:X\rightarrow X$ by

$$fx=\begin{cases}\frac{1}{4}, & \text{if } x\in[0,1),\\ 0, & \text{if } x=1.\end{cases} \tag{3.22}$$

In this case, f is not continuous. Define the mapping $\alpha:X\times X\rightarrow(0,\infty)$ by

$$\alpha(x,y)=\begin{cases}1, & \text{if } (x,y)\in\left(\left[0,\frac{1}{4}\right]\times\left[\frac{1}{4},1\right]\right)\cup\left(\left[\frac{1}{4},1\right]\times\left[0,\frac{1}{4}\right]\right),\\ 0, & \text{otherwise.}\end{cases} \tag{3.23}$$

We will prove that

(A) $f:X\rightarrow X$ is a generalized α-ψ contractive mapping, where $\psi(t)=\frac{t}{2}$ for all $t\geq 0$;

(B) f is α-admissible;

(C) $x_0\in X$ such that $\alpha(x_0,fx_0)\geq 1$;

(D) if $\{x_n\}$ is a sequence in X such that $\alpha(x_n,x_{n+1})\geq 1$ for all n and $x_n\rightarrow x\in X$ as $n\rightarrow+\infty$, there exists a subsequence $\{x_{n(k)}\}$ of $\{x_n\}$ such that $\alpha(x_{n(k)},x)\geq 1$ for all k;

(E) condition (H) is satisfied.

Proof of (A). We have to prove that (3.18) is satisfied for every $x, y \in X$. If $x \in [0, \frac{1}{4}]$ and $y = 1$, we have

$$\alpha(x,y)d(fx,fy) = d(fx,fy) = \left|\frac{1}{4} - 0\right| = \frac{1}{4}d(y,fy) \leq \psi(M(x,y)).$$

Then (3.18) holds. If $x = 1$ and $y \in (0, \frac{1}{4})$, we have

$$\alpha(x,y)d(fx,fy) = d(fx,fy) = \left|0 - \frac{1}{4}\right| = \frac{1}{4}d(x,fx) \leq \psi(M(x,y)).$$

Then (3.18) holds also in this case. The other cases are trivial. Thus (3.18) is satisfied for every $x, y \in X$.

Proof of (B). Let $(x, y) \in X \times X$ such that $\alpha(x, y) \geq 1$. From the definition of α, we have two cases.

Case 1 (if $(x, y) \in [0, 1/4] \times [1/4, 1]$). In this case, we have $(fx, fy) \in [1/4, 1] \times [0, 1/4]$, which implies that $\alpha(fx, fy) = 1$.

Case 2 (if $(x, y) \in [1/4, 1] \times [0, 1/4]$). In this case, we have $(fx, fy) \in [0, 1/4] \times [1/4, 1]$, which implies that $\alpha(fx, fy) = 1$.

In all cases, we have $\alpha(fx, fy) \geq 1$. Thus f is α-admissible.

Proof of (C). Taking $x_0 = 0$, we have $\alpha(x_0, fx_0) = \alpha(0, 1/4) = 1$.

Proof of (D). Let $\{x_n\}$ be a sequence in X such that $\alpha(x_n, x_{n+1}) \geq 1$ for all n and $x_n \to x$ as $n \to +\infty$ for some $x \in X$. From the definition of α, for all n, we have

$$(x_n, x_{n+1}) \in \left(\left[0, \frac{1}{4}\right] \times \left[\frac{1}{4}, 1\right]\right) \cup \left(\left[\frac{1}{4}, 1\right] \times \left[0, \frac{1}{4}\right]\right).$$

Since $([0, 1/4] \times [1/4, 1]) \cup ([1/4, 1] \times [0, 1/4])$ is a closed set with respect to the Euclidean metric, we get

$$(x, x) \in \left(\left[0, \frac{1}{4}\right] \times \left[\frac{1}{4}, 1\right]\right) \cup \left(\left[\frac{1}{4}, 1\right] \times \left[0, \frac{1}{4}\right]\right),$$

which implies that $x = 1/4$. Thus we have $\alpha(x_n, x) \geq 1$ for all n.

Proof of (E). Let $(x, y) \in X \times X$. It is easy to show that, for $z = 1/4$, we have $\alpha(x, z) = \alpha(y, z) = 1$ and condition (H) is satisfied.

Here all the hypotheses of 3.15 are satisfied; thus f has a unique fixed point $u = 1/4$ in X.

3.3 Fixed and common fixed point of Meir-Keeler α-contractive type mappings

In 1969, Meir and Keeler [40] gave another aspect to metrical fixed point theory by introducing a new contractive condition, that is, for given $\epsilon > 0$ there exists a $\delta > 0$ such that $\epsilon \leq d(x,y) < \epsilon + \delta$ implies $d(fx, fy) < \epsilon$. This contraction has been further generalized and studied by various authors [8, 9, 15, 16, 26, 29, 30, 31, 39, 45, 47, 51, 52].

Using the concept of α-admissibility, Abdeljawad [3] (see also [4]) established some new fixed point results for α-contractive type mappings to Meir-Keeler versions of single and a pair of mappings and generalized the results obtained in [47] and others.

Definition 3.6 *[3]* Let $f, g : X \to X$ be self mappings of a set X, and let $\alpha : X \times X \to [0, \infty)$ be a mapping; then the pair (f, g) is called α-admissible if

$$x, y \in X, \quad \alpha(x,y) \geq 1 \Rightarrow \alpha(fx, gy) \geq 1 \text{ and } \alpha(gx, fy) \geq 1.$$

Example 3.10 Let $X = \mathbb{R}$ and

$$\alpha(x,y) = \begin{cases} 1 & \text{if } x, y \in [0,1], \\ 0 & \text{otherwise.} \end{cases}$$

Then the pair $(x^{1/2}, x^{1/3})$ is α-admissible but the pair $(x^{1/2}, x+1)$ is not.

Definition 3.7 Let(X, d) be a metric space and $f : X \to X$ be a self mapping, $\alpha : X \times X \to [0, \infty)$ be a mapping. Then, f is called Meir-Keeler α-contractive if, given $\epsilon > 0$, there exists a $\delta > 0$ such that

$$\epsilon \leq d(x,y) < \epsilon + \delta \quad \Rightarrow \quad \alpha(x,y)d(fx, fy) < \epsilon.$$

Definition 3.8 Let(X, d) be a metric space and $f : X \to X$ be a self mapping, $\alpha : X \times X \to [0, \infty)$ be a mapping. Then, f is called generalized Meir-Keeler α-contractive if, given $\epsilon > 0$, there exists a $\delta > 0$ such that

$$\epsilon \leq M_f(x,y) < \epsilon + \delta \quad \Rightarrow \quad \alpha(x,y)d(fx, fy) < \epsilon,$$

where

$$M_f(x,y) = \max\left\{d(x,y), d(x, fx), d(y, fy), \frac{d(x, fy) + d(y, fx)}{2}\right\}.$$

Definition 3.9 Let(X, d) be a metric space and $f, g : X \to X$ be self mappings, $\alpha : X \times X \to [0, \infty)$ be a mapping. Then, the pair (f, g) is called generalized Meir-Keeler α-contractive if, given $\epsilon > 0$, there exists a $\delta > 0$ such that

$$\epsilon \leq M_{(f,g)}(x,y) < \epsilon + \delta \quad \Rightarrow \quad \alpha(x,y)d(fx, gy) < \epsilon, \tag{3.24}$$

where

$$M_{(f,g)}(x,y) = \max\left\{d(x,y), d(x,fx), d(y,gy), \frac{d(x,gy)+d(y,fx)}{2}\right\}.$$

We write $M_f(x,y) = M_{(f,f)}(x,y)$.

Clearly, f is generalized Meir-Keeler α-contractive if and only if (f,f) is generalized Meir-Keeler α-contractive.

Definition 3.10 Let X be any set, $x_0 \in X$ and f, g be self maps of X. Define $x_{2n+1} = fx_{2n}$ and $x_{2n+2} = gx_{2n}$, $n = 0, 1, 2, ...$ Then $\{x_n\}$ is called the (f,g)-orbit of x_0. If d is a metric on X then (X,d) is called (f,g)-orbitally complete if every Cauchy sequence in the (f,g)-orbit of x_0 is convergent and the map f or g is called orbitally continuous if it is continuous on the orbit.

Lemma 3.3 *Let $f, g : X \to X$ be self mappings of a set X, $\alpha : X \times X \to [0,\infty)$ be a mapping and $\{x_n\}$ be the (f,g)-orbit of x_0 with $\alpha(x_0, fx_0) \geq 1$. If the pair (f,g) is α-admissible, then $\alpha(x_n, x_{n+1}) \geq 1$ for all $n = 0, 1, 2, ...$*

Theorem 3.16 *Let (X,d) be an (f,g)-orbitally complete metric space, where f, g are self mappings of X. Also, let $\alpha : X \times X \to [0,\infty)$ be a mapping. Assume the following:*

(1) (f,g) is α-admissible and there exists an $x_0 \in X$ such that $\alpha(x_0, fx_0) \geq 1$;

(2) the pair (f,g) is generalized Meir-Keeler α-contractive.

Then the sequence $d_n = d(x_n, x_{n+1})$ is monotone decreasing. If, moreover we assume that

(3) on the (f,g)-orbit of x_0, we have $\alpha(x_n, x_j) \geq 1$ for all n even and $j > n$ odd and that f and g are continuous on the (f,g)-orbit of x_0.

Then (1) f or g has a fixed point in the (f,g)-orbit $\{x_n\}$ of x_0 or (2) f and g have a common fixed point p and $\lim x_n = p$. If we assume that the following condition holds (H): For all fixed points x and y of (f,g), $\alpha(x,y) \geq 1$, and the uniqueness of the fixed point is obtained.

Proof. Define $d_n = d(x_n, x_{n+1})$, for $n = 0, 1, 2, ...$ If $d_n = 0$ for some even integer n, then f has a fixed point. If $d_n = 0$ for some odd integer n, then g has a fixed point. Hence, we may assume that $d_n \neq 0$ for each n. That the pair (f,g) is generalized Meir-Keeler α-contractive implies that

$$\alpha(x,y)d(fx,gy) < M_f(x,y) \quad \text{for each} \quad x, y \in X, \; x, y \neq 0. \tag{3.25}$$

Note that assumption (3) implies that $\alpha(x_0, fx_0) \geq 1$. Since (f,g) is α-admissible, Lemma 3.3 implies that $\alpha(x_n, x_{n+1}) \geq 1$ for all $n = 0, 1, 2, ...$

and hence by (3.25) we have

$$\begin{aligned} d_{2n} &= d(fx_{2n}, gx_{2n-1}) \\ &\leq \alpha(x_{2n}, x_{2n-1})d(fx_{2n}, gx_{2n-1}) \\ &< \max\left\{d_{2n-1}, \frac{d(x_{2n-1}, x_{2n+1})}{2}\right\} \\ &\leq \max\left\{d_{2n-1}, \frac{d_{2n-1}+d_{2n}}{2}\right\}, \end{aligned} \tag{3.26}$$

whence $d_{2n} < d_{2n-1}$.

Similarly, it can be shown that $d_{2n+1} < d_{2n}$. Thus $\{d_n\}$ is monotone decreasing in n and converges to a limit, say ϱ.

Suppose $\varrho > 0$. Then, for each $\delta > 0$ there exists a positive integer $N = N(\delta)$ such that $\varrho \leq d_N = d(x_N.x_{N+1}) < \varrho + \delta$, where N can be even. Thus from assumption (1) and Lemma 3.3 we have $d_{N+1} \leq \alpha(x_N, x_{N+1})d(fx_N, gx_{N+1}) < \varrho$, a contradiction. Therefore $\varrho = 0$. To show that $\{x_n\}$ is Cauchy we assume the contrary. Thus $\epsilon' > 0$ such that for each integer N there exist integers $m > n > N$ such that $d(x_m, x_n) \geq \epsilon'$. Define ϵ by $\epsilon' = 2\epsilon$. Choose a number δ, $0 < \delta < \epsilon$ for which (3.24) is satisfied. Since $\varrho = 0$, there exists an integer $N = N(\delta)$ such that $d_i < \frac{\delta}{6}$ for $i \geq N$. With this choice of N, we pick integers $m > n > N$ such that

$$d(x_m, x_n) \geq 2\epsilon > \delta + \epsilon. \tag{3.27}$$

It is clear that $m - n > 6$. Otherwise, $d(x_m.x_n)) \leq \sum_{i=0}^{5} d_{i+n} < \delta < \delta + \epsilon$, contradicts (3.27). Without loss of generality we may assume that n is even since, from (3.27), it follows that $d(x_m, x_{n+1}) > \epsilon + \frac{\delta}{3}$. From (3.27) there exists a smallest odd integer $j > n$ such that

$$d(x_n, x_j) \geq \epsilon + \frac{\delta}{3}. \tag{3.28}$$

Hence, $d(x_n, x_{j-2}) < \epsilon + \frac{\delta}{3}$, and so $d(x_n.x_j) \leq d(x_n, x_{j-2}) + d_{j-1} + d_j < \epsilon + \frac{\delta}{3} + 2(\frac{\delta}{6}) = \epsilon + \frac{2\delta}{3}$. Therefore, we have

$$\begin{aligned} \epsilon &< d(x_n, x_j) \leq M_{(f,g)}(x_n, x_j) \\ &\leq \max\left\{d(x_n.x_j), \frac{d(x_n, x_{j+1}) + d(x_j, x_{n+1})}{2}\right\} \\ &\leq \frac{d(x_n.x_j) + d_j + d(x_j, x_n) + d_n}{2} \\ &\leq d(x_n.x_j) + \frac{\delta}{6} \leq \epsilon + \delta, \end{aligned}$$

so that, by (3.27) and assumption (3), $d(x_{n+1}, x_{j+1}) \leq \alpha(x_n, x_j)d(x_{n+1}, x_{j+1}) < \epsilon$. Then, we have

$$d(x_n.x_j) \leq d_n + d(x_{n+1}, d_{j+1}) + d_j < \frac{\delta}{6} + \epsilon + \frac{\delta}{6} = \epsilon + \frac{\delta}{3}.$$

This contradicts the choice of j in (3.28). Therefore, $\{x_n\}$ is Cauchy.

Since X is (f,g)-orbitally complete, $\{x_n\}$ converges to some point $p \in X$. Since f and g are orbitally continuous, p is a common fixed point of f and g. To prove uniqueness, assume p is the common fixed point obtained as $x_n \to p$ and q is another common fixed point. Then, Equation (3.25) and the condition (H) yield to

$$\begin{aligned} d(p,q) &= d(fp,gq) \\ &\leq \alpha(p,q)d(fp,gq) \\ &< \max\{d(p,q), d(p,fp), d(q,gq), \frac{d(p,gq)+d(q,fp)}{2}\} \\ &= d(p,q). \end{aligned}$$

Thus we reach at $d(p,q) < d(p,q)$ a contradiction which implies that $p = q$.

Corollary 3.1 *Let (X,d) be an f-orbitally complete metric space, where f is a self mapping of X. Also, let $\alpha : X \times X \to [0,\infty]$ be a mapping. Assume the following:*

(1) f is α-admissible and $x_0 \in X$ such that $\alpha(x_0, fx_0) \geq 1$;

(2) f is generalized Meir-Keeler α-contractive.

Then the sequence $d_n = d(x_n, x_{n+1})$ is monotone decreasing. If, moreover we assume that

(3) on the f-orbit of x_0, we have $\alpha(x_n, x_j) \geq 1$ for all n even and $j > n$ odd.

Then either (1) f has a fixed point in the f-orbit $\{x_n\}$ of x_0 or (2) f has a fixed point p and $\lim x_n = p$. Moreover, if condition (H) holds for (f,f) the uniqueness of the fixed point is obtained.

Since generalized Meir-Keeler α-contractions are Meir-Keeler α-contractions, Corollary 3.1 is valid also for Meir-Keeler α-contractions. In the following example the existence and uniqueness of the fixed point cannot be proved in the category of Meir-Keeler contractions, but can be proved by Corollary 3.1.

Example 3.11 Let $X = [0,2]$ with the absolute value metric $d(x,y) = |x-y|$. Define $f : X \to X$ by

$$f(x) = \begin{cases} 0 & \text{if } x = \frac{1}{4}, \\ 1 & \text{if } x \in [0,\frac{1}{2}) - \{\frac{1}{4}\}, \\ \frac{3}{2} & \text{if } x \in [\frac{1}{2}, 2]. \end{cases}$$

Then, for $\epsilon = \frac{1}{2}$, $x = \frac{1}{4}$ and any $\delta > 0$ we have $\frac{1}{2} \leq |\frac{1}{4} - y| < \delta + \frac{1}{2}$ which implies $y \in [\frac{1}{2}, 2]$ and $d(fx, fy) = d(0, \frac{3}{2}) = \frac{3}{2} > \epsilon$. Hence, f is not a Meir-Keeler contraction. However, f is a Meir-Keeler α-contraction where

$$\alpha(x,y) = \begin{cases} 1 & \text{if } x, y \in [\frac{1}{2}, 2], \\ 0 & \text{otherwise.} \end{cases}$$

Indeed, for $0 < \epsilon < 1$ (the case $\epsilon \geq 1$ is trivial, since $|fx - fy| \leq 1$), let $\delta = (1 - \epsilon)$; then $\epsilon \leq \alpha(x, y)d(x, y) < \delta + \epsilon = 1$ implies that $x, y \in [\frac{1}{2}, 2]$ and hence $d(fx, fy) = |\frac{3}{2} - \frac{3}{2}| = 0 < \epsilon$. Also, notice that f is continuous on the orbit of $x_0 = 1$ and that $\alpha(x_n, x_j) \geq 1$ for all n, j. Clearly, $p = \frac{3}{2}$ is the unique fixed point.

The following example shows that we lose the uniqueness if our modified (H) condition is not satisfied.

Example 3.12 Let $X = [0, 2]$ with the absolute value metric $d(x, y) = |x - y|$. Define $f : X \to X$ by

$$f(x) = \begin{cases} 0 & \text{if } x \in \{0, \frac{1}{4}\}, \\ 1 & \text{if } x \in (0, \frac{1}{2}) - \{\frac{1}{4}\}, \\ \frac{3}{2} & \text{if } x \in [\frac{1}{2}, 2]. \end{cases}$$

and define

$$\alpha(x, y) = \begin{cases} 1 & \text{if } x, y \in [\frac{1}{2}, 2], \\ 0 & \text{otherwise.} \end{cases}$$

Notice that f has two common fixed points $x = 0$ and $x = \frac{3}{2}$ because f satisfies all the hypotheses of Corollary 3.1 except condition (H), i.e., $\alpha(0, \frac{3}{2}) = 0 < 1$.

Remark 3.11 Note that the admissibility condition (1) in Theorem 3.16 is not enough to guarantee the existence of the fixed point. However, such an admissibility condition was used to obtain the main result in Theorem 3.2 by Samet *et al.* [58].

Definition 3.11 Let f be a continuous self map of a metric space (X, d), $C_f = \{g : g : X \to X$, such that $fg = gf$ and $gX \subseteq fX\}$. The sequence $\{fx_n\}$ defined by $fx_{n+1} = gx_n, \quad n = 0, 1, 2, ...$, with the understanding that if $fx_n = fx_{n+1}$ for some n, then $fx_{n+j} = fx_n$ for each $j \geq 0$ is called the f-iteration of x_0 under g.

Definition 3.12 Let f be a self map of a metric space (X, d) and $g \in C_f$. Then g is called a Meir-Keeler α-f-contractive map if for each $\epsilon > 0$ there exists a $\delta > 0$ such that for all $x, y \in X$,

$$\epsilon \leq d(fx, fy) < \epsilon + \delta \quad \Rightarrow \quad \alpha(x, y)d(gx, gy) < \epsilon. \tag{3.29}$$

Definition 3.13 Let f be a self map of a metric space (X, d) and $g \in C_f$. Then g is called a generalized Meir-Keeler α-f-contractive map if for each $\epsilon > 0$ there exists a $\delta > 0$ such that for all $x, y \in X$,

$$\epsilon \leq M_g(f)(x, y) < \epsilon + \delta \quad \Rightarrow \quad \alpha(x, y)d(gx, gy) < \epsilon, \tag{3.30}$$

where

$$M_g(f)(x, y) = \max\left\{d(fx, fy), d(fx, gx), d(fy, gy), \frac{d(fx, gy) + d(fy, gx)}{2}\right\}.$$

Lemma 3.4 *Let f, g be continuous self maps of a metric space (X, d) such that $g \in C_f$. Assume g is a generalized Meir-Keeler α-f-contractive map such that $\alpha(x_n, x_{n+1}) \geq 1$ for all n. Then* $\inf\{d(fx_n, fx_{n+1}) : n = 0, 1, 2, ...\} = 0$.

Proof. Let $\sigma = \inf\{d(fx_n, fx_{n+1}) : n = 0, 1, 2, ...\}$, and $\sigma > 0$. From the definition of the f-iteration of x_0 under g and assuming that g is a generalized Meir-Keeler α-f-contractive map, for each n we have

$$\begin{aligned}
d(fx_{n+1}, fx_{n+2}) &= d(gx_n, gx_{n+1}) \leq \alpha(x_n, x_{n+1}) d(gx_n, gx_{n+1}) \\
&< \max\Big\{d(fx_n, fx_{n+1}), d(fx_n, gx_n), d(fx_{n+1}, gx_{n+1}), \\
&\qquad \frac{d(fx_n, gx_{n+1}) + d(fx_{n+1}, gx_n)}{2}\Big\} \\
&= \max\Big\{d(fx_n, fx_{n+1}), d(fx_n, fx_{n+1}), d(fx_{n+1}, fx_{n+2}), \\
&\qquad \frac{d(fx_n, fx_{n+2}) + 0}{2}\Big\} \\
&= \max\left\{d(fx_n, fx_{n+1}), \frac{d(fx_n, fx_{n+2})}{2}\right\} \\
&\leq \max\left\{d(fx_n, fx_{n+1}), \frac{d(fx_n, fx_{n+1}) + d(fx_{n+1}, fx_{n+2})}{2}\right\}.
\end{aligned}$$

Hence, $d(fx_{n+1}, fx_{n+2}) < d(fx_n, fx_{n+1})$ and $\{d(fx_n, fx_{n+1})\}$ is monotone decreasing, so that $\sigma = \lim_{n\to\infty} d(fx_n, fx_{n+1})$. From the assumption that g is a Meir-Keeler α-f-contractive map, for $\epsilon = \sigma$ find $\delta > 0$ such that (3.30) is satisfied. For the chosen δ pick N so that $\sigma \leq d(fx_n, fx_{n+1}) < \sigma + \delta$. Noting that for $x = x_n$ and $y = x_{n+1}$, $M_g(f)(x, y) = d(fx_n, fx_{n+1})$, by (3.30) we conclude that $d(gx_n, gx_{n+1}) \leq \alpha(x_n, x_{n+1}) d(gx_n, gx_{n+1}) < \sigma$. But $d(gx_n, gx_{n+1}) = d(fx_{n+1}, fx_{n+2}) < \sigma$, a contradiction.

Theorem 3.17 *Let f, g be continuous self maps of a metric space (X, d) such that $g \subset C_f$. Assume $\alpha(x_n, x_m) \geq 1$ for all $m > n$. If g is a generalized Meir-Keeler α-f-contractive map such that α satisfies the condition (f-H). If $\{x_n\}$ is a sequence in X such that $\alpha(x_n, x_m) \geq 1$ for all $m > n$ and $fx_n \to z$ then $\alpha(z, fz) \geq 1$ and f and g have a unique common fixed point.*

Proof. Let $x_0 \in X$ for which its f-iteration under g satisfies the assumptions of the theorem. The proof will be divided into four steps.

Step 1: By Lemma 3.4, $\inf\{d(fx_n, fx_{n+1}) : n = 0, 1, 2, ...\} = 0$.

Step 2: We find a coincidence point for f and g, that is we find a $z \in X$ such that $fz = gz$. If there exists an n such that $d(fx_n, fx_{n+1}) = 0$, then $fx_{n+1} = gx_n = fx_n$ and we are finished. Hence, we may assume that $d(fx_n, fx_{n+1}) \neq 0$ for each n. We claim to show that $\{fx_n\}$ is Cauchy. Suppose the converse.

Then there exists an $\epsilon > 0$ and a subsequence $\{fx_{n_i}\}$ of $\{fx_n\}$ such that $d(fx_{n_i}, fx_{n_{i+1}}) > 2\epsilon$. From (3.30), there exists a δ satisfying $0 < \delta < \epsilon$ for which (3.30) is true. Since $\lim_{n\to\infty} d(fx_n, fx_{n+1})=0$, there exists an N such that

$$d(fx_m, fx_{m+1}) < \frac{\delta}{6} \text{ for all } m > N.$$

Let $n_i \geq N$. We shall show an integer j satisfying $n_i < j < n_{i+1}$, such that

$$\epsilon + \frac{\delta}{3} \leq d(fx_{n_i}, fx_j) < \epsilon + \frac{2\delta}{3}. \tag{3.31}$$

First, there exist values of j such that $d(fx_{n_i}, fx_j) \geq \epsilon + \frac{\delta}{3}$. For example, choose $j = n_{i+1}$. The inequality is also true for $j = n_{i+1} - 1$. If not, $d(fx_{n_i}, fx_j) < \epsilon + \frac{\delta}{3}$ and hence

$$\begin{aligned} d(fx_{n_i}, fx_{n_{i+1}}) &\leq d(fx_{n_i}, fx_{n_{i+1}} - 1) + d(fx_{n_{i+1}} - 1, fx_{n_{i+1}}) \\ &< \epsilon + \frac{\delta}{3} + \frac{\delta}{6} < 2\epsilon, \end{aligned}$$

a contradiction. There are also values of j such that $d(fx_{n_i}, fx_j) < \epsilon + \frac{\delta}{3}$. For example choose $j = n_i + 1$ and $j = n_i + 2$. Pick j to be the smallest integer greater than n_i such that $d(fx_{n_i}, fx_j) \geq \epsilon + \frac{\delta}{3}$. Then $d(fx_{n_i}, fx_i - 1) < \epsilon + \frac{\delta}{3}$, and hence

$$d(fx_{n_i}, fx_j) \leq d(fx_{n_i}, fx_j - 1) + d(fx_j - 1, fx_j) < \epsilon + \frac{\delta}{3} + \frac{\delta}{6} < \epsilon + \frac{2\delta}{3}.$$

Thus (3.31) is established. Now note that

$$\begin{aligned} \epsilon + \frac{\delta}{3} \leq d(fx_{n_i}, fx_j) \leq \max\{&d(fx_{n_i}, fx_j), d(fx_{n_i}, gx_{n_i}), d(fx_j, gx_j), \\ &\frac{d(fx_{n_i}, gx_j) + d(fx_j, gx_{n_i})}{2}\}. \end{aligned}$$

Then from the choice of j and since $fx_{n_i} + 1 = gx_{n_i}$, $fx_j + 1 = gx_j$ we reach

$$\epsilon \leq d(fx_{n_i}, fx_j) < \delta + \epsilon.$$

Hence,

$$d(fx_{n_i+1}, fx_{j+1}) = d(gx_{n_i}, gx_j) \leq \alpha(x_{n_i}, x_j) d(gx_{n_i}, gx_j) < \epsilon.$$

On the other hand,

$$\begin{aligned} d(fx_{n_i}, fx_j) &\leq d(fx_{n_i}, f_{n_i+1}) + d(f_{n_i+1}, fx_{j+1}) + d(fx_{j+1}, fx_j) \\ &< \frac{\delta}{6} + \epsilon + \frac{\delta}{6} = \epsilon + \frac{\delta}{3}, \end{aligned}$$

contradicting (3.31). Therefore, $\{fx_n\}$ is Cauchy and convergent to $z \in X$. Since $ffx_n = fgx_{n-1} = gfx_{n-1}$, the continuity of f and g implies that $fz = gz$.

Step 3: We show that $\eta = fz = gz$ is a common fixed point for f and g: Assume $f\eta \neq \eta$. Then $f^2z \neq fz$ and by the help of the (f-H) condition we have

$$\begin{aligned} d(\eta, f\eta) &= d(gz, fgz) = d(gz, gfz) \\ &\leq \alpha(z, fz)d(gz, gfz) \\ &< \max\{d(fz, ffz), d(fz, gz), d(ffz, gfz), \frac{d(fz, gfz) + d(ffz, gz)}{2}\} \\ &= \max\{d(\eta.f\eta), 0, 0, d(\eta, f\eta)\} \end{aligned}$$

Thus $d(\eta, f\eta) < d(\eta, f\eta)$ is a contradiction and therefore $f\eta = \eta$. Moreover, $g\eta = gfz = f\eta = \eta$.

Step 4: Uniqueness of the common fixed point: Assume $\eta = fz = gz$ is our common fixed point for f and g where $fx_n \to z$ and ω is another common fixed point. Then by the (H) condition we have

$$\begin{aligned} d(\eta, \omega) &= d(g\eta, g\omega) \\ &\leq \alpha(\eta, \omega)d(g\eta, g\omega) \\ &< \max\{d(f\eta, f\omega), d(f\eta, g\eta), d(f\omega, g\omega), \frac{d(f\eta, g\omega) + d(f\omega, g\eta)}{2}\}, \end{aligned}$$

which gives $d(\eta, \omega) < d(\eta, \omega)$, a contradiction, and hence $\eta = \omega$.

Remark 3.12 Instead of the condition (f-H), the following condition can be used (s-f-H): If $\{x_n\}$ is a sequence in X such that $\alpha(x_n, x_m) \geq 1$ for all $m > n$ and $fx_n \to z$ then $\alpha(fx_n, z) \geq 1$ and $\alpha(fx_n, fz) \geq K$ for all n, where $K > 1$, and Step 4 will be proved as follows:

We show that $\eta = fz = gz$ is a common fixed point for f and g. Assume $f\eta \neq \eta$. Then $f^2z \neq fz$ and by the help of the (s-f-H) condition we have

$$\begin{aligned} d(\eta, f\eta) &= d(gz, fgz) = d(gz, gfz) \\ &\leq d(gz, gfx_n) + d(gfx_n, gfz) \\ &\leq \alpha(fx_n, z)d(gz, gfx_n) + K^{-1}\alpha(fx_n, fz)d(gfx_n, gfz) \\ &< \max\Big\{d(fz, ffx_n), d(fz, gz), d(ffx_n, gfx_n), \\ &\qquad \frac{d(fz, gfx_n) + d(ffx_n, gz)}{2}\Big\} \\ &\qquad + K^{-1}\max\Big\{d(ffx_n, ffz), d(ffx_n, gfx_n), d(ffz, gfz), \\ &\qquad\qquad \frac{d(ffx_n, gfz) + d(ffz, gfx_n)}{2}\Big\}. \end{aligned}$$

If we let $n \to \infty$ above and use continuity and commutativity of f and g, we reach $d(\eta, f\eta) \leq K^{-1}d(\eta, f\eta) < d(\eta, f\eta)$ and hence $f\eta = \eta$. Moreover, $g\eta = gfz = f\eta = \eta$.

Remark 3.13 Theorem 3.17 has been proved for commuting maps. It would be interesting to extend it for weakly commuting and compatible mappings. For example can we extend the results in [60, 29, 15] to α-type contractions?

Motivated by Abdeljawad [3], Patel *et al.* [48] established some common fixed point theorems for three and four self mappings satisfying generalized Meir-Keeler α-contractions in metric spaces and extended the results of Sessa [60], Jungck [29], Rao and Rao [51] and Cho *et al.* [15] to α-type contractions.

Definition 3.14 Let $f, g, T : X \to X$ be three self mappings of a nonempty set X, and let $\alpha : T(X) \times T(X) \to [0, \infty)$ be a mapping; then the pair (f, g) is called α-admissible with respect to T (in short, (f, g) is α_T-admissible) if for all $x, y \in X$,

$$\alpha(Tx, Ty) \geq 1 \quad \Rightarrow \quad \alpha(fx, gy) \geq 1 \text{ and } \alpha(gx, fy) \geq 1. \tag{3.32}$$

Clearly, if $T = I$ (identity mapping), the above definition implies Definition 3.6.

Definition 3.15 Let (X, d) be a metric space, and $f, g, T : X \to X$ be three self mappings. Then we say that the pair (f, g) is a generalized Meir-Keeler α_T-contractive pair of type m_3 (M_3, respectively) if given $\epsilon > 0$ and a $\delta > 0$ such that

$$\epsilon \leq m_3(x, y) \ (M_3(x, y), \text{ respectively}) < \epsilon + \delta \quad \Rightarrow \quad \alpha(Tx, Ty)d(fx, gy) < \epsilon, \tag{3.33}$$

where

$$m_3(x, y) = \max\left\{d(Tx, Ty),\ \frac{1}{2}[d(Tx, fx) + d(Ty, gy)],\ \frac{1}{2}[d(Tx, gy) + d(Ty, fx)]\right\},$$

and

$$M_3(x, y) = \max\left\{d(Tx, Ty),\ d(Tx, fx),\ d(Ty, gy),\ \frac{1}{2}[d(Tx, gy) + d(Ty, fx)]\right\}.$$

Definition 3.16 Let f, g and T be three self mappings on a metric space (X, d) such that $f(X) \cup g(X) \subseteq T(X)$. If for a point $x_0 \in X$, there exists a sequence $\{x_n\}$ such that $Tx_{2n+1} = fx_{2n}$, $Tx_{2n+2} = gx_{2n+1}$, $n = 0, 1, 2, \ldots$, then $\mathcal{O}(f, g, T, x_0) = \{Tx_n : n = 1, 2, \ldots\}$ is called the orbit for (f, g, T) at x_0. The space (X, d) is called (f, g, T)-orbitally complete at x_0 iff every Cauchy sequence in $\mathcal{O}(f, g, T, x_0)$ converges to a point in X, and it is called (f, g, T)-orbitally complete if it is so at every $x \in X$.

The following theorem shows the common fixed points for three self mappings.

Theorem 3.18 *Let (X, d) be an (f, g, T)-orbitally complete metric space. Suppose that (f, g) is a generalized Meir-Keeler α_T-contractive pair of type m_3 and satisfies the following conditions:*

(1) (f, g) is α_T-admissible;

(2) $x_0 \in X$ such that $\alpha(Tx_0, fx_0) \geq 1$;

(3) on the (f, g, T)-orbit of x_0, we have $\alpha(Tx_n, Tx_j) \geq 1$ for all n even and $j > n$ odd.

Then $\{Tx_n\}$ is a Cauchy sequence. Moreover, if

(4) $\alpha(Tx_n, Tx_{n+1}) \geq 1$ for all n, and $Tx_n \to x$ implies that $\alpha(Tx_n, Tx) \geq 1$ for all n;

(5) one of the pairs (f, T) and (g, T) is absorbing as well as reciprocal continuous.

Then f, g and T have a common fixed point.

Proof. Let $x_0 \in X$ such that $\alpha(Tx_0, fx_0) \geq 1$. Define the sequences $\{x_n\}$ and $\{Tx_n\}$ in X given by the rule

$$Tx_{2n+1} = fx_{2n}, \quad Tx_{2n+2} = gx_{2n+1}, \quad n = 0, 1, 2, \ldots \; .$$

Since (f, g) is α_T-admissible, we have

$$\alpha(Tx_0, fx_0) = \alpha(Tx_0, Tx_1) \geq 1 \;\Rightarrow\; \alpha(fx_0, gx_1) \geq 1 \text{ and } \alpha(gx_0, fx_1) \geq 1,$$

which gives

$$\alpha(Tx_1, Tx_2) \geq 1.$$

Again by (1), we have

$$\alpha(Tx_1, Tx_2) \geq 1 \;\Rightarrow\; \alpha(fx_1, gx_2) \geq 1 \text{ and } \alpha(gx_1, fx_2) \geq 1,$$

which gives

$$\alpha(Tx_2, Tx_3) \geq 1.$$

Inductively, we have

$$\alpha(Tx_n, Tx_{n+1}) \geq 1 \qquad n = 0, 1, 2, \ldots \; . \tag{3.34}$$

The fact that (f, g) is generalized Meir-Keeler α_T-contractive implies that

$$\alpha(Tx, Ty)d(fx, fy) < m_3(x, y) \qquad \text{for each } x, y \in X, \; x \neq y. \tag{3.35}$$

Now, to obtain a common fixed point of f, g and T, we take the following steps:

STEP 1: We show that there exists a point $z \in X$ such that $Tx_n \to z$ as $n \to \infty$. For this, first we claim that $\{Tx_n\}$ is a Cauchy sequence. Two cases arise: either $Tx_n = Tx_{n+1}$ for some n or $Tx_n \neq Tx_{n+1}$ for each n.

CASE I: Suppose that $Tx_n = Tx_{n+1}$ for some n. We first assume that n is even, *i.e.*, $Tx_{2m} = Tx_{2m+1}$ but $Tx_{2m+1} \neq Tx_{2m+2}$, then by (3.35),

$$\begin{aligned}
d(Tx_{2m+1}, Tx_{2m+2}) &= d(fx_{2m}, gx_{2m+1}) \\
&\leq \alpha(Tx_{2m}, Tx_{2m+1}) d(fx_{2m}, gx_{2m+1}) \\
&< \max\left\{ d(Tx_{2m}, Tx_{2m+1}), \frac{1}{2}\left[d(Tx_{2m}, fx_{2m}) + d(Tx_{2m+1}, gx_{2m+1})\right], \right. \\
&\qquad \left. \frac{1}{2}\left[d(Tx_{2m}, gx_{2m+1}) + d(Tx_{2m+1}, fx_{2m})\right] \right\} \\
&= \max\left\{ 0, \ \frac{1}{2} d(Tx_{2m+1}, Tx_{2m+2}), \ \frac{1}{2} d(Tx_{2m}, Tx_{2m+2}) \right\} \\
&= \frac{1}{2} d(Tx_{2m+1}, Tx_{2m+2}),
\end{aligned}$$

which is a contradiction. Hence $Tx_{2m+1} = Tx_{2m+2}$. By proceeding in this way, we obtain $Tx_{2m+k} = Tx_{2m}$ for all $k \in \mathbb{N}$. Similar is the case when n is odd. Thus, we conclude that $\{Tx_n\}$ is a Cauchy sequence.

CASE II: Suppose that $Tx_n \neq Tx_{n+1}$ for all integers n. Applying (3.35), we have

$$\begin{aligned}
d(Tx_{2n+1}, Tx_{2n+2}) &= d(fx_{2n}, gx_{2n+1}) \\
&\leq \alpha(Tx_{2n}, Tx_{2n+1}) d(fx_{2n}, gx_{2n+1}) \\
&< \max\left\{ d(Tx_{2n}, Tx_{2n+1}), \frac{1}{2}\left[d(Tx_{2n}, fx_{2n}) + d(Tx_{2n+1}, gx_{2n+1})\right], \right. \\
&\qquad \left. \frac{1}{2}\left[d(Tx_{2n}, gx_{2n+1}) + d(Tx_{2n+1}, fx_{2n})\right] \right\} \\
&= \max\left\{ d(Tx_{2n}, Tx_{2n+1}), \frac{1}{2}\left[d(Tx_{2n}, Tx_{2n+1}) + d(Tx_{2n+1}, Tx_{2n+2})\right], \right. \\
&\qquad \left. \frac{1}{2}\left[d(Tx_{2n}, Tx_{2n+2}) + d(Tx_{2n+1}, Tx_{2n+1})\right] \right\} \\
&= d(Tx_{2n}, Tx_{2n+1}).
\end{aligned}$$

Similarly, it can be shown that

$$d(Tx_{2n+2}, Tx_{2n+3}) < d(Tx_{2n+1}, Tx_{2n+2}).$$

Thus, $\{d(Tx_n, Tx_{n+1})\}$ is a strictly decreasing sequence of positive real numbers, and therefore, converges to a limit $r \geq 0$. If $r > 0$, given $\delta > 0$, there exists a positive integer $N = N(\delta)$ such that $r \leq d(Tx_n, Tx_{n+1}) < r + \delta$ for all $n \geq N$. Let $2k \geq N$, then

$$r \leq d(Tx_{2k}, Tx_{2k+1}) = m_3(x_{2k}, x_{2k+1}) < r + \delta, \tag{3.36}$$

and by applying (3.33) we have

$$\alpha(Tx_{2k}, Tx_{2k+1})d(fx_{2k}, gx_{2k+1}) < r,$$

$$\Rightarrow \; d(fx_{2k}, gx_{2k+1}) = d(Tx_{2k+1}, Tx_{2k+2}) < r,$$

which is a contradiction since r is the limit. Hence $r = 0$ and thus

$$\lim_{n\to\infty} d(Tx_n, Tx_{n+1}) = 0. \tag{3.37}$$

We now show that $\{Tx_n\}$ is a Cauchy sequence.

Suppose it is not. Then there exists $\epsilon > 0$ such that for each positive integer m, n with $m > n > N$, we have $d(Tx_m, Tx_n) \geq 2\epsilon$. Choose a number δ, $0 < \delta < \epsilon$ for which contractive condition (3.33) is satisfied. Since $d(Tx_n, Tx_{n+1}) \to 0$, there exists integer $N = N(\delta)$ such that $d(Tx_i, Tx_{i+1}) < \frac{\delta}{6}$ for all $i \geq N$. With this choice of N, pick m, n with $m > n > N$ such that

$$d(Tx_n, Tx_m) \geq 2\epsilon > \epsilon + \delta, \tag{3.38}$$

in which it is clear that $m - n > 6$. Otherwise, we have

$$d(Tx_n, Tx_m) \leq \sum_{i=0}^{5} d(Tx_{n+i}, Tx_{n+i+1}) < \delta < \epsilon + \delta,$$

which contradicts (3.38). Also from (3.38), it follows that

$$d(Tx_{n+1}, Tx_m) > \epsilon + \frac{\delta}{3}.$$

Without loss of generality, we may assume that n is even. Suppose that

$$d(Tx_n, Tx_{m-1}) < \epsilon + \frac{\delta}{3},$$

then

$$\begin{aligned} d(Tx_n, Tx_m) &\leq d(Tx_n, Tx_{m-1}) + d(Tx_{m-1}, Tx_m) \\ &< \left(\epsilon + \frac{\delta}{3}\right) + \left(\frac{\delta}{6}\right) \\ &< \epsilon + \delta, \end{aligned}$$

which is a contradiction to (3.38). So we have

$$d(Tx_n, Tx_{m-1}) \geq \epsilon + \frac{\delta}{3}.$$

Similarly, suppose that

$$d(Tx_n, Tx_{m-2}) < \epsilon + \frac{\delta}{3},$$

then

$$\begin{aligned} d(Tx_n, Tx_m) &\leq d(Tx_n, Tx_{m-2}) + d(Tx_{m-2}, Tx_{m-1}) + d(Tx_{m-1}, Tx_m) \\ &< \left(\epsilon + \frac{\delta}{3}\right) + \left(\frac{\delta}{6}\right) + \left(\frac{\delta}{6}\right) \\ &< \epsilon + \delta, \end{aligned}$$

which is a contradiction to (3.38). So we have

$$d(Tx_n, Tx_{m-2}) \geq \epsilon + \frac{\delta}{3}.$$

Thus, there exists a smallest odd integer $j > n$ such that

$$d(Tx_n, Tx_j) \geq \epsilon + \frac{\delta}{3}, \tag{3.39}$$

and hence

$$d(Tx_n, Tx_{j-2}) < \epsilon + \frac{\delta}{3}.$$

Now,

$$\begin{aligned} d(Tx_n, Tx_j) &\leq d(Tx_n, Tx_{j-2}) + d(Tx_{j-2}, Tx_{j-1}) + d(Tx_{j-1}, Tx_j) \\ &< \left(\epsilon + \frac{\delta}{3}\right) + \left(\frac{\delta}{6}\right) + \left(\frac{\delta}{6}\right) \\ &= \epsilon + \left(\frac{2\delta}{3}\right). \end{aligned}$$

Thus, there exists an odd integer $j \in (n, m)$ such that

$$\epsilon + \left(\frac{\delta}{3}\right) \leq d(Tx_n, Tx_j) < \epsilon + \left(\frac{2\delta}{3}\right). \tag{3.40}$$

Since we have

$$\begin{aligned}\epsilon < d(Tx_n, Tx_j) &\le m_3(x_n, x_j)\\ &= \max\left\{d(Tx_n, Tx_j),\ \frac{1}{2}\left[d(Tx_n, fx_n) + d(Tx_j, gx_j)\right],\right.\\ &\qquad\left.\frac{1}{2}\left[d(Tx_n, gx_j) + d(Tx_j, fx_n)\right]\right\}\\ &< d(Tx_n, Tx_j) + \left(\frac{\delta}{6}\right)\\ &< \epsilon + \delta.\end{aligned}$$

Using (3.33) and assumption (3), we get

$$d(fx_n, gx_j) \le \alpha(Tx_n, Tx_j)d(fx_n, gx_j) < \epsilon,$$

that is, $d(Tx_{n+1}, Tx_{j+1}) < \epsilon$. But then

$$\begin{aligned}d(Tx_n, Tx_j) &\le d(Tx_n, Tx_{n+1}) + d(Tx_{n+1}, Tx_{j+1}) + d(Tx_{j+1}, Tx_j)\\ &< \left(\frac{\delta}{6}\right) + \epsilon + \left(\frac{\delta}{6}\right) = \epsilon + \left(\frac{\delta}{3}\right),\end{aligned}$$

which contradicts (3.40). Therefore, $\{Tx_n\}$ is a Cauchy sequence. Since X is (f, g, T)-orbitally complete, there exists a point $z \in X$ such that $Tx_n \to z$ as $n \to \infty$. Consequently, $fx_{2n} \to z$ and $gx_{2n+1} \to z$.

STEP 2: We show that z is common fixed point of (f, g, T). In view of assumption (5), without loss of generality, let the pair (f, T) be absorbing and reciprocal continuous. Then the reciprocal continuity of f and T implies that

$$\lim_{n\to\infty} fTx_{2n} = fz \quad \text{and} \quad \lim_{n\to\infty} Tfx_{2n} = Tz.$$

Since T is f-absorbing, there exists $R > 0$ such that

$$d(fx_{2n}, fTx_{2n}) \le Rd(fx_{2n}, Tx_{2n}).$$

Letting $n \to \infty$, we get $fTx_{2n} \to z$. Similarly, since f is T-absorbing,

$$d(Tx_{2n}, Tfx_{2n}) \le Rd(fx_{2n}, Tx_{2n}),$$

and letting $n \to \infty$, we get $Tfx_{2n} \to z$. By the uniqueness of limits, we have $z = fz = Tz$.

Now suppose that $z \neq gz$, then by assumption (4) and Equation (3.35), we have

$$d(fx_{2n}, gz) \leq \alpha(Tx_{2n}, Tz)d(fx_{2n}, gz)$$
$$< \max\left\{d(Tx_{2n}, Tz),\ \frac{1}{2}\left[d(Tx_{2n}, fx_{2n}) + d(Tz, gz)\right],\right.$$
$$\left.\frac{1}{2}\left[d(Tx_{2n}, gz) + d(Tz, fx_{2n})\right]\right\}.$$

Letting $n \to \infty$, we get $d(z, gz) \leq \frac{1}{2}d(z, gz)$, which implies that $z = gz$. Thus, z is a common fixed point of f, g and T. This completes the proof of the theorem.

If we use $f = g$ and $T = I$ (identity mapping) in Theorem 3.18, we get the following result as a corollary.

Corollary 3.2 Let (X, d) be an f-orbitally complete metric space, where f is a self mapping on X. Also, let $\alpha : X \times X \to [0, \infty)$ be a mapping. Assume the following:

(1) f is α-admissible;

(2) there exists an $x_0 \in X$ such that $\alpha(x_0, fx_0) \geq 1$;

(3) for given $\epsilon > 0$, there exists a $\delta > 0$ such that

$$\epsilon \leq m_1(x, y) < \epsilon + \delta \implies \alpha(x, y)d(fx, fy) < \epsilon,$$

where

$$m_1(x, y) = \max\left\{d(x, y), \frac{1}{2}\left[d(x, fx) + d(y, fy)\right],\ \frac{1}{2}\left[d(x, fy) + d(y, fx)\right]\right\};$$

(4) on the f-orbit of x_0, we have $\alpha(x_n, x_j) \geq 1$ for all n even and $j > n$ odd.

Then f has a fixed point in the f-orbit $\{x_n\}$ of x_0, or f has a fixed point z and $\lim_n x_n = z$.

Example 3.13 Let $X = [0, 2]$ be endowed with the standard metric $d(x, y) = |x - y|$ for all $x, y \in X$. Define $f : X \to X$ by

$$fx = \begin{cases} 0 & \text{if } x \in \{0, \frac{1}{4}\}, \\ 1 & \text{if } x \in (0, \frac{1}{2}) - \{\frac{1}{4}\}, \\ \frac{3}{2} & \text{if } x \in \left[\frac{1}{2}, 2\right]. \end{cases}$$

Then f is not a Meir-Keeler contraction. To see this, consider $\epsilon = \frac{1}{2}$, $x = \frac{1}{4}$, and $y = \frac{3}{4}$; then for any $\delta > 0$, we have $\epsilon \leq m_1(x, y) < \epsilon + \delta$, but $d(fx, fy) = d(0, \frac{3}{2}) = \frac{3}{2} > \epsilon$. However, f is a generalized Meir-Keeler α_T-contraction, where $\alpha : X \times X \to [0, \infty)$ is defined by

$$\alpha(x, y) = \begin{cases} 1 & \text{if } x, y \in [\frac{1}{2}, 2], \\ 0 & \text{otherwise.} \end{cases}$$

Clearly, f has two fixed points, $x = 0$ and $x = \frac{3}{2}$. Notice that $\alpha(\frac{3}{2}, 0) = 0 < 1$.

For the uniqueness of the fixed point of generalized Meir-Keeler α-contractive mappings, we will consider the following hypothesis (H): For all fixed points x and y of (f, g, T), we have $\alpha(Tx, Ty) \geq 1$.

Theorem 3.19 *Adding condition (H) to the hypotheses of Theorem 3.18 (resp., Corollary 3.2), we obtain the uniqueness of the common fixed point of f, g and T.*

Proof. Let z be the common fixed point obtained as $Tx_n \to z$ and u is another common fixed point. Then, (3.35) and condition (H) yield to

$$\begin{aligned} d(z, u) &= d(fz, gu) \\ &\leq \alpha(Tz, Tu)d(fz, gu) \\ &< \max\left\{d(Tz, Tu), \frac{1}{2}\left[d(Tz, fz) + d(Tu, gu)\right], \frac{1}{2}\left[d(Tz, gu) + d(Tu, fz)\right]\right\} \\ &- d(z, u). \end{aligned}$$

Thus, we reach $d(z, u) < d(z, u)$, and hence $z = u$.

The following example illustrates Theorem 3.19.

Example 3.14 Let $X = [2, 20]$ and d be the usual metric on X. Define $f, g, T : X \to X$ as follows:

$$fx = \begin{cases} 3 & \text{if } x \in [2, 4], \\ 2 & \text{if } x > 4, \end{cases} \qquad gx = \begin{cases} 2 & \text{if } x \in [2, 3), \\ 3 & \text{if } x \geq 3, \end{cases}$$

$$\text{and} \qquad Tx = \begin{cases} 3 & \text{if } x = 3, \\ \frac{5}{2} & \text{if } x \in [2, 20] - \{3, 4\}, \\ 2 & \text{if } x = 4. \end{cases}$$

In this example, the mappings f, g and T do not satisfy the general Meir-Keeler contractive condition. To see this, consider $\epsilon = \frac{3}{4}$, $x = 3$ and $y \in [2, 3)$;

then for any $\delta > 0$, we have $\epsilon \leq m(x,y) < \epsilon + \delta$, but $d(fx, gy) = d(3,2) = 1 > \epsilon$. However, f, g and T satisfy the generalized Meir-Keeler α_T-contractive condition (3.33) with the mapping $\alpha : T(X) \times T(X) \to [0, \infty)$ defined by

$$\alpha(u,v) = \begin{cases} 2 & \text{if } u, v \in \{2,3\}, \\ \\ 0 & \text{otherwise.} \end{cases}$$

Also, all the hypotheses of Theorem 3.18 with condition (H) are satisfied, and clearly $x = 3$ is our unique common fixed point. Indeed, hypothesis (2) is satisfied with $x_0 = 3$, and here $Tx_n = 3$ is a sequence, for which hypotheses (3) and (4) are satisfied. Also in view of the sequence $x_n = 3$, here both pairs (f,T) and (g,T) are reciprocal continuous and absorbing. Notice that $x = 3$ is the point of discontinuity of the mappings g and T.

Theorem 3.20 *The conclusion of Theorem 3.18 remains true if the assumption (5) of Theorem 3.18 is replaced by one of the following conditions:*

(a) $d(gx, Ty) \leq \max\{d(y, gx), d(y, Tx)\}$ *for all* $x, y \in X$,

(b) $d(fx, Ty) \leq \max\{d(y, Tx), d(y, fx)\}$ *for all* $x, y \in X$,

with right-hand side positive in both (a) and (b).

Proof. In view of Theorem 3.18, $\{Tx_n\}$ is a Cauchy sequence, and $Tx_n \to z \in X$ as $n \to \infty$, and consequently, fx_{2n} and gx_{2n+1} also converge to z as $n \to \infty$.

Clearly, $Tx_n \neq z$ for infinitely many n. We can as well assume that $Tx_n \neq z$ for all n.

If (a) holds, then

$$d(gx_{2n+1}, Tz) \leq \max\{d(z, gx_{2n+1}),\ d(z, Tx_{2n+1})\}.$$

Letting $n \to \infty$, we get $d(z, Tz) \leq 0$, *i.e.*, $Tz = z$. If (b) holds, then $Tz = z$.

Now suppose $z \neq gz$. Since $Tx_{2n} \neq Tx_{2n+1}$, by assumption (4) and Equation (3.35),

we have

$$\begin{aligned} d(fx_{2n}, gz) &\leq \alpha(Tx_{2n}, Tz) d(fx_{2n}, gz) \\ &< \max\left\{ d(Tx_{2n}, Tz),\ \frac{1}{2}\left[d(Tx_{2n}, fx_{2n}) + d(Tz, gz)\right], \right. \\ &\qquad \left. \frac{1}{2}\left[d(Tx_{2n}, gz) + d(Tz, fx_{2n})\right] \right\}, \end{aligned}$$

and letting $n \to \infty$, we get $d(z, gz) \leq \frac{1}{2} d(z, gz)$, which implies that $z = gz$.

Now letting $fz \neq z = Tz$, by the process above, we have

$$
\begin{aligned}
d(fz, gx_{2n+1}) &\leq \alpha(Tx_{2n+1}, Tz)d(fz, gx_{2n+1}) \\
&< \max\left\{d(Tz, Tx_{2n+1}),\ \frac{1}{2}\left[d(Tz, fz) + d(Tx_{2n+1}, gx_{2n+1})\right],\right. \\
&\qquad \left.\frac{1}{2}\left[d(Tz, gx_{2n+1}) + d(Tx_{2n+1}, fz)\right]\right\},
\end{aligned}
$$

and letting $n \to \infty$, we get $d(fz, z) \leq \frac{1}{2}d(z, fz)$, which implies that $fz = z$. Thus, z is the common fixed point of f, g and T.

The following example demonstrates Theorem 3.20.

Example 3.15 Let $X = [0, 1]$ and d be the usual metric on X. Define $f, g, T : X \to X$ as follows:

$$
fx = \begin{cases} 0 & \text{if } x \in [0, \frac{1}{4}], \\ \frac{1}{20} & \text{if } x \in \left(\frac{1}{4}, \frac{1}{2}\right), \\ \frac{1}{4} & \text{if } x \in \left[\frac{1}{2}, 1\right] \end{cases} \qquad gx = \begin{cases} \frac{x}{3} & \text{if } x \in [0, \frac{1}{4}], \\ x & \text{if } x \in \left(\frac{1}{4}, \frac{1}{2}\right), \\ 0 & \text{if } x \in \left[\frac{1}{2}, 1\right], \end{cases}
$$

$$
\text{and} \qquad Tx = \begin{cases} \frac{x}{3} & \text{if } x \in [0, \frac{1}{4}], \\ \frac{1}{4} & \text{if } x \in \left(\frac{1}{4}, \frac{1}{2}\right), \\ \frac{x}{2} & \text{if } x \in \left[\frac{1}{2}, 1\right]. \end{cases}
$$

Here the mappings f, g and T satisfy all the conditions of Theorem 3.20 with the mapping $\alpha : T(X) \times T(X) \to [0, \infty)$ defined by

$$
\alpha(u, v) = \begin{cases} 1 & \text{if } (u, v) \in \left[0, \frac{1}{12}\right] \times \left[\frac{1}{4}, \frac{1}{2}\right], \\ 0 & \text{otherwise.} \end{cases}
$$

Clearly, no pairs (f, T) and (g, T) are reciprocal continuous. To see this, consider the sequence $x_n = \frac{1}{2} + \frac{1}{n}$, then $\lim_{n\to\infty} fx_n = \lim_{n\to\infty} Tx_n = \frac{1}{4}$, but $\lim_{n\to\infty} fTx_n = \lim_{n\to\infty} f\left(\frac{1}{4} + \frac{1}{2n}\right) = \frac{1}{20} \neq 0 = f\left(\frac{1}{4}\right)$. Therefore, (f, T) is not reciprocal continuous. To see that (g, T) is not reciprocal continuous, one can consider the sequence $y_n = \frac{1}{4} + \frac{1}{n}$ in which the involved mappings satisfy condition (a) of Theorem 3.20, and they have the unique common fixed $x = 0$.

Remark 3.14 Theorem 3.20 generalizes and extends Theorem 1.2 of Rao and Rao [51].

Theorem 3.21 *Theorem 3.18 remains true if we replace $m_3(x, y)$ by $M_3(x, y)$ and condition (4) by the following (4′):*

(4′) $\alpha(Tx_n, Tx_{n+1}) \geq 1$ *for all* n *and* $Tx_n \to x$ *implies that* $\alpha(Tx_n, Tx) \geq K$ *for all* n*, where* $K > 1$.

Proof. The proof of $z = fz = Tz$ follows from Theorem 3.18. Suppose that $z \neq gz$; then by the help of condition (4′), we have

$$\begin{aligned} d(fx_{2n}, gz) &\leq K^{-1}\alpha(Tx_{2n}, Tz)d(fx_{2n}, gz) < K^{-1}M_3(x_{2n}, z) \\ &= K^{-1}\max\{d(Tx_{2n}, Tz),\ d(Tx_{2n}, fx_{2n}),\ d(Tz, gz), \\ &\qquad \frac{1}{2}\left[d(Tx_{2n}, gz) + d(Tz, fx_{2n})\right]\right\}. \end{aligned}$$

By letting $n \to \infty$, we conclude that $d(z, gz) \leq K^{-1}d(z, gz) < d(z, gz)$, and hence $z = gz$. Thus, z is a common fixed point of f, g and T.

Example 3.14 also satisfies Theorem 3.21.

Remark 3.15 Theorem 3.21 generalizes and extends Theorem 1.3 of Rao and Rao [51].

Using $T = I$ (identity) in Theorem 3.21, we derive the following result as a corollary.

Corollary 3.3 *Let* (X, d) *be an* (f, g)*-orbitally complete metric space, where* f, g *are self mappings of* X*. Also, let* $\alpha : X \times X \to [0, \infty)$ *be a mapping. Assume the following:*

(1) (f, g) *is* α*-admissible, and there exists* $x_0 \in X$ *such that* $\alpha(x_0, fx_0) \geq 1$;

(2) for given $\epsilon > 0$, $\delta > 0$ *such that* $\epsilon \leq M(x, y) < \epsilon + \delta$ *implies that* $\alpha(x, y)d(fx, gy) < \epsilon$,
where $M(x, y) = \left\{d(x, y),\ d(x, fx),\ d(y, gy),\ \frac{1}{2}\left[d(x, gy) + d(y, fx)\right]\right\}$;

(3) on the (f, g)*-orbit of* x_0, $\alpha(x_n, x_j) \geq 1$ *for all* n *even and* $j > n$ *odd;*

(4) $\alpha(x_n, x_{n+1}) \geq 1$ *for all* n, *and* $x_n \to x$ *implies that* $\alpha(x_n, x) \geq K$ *for all* n*, where* $K > 1$.

Then pair (f, g) *has a common fixed point provided that it is absorbing and reciprocal continuous.*

Remark 3.16 Corollary 3.3 improves Theorem 3.16.

Similarly the common fixed point of four self mappings can be obtained by defining the α-admissibility for the four self mappings as follows:

Definition 3.17 Let $f, g, S, T : X \to X$ be four self mappings of a nonempty set X, and let $\alpha : S(X) \cup T(X) \times S(X) \cup T(X) \to [0, \infty)$ be a mapping; then the pair (f, g) is called α-admissible with respect to S and T (in short, (f, g) is $\alpha_{S,T}$-admissible) if for all $x, y \in X$,

$$\alpha(Sx, Ty) \geq 1 \text{ or } \alpha(Tx, Sy) \geq 1 \Rightarrow \alpha(fx, gy) \geq 1 \text{ and } \alpha(gx, fy) \geq 1. \tag{3.41}$$

Clearly, if $S = T = I$ (identity), the definition implies Definitions 3.6 and 3.14.

Theorem 3.22 *Let f, g, S and T be four self mappings on a complete metric space (X, d) such that $f(X) \subseteq T(X)$ and $g(X) \subseteq S(X)$, and they satisfy the following conditions:*

(1) the pair (f, g) is $\alpha_{S,T}$-admissible;

(2) $x_0 \in X$ such that $\alpha(Sx_0, fx_0) \geq 1$;

(3) for given $\epsilon > 0$, there exists a $\delta > 0$ such that

$$\epsilon \leq m_4(x, y) < \epsilon + \delta \implies \alpha(Sx, Ty)d(fx, gy) < \epsilon, \tag{3.42}$$

$$\text{where } m_4(x, y) = \max\Big\{d(Sx, Ty), d(fx, Sx), d(gy, Ty), \frac{1}{2}[d(fx, Ty) + d(gy, Sx)]\Big\};$$

(4) there exists a sequence $\{x_n\}$ in X such that $\alpha(Sx_n, Tx_j) \geq 1$ for all n even and $j > n$ odd.

Then f, g, S and T have a common fixed point provided that the pairs (f, S) and (g, T) are absorbing and reciprocal continuous.

Theorem 3.23 *We add condition (H-2) all common fixed points x and y of f, g, S and T, $\alpha(Sx, Ty) \geq 1$, to the hypotheses of Theorem 3.22, for the uniqueness of the fixed point to obtain.*

Remark 3.17 Theorem 3.23 generalizes, extends and improves the results of Jungck (Theorem 3.1, [29]), Cho *et al.* (Theorem 3.2, [15]) and Rao and Rao [51].

3.4 Fixed point of α-type F-contractive mappings

It is well known that Wardowski [65, 66] introduced the F-contraction and F-weak contraction to generalize Banach's contraction principle.

Definition 3.18 *[65]* Let (X, d) be a metric space. A mapping $f : X \to X$ is called an F-contraction on X if there exist $F \in \mathcal{F}$ and $\tau > 0$ such that, for all $x, y \in X$ with $d(fx, fy) > 0$, we have

$$\tau + F(d(fx, fy)) \leq F(d(x, y)).$$

Definition 3.19 *[66]* Let (X, d) be a metric space. A mapping $f : X \to X$ is called an F-weak contraction on X if there exist $F \in \mathcal{F}$ and $\tau > 0$ such that, for all $x, y \in X$ with $d(fx, fy) > 0$, we have

$$\tau + F(d(fx, fy)) \leq F\left(\max\left\{d(x, y), d(x, fx), d(y, fy), \frac{d(x, fy) + d(y, fx)}{2}\right\}\right). \tag{3.43}$$

Remark 3.18 *[66]* Every F-contraction is an F-weak contraction but the converse is not necessarily true.

On the other hand, Hussain *et al.* [25] introduced the α-GF-contraction as a generalization of the F-contraction and obtained some interesting fixed point results.

Definition 3.20 Let Δ_G denote the set of all functions $G : (\mathbb{R}^+)^4 \to \mathbb{R}^+$ satisfying the condition:

(G) for all $t_1, t_2, t_3, t_4 \in \mathbb{R}^+$ with $t_1 t_2 t_3 t_4 = 0$, there exists $\tau > 0$ such that $G(t_1, t_2, t_3, t_4) = \tau$.

Example 3.16 The following function $G : (\mathbb{R}^+)^4 \to \mathbb{R}$ belongs to Δ_G:

(1) $G(t_1, t_2, t_3, t_4) = L \min(t_1, t_2, t_3, t_4) + \tau$;

(2) $G(t_1, t_2, t_3, t_4) = \tau e^{L \min(t_1, t_2, t_3, t_4)}$.

Definition 3.21 *[25]* Let (X, d) be a metric space and $f : X \to X$. Also suppose that $\alpha, \eta : X \times X \to \mathbb{R}^+$ are two functions. We say that f is an α-GF-contraction if for all $x, y \in X$ with $\eta(x, fx) \leq \alpha(x, y)$ and $d(fx, fy) > 0$ we have

$$G(d(x, fx), d(y, fy), d(x, fy), d(y, fx)) + F(d(fx, fy)) \leq F(d(x, y))$$

where $G \in \Delta_G$ and $F \in \mathcal{F}$.

Following this direction of research, Gopal *et al.* [22] introduced new concepts of α-type F-contractive mappings and proved some fixed point and periodic point theorems concerning such contractions.

Definition 3.22 Let (X, d) be a metric space. A mapping $f : X \to X$ is said to be an α-type F-contraction on X if there exist $\tau > 0$ and two functions $F \in \mathcal{F}$ and $\alpha : X \times X \to \{-\infty\} \cup (0, +\infty)$ such that, for all $x, y \in X$ satisfying $d(fx, fy) > 0$, the following inequality holds:

$$\tau + \alpha(x, y) F(d(fx, fy)) \leq F(d(x, y)).$$

Definition 3.23 Let (X, d) be a metric space. A mapping $f : X \to X$ is said to be an α-type F-weak contraction on X if there exists $\tau > 0$ and two functions $F \in \mathcal{F}$ and $\alpha : X \times X \to \{-\infty\} \cup (0, +\infty)$ such that, for all $x, y \in X$ satisfying $d(fx, fy) > 0$, the following inequality holds:

$$\tau+\alpha(x,y)F(d(fx,fy)) \leq F\left(\max\left\{d(x,y), d(xfx), d(y,fy), \frac{d(x,fy)+d(y,fx)}{2}\right\}\right). \tag{3.44}$$

Remark 3.19 Every α-type F-contraction is an α-type F-weak contraction, but the converse is not necessarily true.

Example 3.17 Let $X = [0, \frac{5}{2}]$ and d be the usual metric on X. Define $f : X \to X$ by

$$f(x) = \begin{cases} 0 & \text{if } x \in [0, \frac{2}{5}], \\ \frac{5}{2} & \text{otherwise.} \end{cases}$$

Then, for $x = 0$ and $y = 1$, by putting $F(t) = \ln t$ with $t > 0$, we get

$$\tau + \alpha(0,1)F(d(f(0), f(1))) = \tau + \alpha(0,1)\ln\left(\frac{5}{2}\right)$$

and

$$F(d(0,1)) = \ln(1).$$

Clearly, we have

$$e^{\tau}\left(\frac{5}{2}\right)^{\alpha(0,1)} \nleq e^{\ln(1)}$$

or equivalently

$$\left(\frac{5}{2}\right)^{\alpha(0,1)} \nleq e^{-\tau}$$

for every $\tau > 0$ and $\alpha(0,1) \in (0, +\infty)$. Thus, f is not an α-type F-contraction.

Further, since

$$\max\left\{d(0,1), d(0,f(0)), d(1,f(1)), \frac{d(0,f(1))+d(1,f(0))}{2}\right\} = \frac{7}{4},$$

the inequality (3.43) reduces to

$$\frac{5}{2} \leq \frac{7}{4}e^{-\tau} < \frac{7}{4},$$

a contradiction, and hence f is not an F-weak contraction.

However, since

$$\inf_{x\in[0,\frac{2}{5}], y\in(\frac{2}{5},\frac{5}{2}]}\left\{\max\left\{d(x,y), d(x,fx), d(y,fy), \frac{d(x,fy)+d(y,fx)}{2}\right\}\right\} = \frac{5}{4},$$

f is an α-type F-weak contraction for the choice

$$\alpha(x,y) = \begin{cases} 1 & \text{if } x,y \in [0,\frac{2}{5}] \text{ or } x,y \in (\frac{2}{5},\frac{5}{2}], \\ \frac{\log 10 - \log 9}{\log 5 - \log 2} & \text{otherwise,} \end{cases}$$

and $\tau > 0$ such that $e^{-\tau} = \frac{8}{9}$.

Remark 3.20 Definition 3.22 (resp., Definition 3.23) reduces to F-contraction (resp., F-weak contraction) for $\alpha(x,y) = 1$.

The next two examples demonstrate that α-type F-contractions and α-GF-contractions [25] are independent.

Example 3.18 Let $X = [0,1]$ and d be the usual metric on X. Define $f : X \to X$ by

$$f(x) = \begin{cases} \frac{1}{2} & \text{if } x \in [0,1), \\ 0 & \text{if } x = 1. \end{cases}$$

Also define $F(t) = \ln t$ with $t > 0$.

Then f is an α-type F-weak contraction with $\alpha(x,y) = 1$ for all $x,y \in X$ and $\tau > 0$ such that $e^{-\tau} = \frac{1}{2}$. But f is not an α-GF-contraction [25]. To see this, consider $\eta : X \times X \to \mathbb{R}^+$ such that

$$\eta(x,fx) = \begin{cases} 0 & \text{if } x = \frac{1}{2}, \\ 2 & \text{otherwise,} \end{cases}$$

and

$$G(t_1,t_2,t_3,t_4) = L\min\{t_1,t_2,t_3,t_4\} + \tau.$$

Then, for $x = \frac{1}{2}$ and $y = 1$, we have

$$\eta\left(\frac{1}{2},\frac{1}{2}\right) = 0 \le 1 = \alpha\left(\frac{1}{2},1\right) = \alpha(x,y), \quad d(fx,fy) = \frac{1}{2} > 0$$

and

$$\begin{aligned} &G(d(x,fx),d(y,fy),d(x,fy),d(y,fx)) \\ &\quad = L\min\{d(x,fx),d(y,fy),d(x,fy),d(y,fx)\} + \tau = \tau. \end{aligned}$$

Consequently, we get

$$\begin{aligned} &G(d(x,fx),d(y,fy),d(x,fy),d(y,fx)) + F(d(fx,fy)) \\ &\quad = \tau + \ln\left(\frac{1}{2}\right) \not\le \ln\left(\frac{1}{2}\right) = F(d(x,y)), \end{aligned}$$

and hence f is not an α-GF-contraction.

Example 3.19 Let $X = \mathbb{R}^+$ and d be the usual metric on X. Define $f : X \to X$, $\alpha : X \times X \to \{-\infty\} \cup (0, +\infty)$, $\eta : X \times X \to \mathbb{R}^+$, $G : (\mathbb{R}^+)^4 \to \mathbb{R}^+$ and $F : \mathbb{R}^+ \to \mathbb{R}$ by

$$f(x) = \begin{cases} \frac{1}{2}e^{-\tau}x^2 & \text{if } x \in [0,1], \\ 3x & \text{if } x \in (1, +\infty), \end{cases} \qquad \alpha(x,y) = \begin{cases} 3 & \text{if } x, y \in [0,1], \\ 1 & \text{otherwise}, \end{cases}$$

$\eta(x,y) = 2$ for all $x, y \in X$, $G(t_1, t_2, t_3, t_4) = \tau > 0$ and $F(t) = \ln t$ with $t > 0$.

We can easily verify by simple calculations that f is an α-GF-contraction; but not an α-type F-weak contraction. To see this consider $x = 0$ and $y = 2$; then we get

$$\tau + \alpha(0,2)F(d(f(0), f(2))) = \tau + 1 \cdot F(6) = \tau + \ln(6)$$

whereas

$$F\left(\max\left\{d(0,2), d(0, f(0)), d(2, f(2)), \frac{d(0, f(2)) + d(2, f(0))}{2}\right\}\right) = \ln(4),$$

and so the inequality $6 \leq 4e^{-\tau}$ does not hold for any $\tau > 0$. Hence f is not an α-type F-weak contraction.

Theorem 3.24 *Let (X, d) be a complete metric space and $f : X \to X$ be an α-type-F-weak contraction satisfying the following conditions:*

(1) f is α-admissible;

(2) there exists $x_0 \in X$ such that $\alpha(x_0, fx_0) \geq 1$;

(3) f is continuous.

Then f has a fixed point $x^* \in X$ and for every $x_0 \in X$ the sequence $\{f^n x_0\}_{n\in\mathbb{N}}$ is convergent to x^*.

Proof. Let $x_0 \in X$ be such that $\alpha(x_0, fx_0) \geq 1$. We define a sequence $\{x_n\}$ in X by the rule $x_{n+1} = fx_n$ for all $n \in \mathbb{N}$.

Obviously, if there exists $n_0 \in \mathbb{N}$ for which $x_{n_0+1} = x_{n_0}$, then $fx_{n_0} = x_{n_0}$ and the proof is finished. Thus, we suppose that $x_{n+1} \neq x_n$ for every $n \in \mathbb{N}$. Now from (2) and (1), we get that

$$\alpha(x_0, x_1) = \alpha(x_0, fx_0) \geq 1 \Longrightarrow \alpha(fx_0, fx_1) = \alpha(x_1, x_2) \geq 1.$$

By induction we get

$$\alpha(x_n, x_{n+1}) \geq 1 \text{ for all } n \in \mathbb{N}. \tag{3.45}$$

Since f is an α-type F-weak contraction, then, for every $n \in \mathbb{N}$, we write

$$\begin{aligned} F(d(x_{n+1}, x_n)) &= F(d(fx_n, fx_{n-1})) \\ &\leq \alpha(x_n, x_{n-1})F(d(fx_n, fx_{n-1})). \end{aligned}$$

Consequently, we have

$$\begin{aligned}\tau + F(d(x_{n+1}, x_n)) &\leq \tau + \alpha(x_n, x_{n-1})F(d(fx_n, fx_{n-1}))\\ &\leq F\left(\max\left\{d(x_n, x_{n-1}), d(x_n, fx_n), d(x_{n-1}, fx_{n-1}),\right.\right.\\ &\qquad \left.\left.\frac{d(x_n, fx_{n-1}) + d(x_{n-1}, fx_n)}{2}\right\}\right)\\ &= F\left(\max\left\{d(x_{n-1}, x_n), d(x_n, x_{n+1}), \frac{d(x_{n-1}, x_{n+1})}{2}\right\}\right)\\ &= F(\max\{d(x_{n-1}, x_n), d(x_n, x_{n+1})\}). \qquad (3.46)\end{aligned}$$

If $n \in \mathbb{N}$ such that $\max\{d(x_{n-1}, x_n), d(x_n, x_{n+1})\} = d(x_n, x_{n+1})$, then (3.46) becomes

$$F(d(x_n, x_{n+1})) \leq F(d(x_n, x_{n+1})) - \tau,$$

a contradiction. Therefore $\max\{d(x_{n-1}, x_n), d(x_n, x_{n+1})\} = d(x_{n-1}, x_n)$ for all $n \in \mathbb{N}$. Thus, from (3.46), we have

$$F(d(x_n, x_{n+1})) \leq F(d(x_{n-1}, x_n)) - \tau \text{ for all } n \in \mathbb{N}.$$

This implies that

$$F(d(x_n, x_{n+1})) \leq F(d(x_0, x_1)) - n\tau \text{ for all } n \in \mathbb{N}. \qquad (3.47)$$

Taking the limit as $n \to +\infty$ in (3.47), we get

$$F(d(x_n, x_{n+1})) = -\infty$$

that together with (F_2) gives us

$$\lim_{n\to\infty} d(x_n, x_{n+1}) = 0. \qquad (3.48)$$

Now, from (F_3), $k \in (0,1)$ such that

$$\lim_{n\to+\infty} \left((d(x_n, x_{n+1}))^k F(d(x_n, x_{n+1}))\right) = 0. \qquad (3.49)$$

From (3.47), for all $n \in \mathbb{N}$, we deduce that

$$d(x_n, x_{n+1})^k \left(F(d(x_n, x_{n+1}) - F(d(x_0, x_1))\right) \leq -\left(d(x_n, x_{n+1})\right)^k n\tau \leq 0. \qquad (3.50)$$

Next using (3.48) and (3.49) and taking the limit as $n \to +\infty$ in (3.50), we get

$$\lim_{n\to+\infty} \left(n(d(x_n, x_{n+1}))^k\right) = 0.$$

Then $n_1 \in \mathbb{N}$ such that $n(d(x_n, x_{n+1}))^k \leq 1$ for all $n \geq n_1$, that is,

$$d(x_n, x_{n+1}) \leq \frac{1}{n^{1/k}} \text{ for all } n \geq n_1. \qquad (3.51)$$

For all $m > n > n_1$, by using (3.51) and the triangle inequality, we get

$$d(x_n, x_m) \leq d(x_n, x_{n+1}) + \cdots + d(x_m, x_{m+1})$$
$$< \sum_{i=1}^{\infty} d(x_i, x_{i+1}) \leq \sum_{i=1}^{\infty} \frac{1}{i^{1/k}}.$$

Since the series $\sum_{i=1}^{\infty} \frac{1}{i^{1/k}}$ is convergent, taking the limit as $n \to +\infty$, we get

$$\lim_{n,m \to +\infty} d(x_n, x_m) = 0.$$

This shows that $\{x_n\}$ is a Cauchy sequence.

Since X is complete, there exists $x^* \in X$ such that $\lim_{n \to +\infty} x_n = x^*$. Finally, the continuity of f yields

$$d(x^*, fx^*) = \lim_{n \to +\infty} d(x_n, fx_n) = \lim_{n \to +\infty} d(x_n, x_{n+1}) = 0.$$

i.e., x^* is a fixed point of f.

The next theorem does not require the continuity of f.

Theorem 3.25 *Let (X, d) be a metric space and $f : X \to X$ be an α-type F-weak contraction satisfying the following conditions:*

(1) there exists $x_0 \in X$ such that $\alpha(x_0, fx_0) \geq 1$;

(2) f is α-admissible;

(3) if $\{x_n\}$ is a sequence in X such that $\alpha(x_n, x_{n+1}) \geq 1$ for all $n \in \mathbb{N}$ and $x_n \to x$ as $n \to +\infty$, then $\alpha(x_n, x) \geq 1$ for all $n \in \mathbb{N}$;

(4) F is continuous.

Then f has a fixed point $x^* \in X$ and for every $x_0 \in X$ the sequence $\{f^n x_0\}_{n \in \mathbb{N}}$ is convergent to x^*.

Proof. Let $x_0 \in X$ be such that $\alpha(x_0, fx_0) \geq 1$ and let $x_n = fx_{n-1}$ for all $n \in \mathbb{N}$. Following the proof of Theorem 3.24, $\{x_n\}$ is a Cauchy sequence in the complete metric space (X, d) and $x^* \in X$ such that $x_n \to x^*$ as $n \to +\infty$. On the other hand, from (3.45) and hypothesis (3),

$$\alpha(x_n, x^*) \geq 1 \quad \text{for all} \quad n \in \mathbb{N}.$$

Case I: Suppose, for each $n \in \mathbb{N}$, there exists $i_n \in \mathbb{N}$ such that $x_{i_n+1} = fx^*$ and $i_n > i_{n-1}$. Then we have

$$x^* = \lim_{n \to +\infty} x_{i_n+1} = \lim_{n \to +\infty} fx^* = fx^*,$$

that is x^* is a fixed point of f.
Case II: Suppose $n_0 \in \mathbb{N}$ such that $x_{n+1} \neq fx^*$ for all $n \geq n_0$, i.e., $d(fx_n, fx^*) > 0$ for all $n \geq n_0$. It follows from (3.44) and (F_1) that

$$\begin{aligned}\tau + F(d(x_{n+1}, fx^*)) &= \tau + F(d(fx_n, fx^*)) \\ &\leq \tau + \alpha(x_n, x^*)F(d(fx_n, fx^*)) \\ &\leq F\left(\max\left\{d(x_n, x^*), d(x_n, fx_n), d(x^*, fx^*),\right.\right. \\ &\qquad \left.\left.\frac{d(x_n, fx^*) + d(x^*, fx_n)}{2}\right\}\right) \\ &= F\left(\max\left\{d(x_n, x^*), d(x_n, x_{n+1}, d(x^*, fx^*),\right.\right. \\ &\qquad \left.\left.\frac{d(x_n, x^*) + d(x^*, fx^*) + d(x^*, x_{n+1})}{2}\right\}\right).\end{aligned} \tag{3.52}$$

If $d(x^*, fx^*) > 0$ and because

$$\lim_{n\to+\infty} d(x_n, x^*) = \lim_{n\to+\infty} d(x_{n+1}, x^*) = 0,$$

there exists $n_1 \in \mathbb{N}$ such that for all $n \geq n_1$, we have

$$\begin{aligned}&\max\left\{d(x_n, x^*), d(x_n, x_{n+1}), d(x^*, fx^*),\right. \\ &\quad \left.\frac{d(x_n, fx^*) + d(x^*, fx^*) + d(x^*, x_{n+1})}{2}\right\} \\ &\quad = d(x^*, fx^*).\end{aligned}$$

By (3.52), we get

$$\tau + F(d(x_{n+1}, fx^*)) \leq F(d(x^*, fx^*)) \tag{3.53}$$

for all $n \geq \max\{n_0, n_1\}$. Since F is continuous, taking the limit as $n \to +\infty$ in (3.53), we obtain

$$\tau + F(d(x^*, fx^*)) \leq F(d(x^*, fx^*)),$$

a contradiction. Therefore $d(x^*, fx^*) = 0$ and hence x^* is a fixed point of f.

Example 3.17 satisfies all the hypotheses of Theorem 3.25; consequently f has at least a fixed point. Here $x = 0$ and $x = \frac{5}{2}$ are two fixed points of f.

Adding the following hypothesis to the above theorems ensures the uniqueness of the fixed point: (H): for all $x, y \in Fix(f)$, $\alpha(x, y) \geq 1$.

Theorem 3.26 *Adding condition (H) to the hypotheses of Theorem 3.24 (resp., Theorem 3.25) the uniqueness of the fixed point is obtained.*

Proof. Suppose that y^* is an another fixed point of f, so $d(fx^*, fy^*) = d(x^*, y^*) > 0$. Then,

$$\begin{aligned}\tau + F(d(x^*, y^*)) &= \tau + F(d(fx^*, fy^*)) \\ &\le \tau + \alpha(x^*, y^*)F(d(fx^*, fy^*)) \\ &\le F(d(x^*, y^*)),\end{aligned}$$

a contradiction which implies that $x^* = y^*$.

Example 3.18 satisfies all the hypotheses of Theorem 3.26; hence f has unique fixed point $x = \frac{1}{2}$.

From Remark 3.19, the following corollary is straightforward.

Corollary 3.4 *Let (X, d) be a complete metric space and $f : X \to X$ be an α-type F-contraction satisfying the hypotheses of Theorem 3.26; then f has unique fixed point.*

Also, many results in the literature can be deduced easily from Theorem 3.26. In fact, taking $\alpha(x, y) = 1$ for all $x, y \in X$ in Theorem 3.26, the following fixed point result can be derived.

Corollary 3.5 *[66] Let (X, d) be a complete metric space and $f : X \to X$ be an F-weak contraction. If F is continuous, f has a unique fixed point x^* in X.*

Since Corollary 3.5 implies the corresponding theorems in Wardowski [65], Ćirić [18] and Hardy and Rogers [24], these results are consequences of Theorem 3.26.

Gopal *et al.* [22] also obtained periodic point results for self mappings on a complete metric space. It is obvious that, if f is a map which has a fixed point x, then x is also a fixed point of f^n for every $n \in \mathbb{N}$. However, the converse is false. Indeed, let $X = [0, 1]$ and $f : X \to X$ be given by $fx = 1 - x$ for all $x \in X$. Then f has a unique fixed point at $x = \frac{1}{2}$, but $f^n = I$ (identity map on X) for each even $n > 1$, has every point of X as a fixed point. On the other hand, if $X = [0, \pi]$, the function $f : X \to X$ given by $fx = \cos x$ for all $x \in X$, is nonexpansive and every iterative of f has the same fixed point as f.

Definition 3.24 A mapping $f : X \to X$ is said to have the property (P) if $Fix(f^n) = Fix(f)$ for every $n \in \mathbb{N}$, where $Fix(f) := \{x \in X : fx = x\}$.

For further details on these properties, we refer to [28].

Theorem 3.27 *Let (X, d) be a complete metric space and $f : X \to X$ be a mapping satisfying the following conditions:*

(1) there exists $\tau > 0$ *and two functions* $F \in \mathcal{F}$ *and* $\alpha : X \times X \to \{-\infty\} \cup (0, +\infty)$ *such that*

$$\tau + \alpha(x, fx)F(d(fx, f^2x)) \leq F(d(x, fx))$$

holds for all $x \in X$ *with* $d(fx, f^2x) > 0$*;*

(2) there exists $x_0 \in X$ *such that* $\alpha(x_0, fx_0) \geq 1$*;*

(3) f *is* α*-admissible;*

(4) if $\{x_n\}$ *is a sequence in* X *such that* $\alpha(x_n, x_{n+1}) \geq 1$ *for all* $n \in \mathbb{N}$ *and* $x_n \to x$ *as* $n \to +\infty$*, then* $fx_n \to fx$ *as* $n \to +\infty$*;*

(5) if $w \in Fix(f^n)$ *and* $w \notin Fix(f)$*, then* $\alpha(f^{n-1}w, f^n w) \geq 1$*.*

Then f *has the property (P).*

Proof. Let $x_0 \in X$ be such that $\alpha(x_0, fx_0) \geq 1$. Now, for $x_0 \in X$, we define the sequence $\{x_n\}$ by the rule $x_n = f^n x_n = fx_{n-1}$. By (3), we get

$$\alpha(x_1, x_2) = \alpha(fx_0, fx_1) \geq 1$$

and by induction we write

$$\alpha(x_n, x_{n+1}) \geq 1 \text{ for all } n \in \mathbb{N}. \tag{3.54}$$

If $n_0 \in \mathbb{N}$ such that $x_{n_0} = x_{n_0+1} = fx_{n_0}$, then x_{n_0} is a fixed point of f and the proof is finished. Hence, we assume $x_n \neq x_{n+1}$ or $d(fx_{n-1}, f^2x_{n-1}) > 0$ for all $n \in \mathbb{N}$.

From (3.54) and (1), we have

$$\begin{aligned} \tau + F(d(x_n, x_{n+1})) &= \tau + F(d(fx_{n-1}, f^2x_{n-1})) \\ &\leq \tau + \alpha(x_{n-1}, fx_{n-1})F(d(fx_{n-1}, f^2x_{n-1})) \\ &\leq F(d(x_{n-1}, fx_{n-1})) \end{aligned}$$

or equivalently

$$F(d(x_n, x_{n+1}) \leq F(d(x_{n-1}, x_n)) - \tau.$$

By using a similar reasoning as in the proof of Theorem 3.24, $\{x_n\}$ is a Cauchy sequence and hence the completeness of (X, d) ensures that $x^* \in X$ such that $x_n \to x^*$ as $n \to +\infty$.

From (4), we get $x_{n+1} = fx_n \to fx^*$ as $n \to \infty$, that is $x^* = fx^*$. Thus, f has a fixed point and $Fix(f^n) = Fix(f)$ is true for $n = 1$. Let $n > 1$ and assume, by contradiction, that $w \in Fix(f^n)$ and $w \notin Fix(f)$, then $d(w, fw) > 0$. Now applying (5) and (1), we get

$$\begin{aligned} F(d(w, fw)) &= F(d(f(f^{n-1}w), f^2(f^{n-1}w)) \\ &\leq \alpha(f^{n-1}w, f(f^n w))F(d(f(f^{n-1}w), f^2(f^{n-1}w))) \end{aligned}$$

and

$$\tau + F(d(w, fw)) \leq \tau + \alpha(f^{n-1}w, f^n w)F(d(f(f^{n-1}w), f^2(f^{n-1}w))) \\ \leq F(d(f^{n-1}w, f^n w)).$$

Consequently, we write

$$\begin{aligned} F(d(w, fw)) &\leq F(d(f^{n-1}w, f^n w)) - \tau \\ &\leq F(d(f^{n-2}w, f^{n-1}w)) - 2\tau \\ &\leq \cdots \\ &\leq F(d(w, fw) - n\tau. \end{aligned}$$

By taking limit as $n \to +\infty$ in the above inequality, $F(d(w, fw)) = -\infty$, which is a contradiction and hence we deduce that $d(w, fw) = 0$. Therefore, $Fix(f^n) = Fix(f)$ for all $n \in \mathbb{N}$.

Using in Theorem 3.27, $\alpha(x, y) = 1$ for all $x, y \in X$, we get the following result which is partial generalization of Theorem 4 of Abbas *et al.* [1].

Corollary 3.6 *Let (X, d) be a complete metric space and $f : X \to X$ be a continuous mapping satisfying*

$$\tau + F(d(fx, f^2x)) \leq F(d(x, fx))$$

for some $\tau > 0$ and for all $x \in X$ such that $d(fx, f^2x) > 0$. Then f has property (P).

Most recently, Piri and Kumam [49] (resp. Minak *et al.* [41]) extended the results of Wardowski [65] by introducing the F-Suzuki contraction (resp. almost F-contraction) and obtained some interesting fixed point results.

Definition 3.25 *[41]* Let (X, d) be a metric space, $f : X \to X$ be a mapping. Then the mapping f is said to be an almost F-contraction if $F \in \mathcal{F}$ and $\tau > 0$, $L \geq 0$ such that

$$d(fx, fy) > 0 \implies \tau + F(d(fx, fy)) \leq F(d(x, y) + Ld(y, fx))$$

$$\text{and} \quad d(fx, fy) > 0 \implies \tau + F(d(fx, fy)) \leq F(d(x, y) + Ld(x, fy))$$

for all $x, y \in X$.

Definition 3.26 *[59, 49]* Let us denote by $\mathcal{G}$ the set of all functions $F : \mathbb{R}^+ \to \mathbb{R}$ satisfying the following conditions

(G1) F is strictly increasing, that is $\alpha < \beta \Rightarrow F(\alpha) < F(\beta)$ *for all* $\alpha, \beta \in \mathbb{R}^+$.

(G2) There is a sequence $\{\alpha_n\}$ of positive real numbers such that $\lim_{n\to\infty} F(\alpha_n) = -\infty$, or $\inf F = -\infty$.

(G3) F is continuous on $(0, \infty)$.

Example 3.20 The following function $F : \mathbb{R}^+ \to \mathbb{R}$ belongs to $\mathcal{G}$:

$$F(\alpha) = -\frac{1}{\alpha}, \; F(\alpha) = -\frac{1}{\alpha} + \alpha, \; F(\alpha) = \ln \alpha \text{ where } \alpha > 0.$$

Lemma 3.5 *[49] Let $F : \mathbb{R}^+ \to \mathbb{R}$ be an increasing function and $\{\alpha_n\}$ be a sequence of positive real numbers. Then the following holds*

(1) if $\lim_{n\to\infty} F(\alpha_n) = -\infty$, then $\lim_{n\to\infty} \alpha_n = 0$;

(2) if $\inf F = -\infty$, and $\lim_{n\to\infty} \alpha_n = 0$, then $\lim_{n\to\infty} F(\alpha_n) = -\infty$.

Definition 3.27 *[49]* Let (X, d) be a metric space. A mapping $f : X \to X$ is said to be an F-Suzuki contraction if $\tau > 0$ such that for all $x, y \in X$ with $fx \neq fy$

$$\frac{1}{2}d(x, fx) < d(x, y) \implies \tau + F(d(fx, fy)) \leq F(d(x, y))$$

where $F \in \mathcal{G}$.

Following this direction, Budhia *et al.* [13] introduced two new concepts: α-type almost-F-contraction and α-type F-Suzuki contraction and prove fixed point theorems for such mappings in complete metric space.

Definition 3.28 Let (X, d) be a metric space, $f : X \to X$ be a mapping and $\alpha : X \times X \to \{-\infty\} \cup (0, \infty)$ be a symmetric function. Then the mapping f is said to be an α-type almost F-contraction if $F \in \mathcal{F}$ and there exist $\tau > 0$ and $L \geq 0$ such that

$$d(fx, fy) > 0 \implies \tau + \alpha(x, y)F(d(fx, fy)) \leq F(d(x, y) + Ld(y, fx))$$

$$\text{and } d(fx, fy) > 0 \implies \tau + \alpha(x, y)F(d(fx, fy)) \leq F(d(x, y) + Ld(x, fy))$$

for all $x, y \in X$.

Example 3.21 Let $X = [0, 3] \cup [5, 6]$ with usual metric and $f : X \to X$ be defined as

$$fx = \begin{cases} 0 & \text{if } x \in [0, 3], \\ 3 & \text{if } x \in [5, 6] \end{cases}$$

and $F(\alpha) = \ln \alpha$.
Then f is not an almost F-contraction. Since at $x = 3$ and $y = 5$, $d(fx, fy) > 0$ but $\tau + F(d(fx, fy)) = \tau + F(3)$ whereas $F(d(x, y) + Ld(x, fy)) = F(2)$.
Define

$$\alpha(x, y) = \begin{cases} 1 & \text{if } x, y \in [0, 3] \text{ or } x, y \in [5, 6], \\ 0.1 & \text{otherwise.} \end{cases}$$

Then f is α-type-almost F-contraction with $\tau = 0.5$ and $L = 3$.

Definition 3.29 Let (X, d) be a metric space $f : X \to X$ be a mapping and $\alpha : X \times X \to \{-\infty\} \cup (0, \infty)$ be a symmetric function. A map $f : X \to X$ is said to be an α-type F-Suzuki contraction if $\tau > 0$ such that for all $x, y \in X$ with $fx \neq Ty$

$$\frac{1}{2}d(x, fx) < d(x, y) \implies \tau + \alpha(x, y)F(d(fx, fy)) \leq F(d(x, y))$$

where $F \in \mathcal{G}$.

Example 3.22 Let $X = [1, 3] \cup [5, 9]$ with usual metric and $F : \mathbb{R}^+ \to \mathbb{R}$ be defined as $F(\alpha) = -\frac{1}{\alpha}$. Define a mapping $T : X \to X$ as

$$fx = \begin{cases} 5 & \text{if } x \in [1, 3], \\ 8 & \text{if } x \in [5, 9]. \end{cases}$$

Then f is not F-Suzuki contraction as the condition

$$\frac{1}{2}d(x, fx) < d(x, y) \implies \tau + F(d(fx, fy)) \leq F(d(x, y))$$

fails for $x = 3$ and $y = 5$.
Define $\alpha : X \times X \to \{-\infty\} \cup (0, \infty)$ as

$$\alpha(x, y) = 2, \text{ for all } x, y \in X.$$

Then f is an α-type F-Suzuki contraction, i.e.,

$$\frac{1}{2}d(x, fx) < d(x, y) \implies \tau + \alpha(x, y)F(d(fx, fy)) \leq F(d(x, y))$$

holds for all $x, y \in X$ with $\tau = \frac{1}{6}$.

Theorem 3.28 *Let (X, d) be a complete metric space and $f : X \to X$ be an α-type-almost F-contraction where $F \in \mathcal{F}$, satisfying the following conditions:*

(1) f is α-admissible;

(2) there exists $x_0 \in X$ such that $\alpha(x_0, fx_0) \geq 1$;

(3) if $\{x_n\}$ is a sequence in X such that $\alpha(x_n, x_{n+1}) \geq 1$ for all $n \in \mathbb{N} \cup \{0\}$ and $x_n \to x \in X$ as $n \to \infty$, then $\alpha(x_n, x) \geq 1$ for all $n \in \mathbb{N} \cup \{0\}$.

Then f has a fixed point $x^* \in X$.

Proof. Let $x_0 \in X$ be such that $\alpha(x_0, fx_0) \geq 1$. Define the sequence $\{x_n\}$ in X by $x_{n+1} = fx_n$, for all $n \in \mathbb{N} \cup \{0\}$. If $x_{n+1} = x_n$ for some $n \in \mathbb{N}$, then $x^* = x_n$ is a fixed point of f. Let us assume that $x_{n+1} \neq x_n$ for all $n \in \mathbb{N} \cup \{0\}$. Since f is α-admissible, $\alpha(x_0, x_1) = \alpha(x_0, fx_0) \geq 1$ which implies

$$\alpha(fx_0, fx_1) = \alpha(x_1, x_2) \geq 1.$$

Continuing in this way,

$$\alpha(x_n, x_{n+1}) \geq 1 \text{ for all } n \in \mathbb{N}. \tag{3.55}$$

$F(d(x_{n+1}, x_n)) = F(d(fx_n, fx_{n-1})) \leq \alpha(x_n, x_{n-1})F(d(fx_n, fx_{n-1}))$.
Therefore,

$$\begin{aligned}\tau + F(d(fx_n, fx_{n-1})) &\leq \tau + \alpha(x_n, x_{n-1})F(d(fx_n, fx_{n-1})) \\ &\leq F(d(x_n, x_{n-1}) + Ld(x_n, fx_{n-1})).\end{aligned}$$

$F(d(x_{n+1}, x_n)) = F(d(fx_n, fx_{n-1})) \leq F(d(x_n, x_{n-1})) - \tau$. In general

$$F(d(x_{n+1}, x_n)) = F(d(fx_n, fx_{n-1})) \leq F(d(x_1, x_0) - n\tau. \tag{3.56}$$

Thus as $n \to \infty$, we have $\lim_{n\to\infty} F(d(x_{n+1}, x_n)) = -\infty$, then by (F2) $\lim_{n\to\infty} d(x_{n+1}, x_n) = 0$. Now, from (F3), $k \in (0, 1)$ such that

$$\lim_{n\to\infty} (d(x_{n+1}, x_n))^k F(d(x_{n+1}, x_n)) = 0.$$

From (3.56) it follows that

$$(d(x_{n+1}, x_n))^k F(d(x_{n+1}, x_n)) \leq (d(x_{n+1}, x_n))^k (F(d(x_1, x_0)) - n\tau).$$

Then as $n \to \infty$ we get

$$\lim_{n\to\infty} n(d(x_{n+1}, x_n))^k = 0.$$

Therefore, $n_0 \in \mathbb{N}$ such that

$$n(d(x_{n+1}, x_n))^k \leq 1, \ \forall\, n \geq n_0$$

$$i.e. \quad d(x_{n+1}, x_n) \leq \frac{1}{n^{1/k}}, \ \forall\, n \geq n_0.$$

Now, for $m > n > n_0$,

$$\begin{aligned}d(x_n, x_m) &\leq d(x_n, x_{n+1}) + d(x_{n+1}, x_{n+2}) + \ldots + d(x_{m-1}, x_m) \\ &\leq \sum_{n\geq n_0} \frac{1}{n^{1/k}}\end{aligned}$$

which is convergent as $k \in (0, 1)$. Therefore as $m, n \to \infty$ we get $d(x_n, x_m) \to 0$. Hence $\{x_n\}$ is a Cauchy sequence. From completeness of X, $x^* \in X$ such that $x_n \to x^*$.
Now we claim that $d(x_{n+1}, fx^*) = d(fx_n, fx^*) \to 0$ as $n \to \infty$. If $x^* = fx^*$, the proof is finished. Assume that $x^* \neq fx^*$. If $x_{n+1} = fx_n = fx^*$ for infinite values of $n \in \mathbb{N} \cup \{0\}$; then the sequence has a subsequence that converges

to fx^* and the uniqueness of limit implies $x^* = fx^*$. We can assume that $fx_n \neq fx^*$ for all $n \in \mathbb{N} \cup \{0\}$. Now using (3),

$$\tau + F(d(fx_n, fx^*)) \leq \tau + \alpha(x_n, x^*)F(d(fx_n, fx^*)) \leq F(d(x_n, x^*) + Ld(fx_n, x^*)).$$

Then, as $n \to \infty$ we get

$$\tau + \lim_{n\to\infty} F(d(fx_n, fx^*)) \leq -\infty$$

which leads to a contradiction of the assumption that $\lim_{n\to\infty} d(fx_n, fx^*) > 0$ (in respect of (F2)). Thus $x_{n+1} = fx_n \to fx^*$ as $n \to \infty$ and hence $fx^* = x^*$.

Theorem 3.29 *If we further assume that $\alpha(x, y) \geq 1$ for all $x, y \in Fix(f)$ and f satisfies the condition: $G \in \mathcal{F}$ and some $L \geq 0, \tau > 0$ such that for all $x, y \in X$*

$$\tau + \alpha(x, y)G(d(fx, fy)) \leq G(d(x, y) + Ld(x, fx))$$

holds. Then the fixed point result is above unique.

Proof. Let $y^* \in X$, $y^* \neq x^*$ such that $fy^* = y^*$. Then $d(fx^*, fy^*) > 0$ which implies

$$\tau + G(d(fx^*, fy^*)) \leq \tau + \alpha(x^*, y^*)G(d(fx^*, fy^*)) \leq G(d(x^*, y^*) + L(x^*, fx^*)).$$

Therefore we have

$$\tau + G(d(x^*, y^*)) = \tau + G(d(fx^*, fy^*)) \leq G(d(x^*, y^*))$$

which is a contradiction as $\tau > 0$.

The following example illustrates Theorem 3.29.

Example 3.23 Let $X = [1, 3] \cup [5, 9]$ with the usual metric and $F : \mathbb{R}^+ \to \mathbb{R}$ be defined as $F(\alpha) = \ln \alpha$ Define a continuous map $f : X \to X$ as

$$fx = \begin{cases} 2x + 3 & x \in [1, 3], \\ 9 & x \in [5, 9]. \end{cases}$$

Then f satisfies all the conditions of Theorem 3.2 for $\tau \leq 1.9408$ and $L = 1$ and hence f has a unique fixed point $x^* = 9$.

If $\alpha(x, y) = 1$ for all $x, y \in X$ then we have following result as in [41].

Corollary 3.7 *Let (X, d) be a complete metric space and $f : X \to X$ be an almost F-contraction. Then f has a fixed point x^* in X.*

If $\alpha(x, y) = 1$ for all $x, y \in X$ and $L = 0$, we have the following result of Wardowski [65].

Corollary 3.8 *Let (X, d) be a complete metric space and $f : X \to X$ be an F-contraction. Then f has a unique fixed point x^* in X.*

To prove the next result Budhia *et al.* [13] used the following property.

Definition 3.30 *[5]* An α-admissible map f is said to have K-property whenever for each sequence $\{x_n\} \subseteq X$ with $\alpha(x_n, x_{n+1}) \geq 1$ for all $n \in \mathbb{N} \cup \{0\}$ and there exists natural number k such that $\alpha(fx_m, fx_n) \geq 1$ for all $m > n \geq k$.

Theorem 3.30 *Let (X, d) be a complete metric space and $f : X \to X$ be an α-type F-Suzuki contraction satisfying following conditions:*

(1) f is α-admissible;

(2) there exists $x_0 \in X$ such that $\alpha(x_0, fx_0) \geq 1$;

(3) f has K-property;

(4) if $\{x_n\}$ is a sequence in X such that $\alpha(x_n, x_{n+1}) \geq 1$ for all $n \in \mathbb{N} \cup \{0\}$ and $x_n \to x \in X$ as $n \to \infty$, then $\alpha(x_n, x) \geq 1$ for all $n \in \mathbb{N} \cup \{0\}$.

Then f has a fixed point $x^* \in X$.

Proof. Let $x_0 \in X$ be such that $\alpha(x_0, fx_0) \geq 1$. Define the sequence $\{x_n\} \subseteq X$ by $x_{n+1} = fx_n$, for all $n \in \mathbb{N} \cup \{0\}$. Since f is α-admissible we have $\alpha(x_0, x_1) = \alpha(x_0, fx_0) \geq 1$ which implies $\alpha(fx_0, fx_1) = \alpha(x_1, x_2) \geq 1$. Continuing in this way we have in general

$$\alpha(x_n, x_{n+1}) \geq 1 \text{ for all } n \in \mathbb{N} \cup \{0\}. \tag{3.57}$$

If $x_{n+1} = x_n$ for some $n \in \mathbb{N} \cup \{0\}$, then $x^* = x_n$ is a fixed point of f. Let us assume that $x_{n+1} \neq x_n$ for all $n \in \mathbb{N} \cup \{0\}$. Therefore $\frac{1}{2}d(x_n, fx_n) < d(x_n, fx_n)$ for all $n \in \mathbb{N} \cup \{0\}$ and hence

$$\begin{aligned} \tau + F(d(fx_n, f^2x_n)) &\leq \tau + \alpha(x_n, fx_n)F(d(fx_n, f^2x_n)) \\ &\leq F(d(x_n, fx_n)) \end{aligned}$$

for all $n \in \mathbb{N} \cup \{0\}$. So $F(d(fx_n, f^2x_n)) \leq F(d(x_n, fx_n)) - \tau$. Repeating this process in general we get

$$F(d(fx_n, f^2x_n)) \leq F(d(x_0, x_1)) - n\tau.$$

As $n \to \infty$, we obtain

$$\lim_{n\to\infty} F(d(x_{n+1}, x_{n+2})) = -\infty,$$

which together with (G2) and Lemma 3.5, gives

$$\lim_{n\to\infty} d(x_{n+1}, x_{n+2}) = 0.$$

Suppose $\{x_n\}$ is not a Cauchy sequence. Then $\varepsilon > 0$ and $p(n) > q(n) > n \geq k$ such that

$$d(x_{p(n)}, x_{q(n)}) \geq \varepsilon \text{ and } d(x_{(p(n)-1)}, x_{q(n)}) < \varepsilon.$$

Now

$$\varepsilon \le d(x_{p(n)}, x_{q(n)}) \le d(x_{p(n)}, x_{p(n)-1}) + d(x_{p(n)-1}, x_{q(n)}) < d(x_{p(n)}, x_{p(n)-1}) + \varepsilon.$$

Therefore

$$\lim_{n\to\infty} d(x_{p(n)}, x_{q(n)}) = \varepsilon. \tag{3.58}$$

Again we have

$$d(x_{p(n)}, x_{q(n)}) \le d(x_{p(n)}, x_{p(n)+1}) + d(x_{p(n)+1}, x_{q(n)+1}) + d(x_{q(n)+1}, x_{q(n)})$$

and

$$d(x_{p(n)+1}, x_{q(n)+1}) \le d(x_{p(n)+1}, x_{p(n)}) + d(x_{p(n)}, x_{q(n)}) + d(x_{q(n)}, x_{q(n)+1})$$

So as $n \to \infty$, from the above two inequalities we have

$$\lim_{n\to\infty} d(x_{p(n)+1}, x_{q(n)+1}) = \varepsilon. \tag{3.59}$$

Therefore $k \in \mathbb{N}$ such that $\frac{1}{2}d(x_{p(n)}, x_{q(n)}) < d(x_{p(n)}, x_{q(n)})$ for all $n \ge k$. Using K-property we have

$$\begin{aligned}\tau + F(d(fx_{p(n)}, fx_{q(n)})) &\le \tau + \alpha(x_{p(n)}, x_{q(n)})F(d(fx_{p(n)}, fx_{q(n)}))\\ &\le F(d(x_{p(n)}, x_{q(n)}).\end{aligned}$$

Since as $n \to \infty$ and using (G3), we get $\tau + F(\varepsilon) \le F(\varepsilon)$, which is a contradiction. Hence $\{x_n\}$ is a Cauchy sequence in X and converges to some x^* in X.

Next, we claim that

$$\frac{1}{2}d(x_n, fx_n) < d(x_n, x^*) \quad \textit{or}$$

$$\frac{1}{2}d(fx_n, f^2x_n) < d(fx_n, x^*) \text{ for all } n \in \mathbb{N}.$$

If $m \in \mathbb{N}$ such that

$$\frac{1}{2}d(x_m, fx_m) \ge d(x_m, x^*) \text{ and } \frac{1}{2}d(fx_m, f^2x_m) \ge d(fx_m, x^*),$$

then,

$$2d(x_m, x^*) \le d(x_m, fx_m) \le d(x_m, x^*) + d(x^*, fx_m)$$

and hence

$$d(x_m, x^*) \le d(x^*, fx_m) \le \frac{1}{2}d(fx_m, f^2x_m). \tag{3.60}$$

Since $\frac{1}{2}d(x_m, fx_m) < d(x_m, fx_m)$, we have

$$\tau + F(d(fx_m, f^2x_m)) \le \tau + \alpha(x_m, fx_m)F(d(fx_m, f^2x_m)) \le F(d(x_m, fx_m))$$

which implies $F(d(fx_m, f^2x_m)) < F(d(x_m, fx_m))$ and so $d(fx_m, f^2x_m) < d(x_m, fx_m)$. Now

$$\begin{aligned} d(fx_m, f^2x_m) &< d(x_m, fx_m) \\ &\leq d(x_m, x^*) + d(x^*, fx_m) \\ &\leq \frac{1}{2}d(fx_m, f^2x_m) + \frac{1}{2}d(fx_m, f^2x_m) = d(fx_m, f^2x_m). \end{aligned}$$

Thus, for every $n \in \mathbb{N}$ either

$$\tau + F(d(fx_n, fx^*)) \leq \tau + \alpha(x_n, x^*)F(d(fx_n, fx^*)) \leq F(d(x_n, x^*)) \qquad \text{or}$$

$$\tau + F(d(f^2x_n, fx^*)) \leq \tau + \alpha(x_{n+1}, x^*)F(d(f^2x_n, fx^*)) \leq F(d(x_{n+1}, x^*)).$$

As $n \to \infty$, we get from above

$$\lim_{n\to\infty} F(d(fx_n, fx^*)) = -\infty \qquad \text{and}$$

$$\lim_{n\to\infty} F(d(f^2x_n, fx^*)) = -\infty$$

respectively. This further gives us

$$\lim_{n\to\infty} d(fx_n, fx^*) = 0 \;\text{ and }\; \lim_{n\to\infty} d(f^2x_n, fx^*) = 0.$$

Then

$$0 \leq d(x^*, fx^*) \leq d(x^*, fx_n) + d(fx_n, fx^*),$$

which as $n \to \infty$ gives us $d(x^*, fx^*) = 0$ and hence $fx^* = x^*$.

Theorem 3.31 *If we further assume that $\alpha(x, y) \geq 1$ for all $x, y \in Fix(f)$, the fixed point is unique in the above result.*

Proof. Let $y^* \in X$, $y^* \neq x^*$ such that $fy^* = y^*$. Then $\frac{1}{2}d(fx^*, x^*) = 0 < d(x^*, y^*)$ which implies

$$\tau + F(d(fx^*, fy^*)) \leq \tau + \alpha(x^*, y^*)F(d(fx^*, fy^*)) \leq F(d(x^*, y^*)).$$

Therefore, we have

$$\tau + F(d(x^*, y^*)) = \tau + F(d(fx^*, fy^*)) \leq F(d(x^*, y^*)),$$

which is a contradiction as $\tau > 0$.

The following example illustrates Theorem 3.31.

Example 3.24 Let $X = [1, 3] \cup [5, 9]$ with the usual metric and $F : \mathbb{R}^+ \to \mathbb{R}$ be defined as $F(\alpha) = -\frac{1}{\alpha}$. Define a continuous map $f : X \to X$ as

$$fx = \begin{cases} 2x + 3 & \text{if } x \in [1, 3] \\ 9 & \text{if } x \in [5, 9]. \end{cases}$$

The f is not a F-Suzuki contraction as the condition

$$\frac{1}{2}d(x, fx) < d(x, y) \implies \tau + F(d(fx, fy)) \leq F(d(x, y))$$

fails for $x = 1$ and $y = 5$.
Now, we distinguish following cases:
CASE I: If $x, y \in [1, 3]$, there are no points for which $\frac{1}{2}d(x, fx) < d(x, y)$ holds, so we are through.
CASE II: If $x, y \in [5, 9]$, $fx = fy$ so we are done.
CASE III: Let $x \in [1, 3]$ and $y \in [5, 9]$. In this case we have $d(fx, fy) \leq 4$ and $2 \leq d(x, y)$. Therefore $\tau + \alpha(x, y)F(d(fx, fy)) \leq \tau + \alpha(x, y)F(4) = \tau - \frac{\alpha(x,y)}{4}$ and $F(2) \leq F(d(x, y))$ by (F2). So for given $\alpha(x, y)$ we can choose τ for which $\tau \leq \frac{\alpha(x,y)-2}{4}$ holds. Then for $\alpha(x, y)$ and $\tau > 0$ we are done.
In particular, if we define $\alpha : X \times X \to (0, \infty) \cup \{-\infty\}$ as

$$\alpha(x, y) = 3 \text{ for all } x, y \in X.$$

Then f satisfies all the conditions of Theorem 3.31 with $\tau = \frac{1}{4}$ and hence f has a unique fixed point $x^* = 9$.

If $\alpha(x, y) = 1$ for all $x, y \in X$, we have following result:

Corollary 3.9 *Let (X, d) be a complete metric space and $f : X \to X$ be an F-Suzuki contraction. Then f has a unique fixed point x^* in X.*

3.5 Fixed point of α-ψ-contractive type mappings along with weakly α-admissible mappings

Inspired by the concept of α-admissible mapping, Sintunavarat [62] introduced weakly α-admissible mapping as follows.

Definition 3.31 *[62]* Let f be a self mapping on a nonempty set X and $\alpha : X \times X \to [0, \infty)$ be a mapping. We say that f is weakly α-admissible if the following condition holds:

$$\text{for } x \in X \text{ with } \alpha(x, fx) \geq 1 \Longrightarrow \alpha(fx, ffx) \geq 1.$$

Remark 3.21 If f is an α-admissible mapping, then f is also a weakly α-admissible mapping. In general, the converse of the previous statement is not true.

Next, we give an example to show the generalization of a weakly α-admissible mapping.

Example 3.25 Let $X = \{1, 2, 3, ...\}$. Define $f : X \to X$ and $\alpha : X \times X \to [0, \infty)$ by

$$fx = \begin{cases} 3 & \text{if } x = 1, \\ 2 & \text{if } x = 2, \\ 1 & \text{if } x = 3, \\ x - 1 & \text{if } x = 4, 5, 6, ..., \end{cases}$$

and

$$\alpha(x, y) = \begin{cases} x + y & \text{if } x, y \in \{1, 2\}, \\ \dfrac{|x - y|}{x + y} & \text{otherwise.} \end{cases}$$

It is easy to see that f is not an α-admissible mapping. Indeed, for $x = 1, y = 2$, we see that

$$\alpha(x, y) = \alpha(1, 2) = 3$$

but

$$\alpha(fx, fy) = \alpha(f1, f2) = \alpha(3, 2) = \frac{1}{5} < 1.$$

Next, we show that f is weakly α-admissible. Suppose that $x \in X$ such that $\alpha(x, fx) \geq 1$ and so $x = 2$. Now we obtain

$$\alpha(fx, ffx) = \alpha(f2, ff2) = \alpha(2, 2) = 4.$$

Here we give the fixed point results for weakly α-admissible mapping along with the α-ψ-contractive condition.

Theorem 3.32 *[62] Let (X, d) be a complete metric space and $f : X \to X$ be an α-ψ-contractive mapping satisfying the following conditions:*

(1) f is weakly α-admissible;

(2) $x_0 \in X$ such that $\alpha(x_0, fx_0) \geq 1$;

(3) f is continuous.

Then the fixed point problem of f has a solution, that is, $x^* \in X$ such that $fx^* = x^*$.

Proof. Starting from x_0 in (2), we have $(x_0, fx_0) \geq 1$. Define the sequence $\{x_n\}$ in X by $x_n = fx_{n-1}$ for all $n \in \mathbb{N}$. If $x_{n^\star} = x_{n^\star-1}$ for some $n^\star \in \mathbb{N}$, then $x_{n^\star}$ is a fixed point for f. We have nothing to prove and may assume that $x_n \neq x_{n-1}$ for all $n \in \mathbb{N}$. Since f is weakly α-admissible, $\alpha(fx_0, ffx_0) = \alpha(x_1, x_2) \geq 1$. By induction, we get

$$\alpha(x_{n-1}, x_n) \geq 1 \tag{3.61}$$

for all $n \in \mathbb{N}$. By α-ψ-contractive condition of f, we get

$$d(x_n, x_{n+1}) = d(fx_{n-1}, fx_n) \leq \alpha(x_{n-1}, x_n)d(fx_{n-1}, fx_n) \leq \psi(d(x_{n-1}, x_n))$$

for all $n \in \mathbb{N}$. By induction, we have

$$d(x_n, x_{n+1}) \leq \psi^n(d(x_0, x_1))$$

for all $n \in \mathbb{N}$.

Next, we show that $\{x_n\}$ is a Cauchy sequence in X. Let $n, m \in \mathbb{N}$ with $m > n$. By using the triangular inequality, we obtain

$$\begin{aligned} d(x_n, x_m) &\leq \sum_{k=n}^{m-1} d(x_k, x_{k+1}) \\ &\leq \sum_{k=n}^{m-1} \psi^k(d(x_0, x_1)) \\ &= \sum_{k=1}^{n-1} \psi^k(d(x_0, x_1)) - \sum_{k=1}^{m-1} \psi^k(d(x_0, x_1)). \end{aligned}$$

This implies that $d(x_n, x_m) \to 0$ as $n, m \to \infty$ because $\sum_{n=1}^{\infty} \psi^n(d(x_0, x_1)) < \infty$. This shows that $\{x_n\}$ is a Cauchy sequence in (X, d). By the completeness of X, there exists $x^* \in X$ such that $x_n \to x^*$ as $n \to \infty$. From the continuity of f, it follows that

$$x^* = \lim_{n\to\infty} x_{n+1} = \lim_{n\to\infty} fx_n = f(\lim_{n\to\infty} x_n) = fx^*,$$

that is, x^* is a fixed point of f. This completes the proof.

In the next theorem, we replace the continuity hypothesis of f by a condition of α-regularity of metric space X.

Theorem 3.33 *[62] Let (X, d) be a complete metric space and $f : X \to X$ be an α-ψ-contractive mapping satisfying the following conditions:*

(1) f is weakly α-admissible;

(2) $x_0 \in X$ such that $\alpha(x_0, fx_0) \geq 1$;

(3) X is α-regular, i.e., if $\{x_n\}$ is a sequence in X such that $\alpha(x_n, x_{n+1}) \geq 1$ for all $n \in \mathbb{N}$ and $x_n \to x \in X$ as $n \to \infty$, then $\alpha(x_n, x) \geq 1$ for all $n \in \mathbb{N}$.

Then the fixed point problem of f has a solution, that is, $x^* \in X$ such that $fx^* = x^*$.

Proof. Following the proof of Theorem 3.32, we know that $\{x_n\}$ is a Cauchy sequence in the complete metric space (X, d). Then $x^* \in X$ such that $x_n \to x^*$ as $n \to \infty$. From (3.61) and (3), we get

$$\alpha(x_n, x^*) \geq 1 \tag{3.62}$$

for all $n \in \mathbb{N}$. By (3.62), the α-ψ-contractive condition of f and the triangle inequality, we get

$$\begin{aligned} d(x^*, fx^*) &\leq d(x^*, x_{n+1}) + d(x_{n+1}, fx^*) \\ &= d(x^*, x_{n+1}) + d(fx_n, fx^*) \\ &\leq d(x^*, x_{n+1}) + \alpha(x_n, x^*)d(fx_n, fx^*) \\ &\leq d(x^*, x_{n+1}) + \psi(d(x_n, x^*)) \end{aligned}$$

for all $n \in \mathbb{N}$. Letting $n \to \infty$ in the above inequality, since ψ is continuous at $t = 0$, we obtain $d(x^*, fx^*) = 0$, that is, $fx^* = x^*$. Therefore, x^* is a fixed point of f. This completes the proof.

Remark 3.22 From Remark 3.21, Theorems 3.32 and 3.33 are generalized and complementary of Theorems 2.1 and 2.2 of Samet *et al.* [58].

Next, we give some interesting examples which support our main theorems although the results of Samet et al. [58] are not applicable.

Example 3.26 Let $X = [0, \infty)$ with the usual metric d. Define $f : X \to X$ and $\alpha : X \times X \to [0, \infty)$ by

$$fx = \begin{cases} 1.9 & \text{if } x = 1, \\ 2 & \text{if } x = 2, \\ x^2 & \text{if } x \notin \{1, 2\}, \end{cases}$$

and

$$\alpha(x, y) = \begin{cases} x + y & \text{if } x, y \in \{1, 2\}, \\ 0 & \text{otherwise.} \end{cases}$$

It is easy to see that f is not an α-admissible mapping. Therefore, the main result of Samet *et al.* [58] is not applicable here.

Next, we show that Theorem 3.33 can guarantee the fixed point of f. First, we can easily to see that f is weakly α-admissible. Clearly f is an α-ψ-contractive mapping with $\psi(t) = \frac{t}{2}$ for all $t \geq 0$. Moreover, there exists $x_0 \in X$ such that $\alpha(x_0, fx_0) \geq 1$. Finally, let $\{x_n\}$ be a sequence in X such that $\alpha(x_n, x_{n+1}) \geq 1$ for all $n \in \mathbb{N}$ and $x_n \to x \in X$ as $n \to \infty$.

Since $\alpha(x_n, x_{n+1}) \geq 1$ for all $n \in \mathbb{N}$, we get $x_n \in \{1, 2\}$ for all $n \in \mathbb{N}$ and $x \in \{1, 2\}$. Then $\alpha(x_n, x) \geq 1$ for all $n \in \mathbb{N}$.

Therefore, all the required hypotheses of Theorem 3.33 are satisfied, and f has a fixed point. Here, 0 and 2 are fixed points of f.

Theorem 3.32 and 3.33 do not claim the uniqueness of fixed point. To assure the uniqueness of the fixed point, we add the following properties:

$$(H_0) : \alpha(a, b) \geq 1 \text{ for all } a, b \in X, \text{ where } a, b \text{ are fixed points of } f.$$

Theorem 3.34 *[62] Adding condition (H_0) to the hypotheses of Theorem 3.32 (resp. Theorem 3.33) we obtain the uniqueness of the fixed point of f.*

Proof. Suppose that x^* and y^* are two fixed points of f. From the α-ψ-contractive condition, we can conclude that $d(x^*, y^*) = 0$ and we get uniqueness of the fixed point of f.

Moreover, Sintunavarat [62] investigated the generalized Ulam-Hyers stability, the well posedness and the limit shadowing for fixed point problems that satisfy some conditions.

Here, we give the definitions of generalized Ulam-Hyers stability, well posedness and limit shadowing property in the context of a fixed point problem.

Definition 3.32 Let (X, d) be a metric space and $f : X \to X$ be a mapping. The fixed point problem

$$fx = x \tag{3.63}$$

is called generalized Ulam-Hyers stable if $\xi : [0, \infty) \to [0, \infty)$ which is increasing continuous at 0 and $\xi(0) = 0$ such that for each $\varepsilon > 0$ and for each $w^* \in X$ which is an ε-solution of the fixed point equation (3.63), i.e., w^* satisfies the inequality

$$d(w^*, fw^*) \leq \varepsilon, \tag{3.64}$$

there exists a solution $x^* \in X$ of the equation (3.63) such that

$$d(x^*, w^*) \leq \xi(\varepsilon).$$

Remark 3.23 If the function ξ is defined by $\xi(t) = ct$ for all $t \geq 0$ where $c > 0$, the fixed point equation (3.63) is said to be Ulam-Hyers stable.

Definition 3.33 Let (X, d) be a metric space and $f : X \to X$ be a mapping. The fixed point problem of f is said to be well posed if satisfies the following conditions:

(1) f has a unique fixed point x^* in X;

(2) for any sequence $\{x_n\}$ in X with $\lim_{n\to\infty} d(x_n, fx_n) = 0$, $\lim_{n\to\infty} d(x_n, x^*) = 0$.

Definition 3.34 Let (X, d) be a metric space and $f : X \to X$ be a mapping. We say that the fixed point problem of f has a limit shadowing property in X if for any sequence $\{x_n\}$ in X satisfying $\lim_{n\to\infty} d(x_n, fx_n) = 0$, there exists $z \in X$ such that $\lim_{n\to\infty} d(f^n z, x_n) = 0$.

Concerning the generalized Ulam-Hyers stable, well posedness and limit shadowing property of the fixed point problem for a self map of a complete metric space satisfying the conditions of Theorem 3.34, we have the following results.

Theorem 3.35 *[62] Let (X, d) be a complete metric space. Suppose that all the hypotheses of Theorem 3.34 hold and that the function $\xi : [0,\infty) \to [0,\infty)$ defined by $\xi(t) := t - \psi(t)$ is strictly increasing and onto. Then the following assertions holds:*

(1) If $\alpha(a, b) \geq 1$ for all a, b for which are an ε-solution of the fixed point equation (3.63), the fixed point problem of f is generalized Ulam-Hyers stable.

(2) If ψ is a continuous function and $\alpha(x_n, x^*) \geq 1$ for all $n \in \mathbb{N}$ where $x_n \in X$ with $\lim_{n\to\infty} d(x_n, fx_n) = 0$ and x^* is a fixed point of f, the fixed point problem of f is well-posed.

(3) If ψ is continuous function and $\alpha(x_n, x^*) \geq 1$ for all $n \in \mathbb{N}$ where $x_n \in X$ with $\lim_{n\to\infty} d(x_n, fx_n) = 0$ and x^* is a fixed point of f, the fixed point problem of f has the limit shadowing property in X.

Proof. From the proof of Theorem (3.34), we see that f has a unique fixed point and so let x^* be a unique fixed point of f.

From the hypothesis in (1), we claim that the fixed point problem of f is generalized Ulam-Hyers stable. Let $\varepsilon > 0$ and $w^* \in X$ be a solution of (3.64), i.e,

$$d(w^*, fw^*) \leq \varepsilon.$$

It is obvious that the fixed point x^* of f satisfies inequality (3.64). From the hypothesis in (1), we get $\alpha(x^*, w^*) \geq 1$. Now we have

$$\begin{aligned} d(x^*, w^*) &= d(fx^*, w^*) \\ &\leq d(fx^*, fw^*) + d(fw^*, w^*) \\ &\leq \alpha(x^*, w^*)d(fx^*, fw^*) + d(fw^*, w^*) \\ &\leq \psi(d(x^*, w^*)) + \varepsilon. \end{aligned}$$

This implies that $d(x^*, w^*) - \psi(d(x^*, w^*)) \leq \varepsilon$, that is, $\xi(d(x^*, w^*)) \leq \varepsilon$ Therefore,

$$d(x^*, w^*) \leq \xi^{-1}(\varepsilon).$$

Since ξ^{-1} is increasing continuous at 0 and $\xi^{-1}(0) = 0$, the fixed point problem of f is generalized Ulam-Hyers stable.

Next, we prove that the fixed point problem of f is well posed under the assumption in (2). Let $\{x_n\}$ be sequence in X such that $\lim_{n\to\infty} d(x_n, fx_n) = 0$. From assumption, we get $\alpha(x_n, x^*) \geq 1$ for all $n \in \mathbb{N}$. Now, we obtain that

$$\begin{aligned} d(x_n, x^*) &\leq d(x_n, fx_n) + d(fx_n, fx^*) \\ &\leq d(x_n, fx_n) + \alpha(x_n, x^*)d(fx_n, fx^*) \\ &\leq d(x_n, fx_n) + \psi(d(x_n, x^*)) \end{aligned}$$

for all $n \in \mathbb{N}$. Letting $n \to \infty$ in the above inequality, we get $\lim_{n\to\infty} d(x_n, x^*) = 0$ and so the fixed point problem of f is well posed.

Finally, we prove that f has a limit shadowing under the assumption (3). Let $\{x_n\}$ be a sequence in X such that $\lim_{n\to\infty} d(x_n, fx_n) = 0$. Similarly to case (2), we get $\lim_{n\to\infty} d(x_n, x^*) = 0$. Since x^* is a fixed point of f, we have $\lim_{n\to\infty} d(x_n, f^n x^*) = \lim_{n\to\infty} d(x_n, x^*) = 0$. Therefore, f has the limit shadowing property.

3.6 Conclusions and future investigations

In summary, we investigated many fixed point results along with the concepts of α-admissible mappings and weakly α-admissible mappings. We noted that the following results can be concluded from theoretical results with α-admissible mappings and weakly α-admissible mappings:

(1) Fixed point results in metric spaces endowed with a binary relation such as strict order (or sharp order), near order, pseudo-order, quasi-order (or preorder), partial order, simple order, weak order, total order (or linear order or chain), tolerance, equivalence

(2) Fixed point results in metric spaces endowed with graph

(3) Fixed point results for cyclic mappings.

We present some problems for further investigation.

- Can α-admissible mappings and weakly α-admissible mappings be extended to the new notions?
- Can the ideas of α-admissible mappings and weakly α-admissible mappings be applied to the fixed point results for mappings satisfying new contractive conditions?
- Can the results involving α-admissible mappings and weakly α-admissible mappings be applied to other applications with (1) through (3) above?
- Can the ideas of α-admissible mappings and weakly α-admissible mappings be applied to the fixed point results in generalization of metric spaces such as b-metric spaces, partial metric spaces, m-metric spaces, complex-valued metric spaces, modular spaces and fuzzy metric spaces?

References

[1] Abbas M., Ali B., Romaguera S., Fixed and periodic points of generalized contractions in metric space. Fixed Point Theory Appl. 2013; Art. 243.

[2] Abbas M., Khan M.A., Radenović S., Common coupled fixed point theorems in cone metric spaces for w-compatible mappings. Appl. Math. Comput. 2010; 217:195-202.

[3] Abdeljawad T., Meir-Keeler α-contractive fixed and common fixed point theorems, Fixed Point Theory Appl. 2013; Art. 19.

[4] Abdeljawad T., Gopal D., Erratum to 'Meir-Keeler α-contractive fixed and common fixed point theorems', Fixed Point Theory Appl. 2013; Art. 110.

[5] Alikhani H., Gopal D., Miandaragh M.A., Rezapour S., Shahzad N., Some endpoint results for β-generalized weak contractive multifunctions. The Scientific World Journal. 2013; 2013: Art. ID 948472, 7 pages.

[6] Alsulami H.H., Karap?nar E., Piri H., Fixed points of generalized F-Suzuki type contraction in complete b-metric spaces. Discrete Dyn. Nat. Soc. 2015; Art. 969726.

[7] Alsulami H.H., Karap?nar E., Piri H., Fixed Points of modified F-contractive mappings in Complete metric-like spaces. Journal of Function Spaces. 2015; Art. 270971.

[8] Aydi H., Karapinar E., A Meir-Keeler common type fixed point theorem on partial metric space, Fixed Point Theory Appl. 2012; Art. 26.

[9] Aydi H., Karapinar E., New Meir-Keeler type tripled fixed point theorems on ordered partial metric spaces. Math. Prob. Eng. 2012; Art. 409872.

[10] Banach S., Sur les opérations dans les ensembles abstraits et leur application aux équations intégrales. Fund. Math. 1922; 3:133-181.

[11] Batra R., Vashistha S., Fixed points of an F-contraction on metric spaces with a graph. Intl. J. Computer Math. 2014; http://dx.doi.org/10.1080/00207160.2014.887700.

[12] Bhaskar T.G., Lakshmikantham V., Fixed point theorems in partially ordered metric spaces and applications. Nonlinear Anal. 2006; 65:1379-1393.

[13] Budhia L., Kumam P., Moreno J.M., Gopal D., Extension of almost-F and F-Suzuki contractions with graph and some applications to fractional calculus. Fixed Point Theory Appl. 2016; 2.

[14] Chandok S., Some fixed point theorems for α-β-admissible Geraghty type contractive mappings and related results. Math. Sci., 2015; 9:127-135.

[15] Cho Y.J., Murthy P.P., Jungck G., A common fixed point theorem of Meir and Keeler type, Int. J. Math. Sci. 1993; 16(4):669-674.

[16] Cho Y.J., Murthy P.P., Jungck G., A theorem of Meir-Keeler type revisited, Int. J. Math. Math. Sci. 2000; 23(7):507-511.

[17] Choudhury B.S., Kundu A., A coupled coincidence point result in partially ordered metric spaces for compatible mappings. Nonlinear Anal. 2010; 73:2524-2531.

[18] Ćirić L.B., A generalization of Banach's contraction principle. Proc. Amer. Math. Soc. 1974; 45:267-273.

[19] Ćirić L.B., Lakshmikantham V. , Coupled random fixed point theorems for nonlinear contractions in partially ordered metric spaces, Stoch. Anal. Appl. 2009; 27:1246-1259.

[20] Cosentino M., Vetro P., Fixed point results for F-contractive mappings of Hardy-Rogers-type. Filomat. 2014; 28(4):715-722.

[21] Du W.S., Coupled fixed point theorems for nonlinear contractions satisfied Mizoguchi-Takahashi's condition in quasiordered metric spaces, Fixed Point Theory Appl. 2010; 2010: Art. 876372.

[22] Gopal D., Abbas M., Patel D.K., Vetro C., Fixed points of α-type F-contractive mappings with an application to nonlinear fractional differential equation. Acta Math. Sci. 2016; 36B(3):1-14.

[23] Gopal D., Abbas M., Vetro C., Patel D.K., Some coincidence and periodic points results in a metric space endowed with a graph and applications. Banach J. Math. Anal. 2015; 9(3):128-139.

[24] Hardy G.E., Rogers T.D., A generalization of a fixed point theorem of Reich. Canadian Math. Bull. 1973; 16:201-206.

[25] Hussain N., Salimi P., Suzuki-Wardowski type fixed point theorems for α-GF-contractions. Taiwanese J. Math. 2014; 18:1879-1895.

[26] Jachymski J., Equivalent conditions and the Meir-Keeler type theorems. J. Math. Anal. Appl. 1995; 194(1):293-303.

[27] Jachymski J., The contraction principle for mapping on a metric space with a graph. Proc. Amer. Math. Soc. 2008; 136:1359-1373.

[28] Jeong G.S., Rhoades B.E., Maps for which $F(T) = F(T^n)$. Fixed Point Theory Appl. 2005; 6:87-131.

[29] Jungck G., Compatible mappings and common fixed points. Int. J. Math. Math. Sci. 1986; 9(4):771-779.

[30] Jungck G., Moon K.B., Park S., Rhoades B.E., On generalizations of the Meir-Keeler type contractive maps: Corrections. J. Math. Anal. Appl. 1993; 180(1):221-222.

[31] Jungck G., Pathak H.K., Fixed points via "biased maps". Proc. Amer. Math. Soc. 1995; 123(7):2049-2060.

[32] Karapinar E., Couple fixed point theorems for nonlinear contractions in cone metric spaces. Comput. Math. Appl. 2010; 59:3656-3668.

[33] Karapinar E., Kutbi M.A., Piri H., O?Regan D., Fixed points of conditionally F-contractions in complete metric-like spaces. Fixed Point Theory Appl. 2015; doi:10.1186/s13663-015-0377-3.

[34] Karapinar E., Samet B., Generalized $(\alpha - \psi)$ contractive type mappings and related fixed point theorems with applications. Abstr. Appl. Anal. 2012:Art. 793486.

[35] Kirk W.A., Srinivasan P.S., Veeramani P., Fixed points for mappings satisfying cyclical contractive conditions. Fixed Point Theory. 2003; 4:79-89.

[36] Kutbi M.A., Sintunavarat W., On new fixed point results for (α, ψ, ξ)-contractive multi-valued mappings on α-complete metric spaces and their consequences, Fixed Point Theory Appl. 2015:2.

[37] Lakshmikantham V., Ćirić L.B., Coupled fixed point theorems for nonlinear contractions in partially ordered metric spaces. Nonlinear Anal. 2009; 70 :4341-4349.

[38] Latif A., Sintunavarat W., Ninsri A., Approximate fixed point theorems for partial generalized convex contraction mappings in α-complete metric spaces. Taiwanese J. Math. 2015; 19:315-333.

[39] Maiti M., Pal T.K., Generalizations on two fixed point theorems. Bull. Calcutta Math. Soc. 1978; 70:57-61.

[40] Meir A., Keeler v, A theorem on contraction mappings. J. Math. Anal. Appl. 1969; 28:326-329.

[41] Minak G., Helvaci A., Altun I., Ćirić type generalized F-contractions on complete metric spaces and fixed point results. Filomat. 2014; 28(6):1143-1151.

[42] Mohammadi B., Rezapour S., Shahzad N., Some results on fixed points of α-ψ-ciric generalized multifunctions. Fixed Point Theory Appl. 2013; doi:10.1186/1687-1812-2013-24.

[43] Nieto J.J., Rodriguez-López R., Existence and uniqueness of fixed point in partially ordered sets and applications to ordinary differential equations. Acta Math. Sin. (Engl. Ser.). 2007; 23:2205-2212.

[44] Ninsri A., Sintunavarat W., Fixed point theorems for partial α-ψ contractive mappings in generalized metric spaces. J. Nonlinear Sci. Appl. 2016; 9:83-91.

[45] Pant R.P., Joshi P.C., Gupta V., A Meir-Keeler type fixed point theorem. Indian J. Pure Appl. Math. 2001; 32(6):779-787.

[46] Park S., Bae J.S., Extensions of fixed point theorem of Meir-Keeler, Ark. Mat. 1981; 19:223-228.

[47] Park S., Rhoades B.E., Meir-Keeler type contractive condition, Math. Japonica 1981; 26(1):13-20.

[48] Patel D.K., Abdeljawad T., Gopal D., Common fixed point of generalized Meir-Keeler-contractions. Fixed Point Theory Appl. 2013; 2013:Art. ID 260.

[49] Piri H., Kumam P., Some fixed point theorems concerning F-contraction in complete metric spaces, Fixed Point Theory Appl. 2014: doi:10.1186/1687-1812-2014-210.

[50] Ran A.C.M., Reurings M.C.B., A fixed point theorem in partially ordered sets and some applications to matrix equations. Proc. Amer. Math. Soc. 2004; 132:1435-1443.

[51] Rao J.H.N., Rao K.P.R., Generalizations of fixed point theorems of Meir and Keeler type. Indian J. Pure Appl. Math. 1985; 16(1):1249-1262.

[52] Rhoades B.E., Park S., Moon K.B., On generalizations of the Meir-Keeler type contraction maps. J. Math. Anal. Appl. 1990; 146(2):482-494.

[53] Salimi P., Hussain N., Shukla S., Fathollahi S., Radenovic S., Fixed point results for cyclic α-$\psi\varphi$-contractions with application to integral equations. J. Comput. Appl. Math. 2015; 290:445-458.

[54] Salimi P., Latif A., Hussain N., Modified (α, ψ)-contractive mappings with applications. Fixed Point Theory Appl. 2013; doi:10.1186/1687-1812-2013-151.

[55] Samet B., Coupled fixed point theorems for a generalized Meir-Keeler contraction in partially ordered metric spaces. Nonlinear Anal. 2010; 72: 4508-4517.

[56] Samet B., Vetro C., Coupled fixed point, F -invariant set and fixed point of N-order. Ann. Funct. Anal. 2010; 1:46-56.

[57] Samet B., Vetro C., Coupled fixed point theorems for multi-valued nonlinear contraction mappings in partially ordered metric spaces. Nonlinear Anal. 2011; 74:4260-4268.

[58] Samet B., Vetro C., Vetro P., Fixed point theorems for α-ψ-contractive type mappings. Nonlinear Anal. 2012; 75:2154-2165.

[59] Secelean N.A., Iterated function systems consisting of F-contractions. Fixed Point Theory Appl. 2013; doi:10.1186/1687-1812-2013-277.

[60] Sessa S., On a weak commutative condition of maps in fixed point consideration. Publ. Inst. Math. (Beograd). 1982; 32(46):149-153.

[61] Shatanawi W., Nashine H.K., A generalization of Banach's contraction principle for nonlinear contraction in a partial metric space. J. Nonlinear Sci. Appl. 2012; 5:37-43.

[62] Sintunavarat W., A new approach to α-ψ-contractive mappings and generalized Ulam-Hyers stability, well posedness and limit shadowing results. Carpathian J. Math. 2015; 31:395-401.

[63] Sintunavarat W., Fixed point results in b-metric spaces approach to the existence of a solution for nonlinear integral equations, RACSAM, 2016; 110:585-600.

[64] Sintunavarat W., Nonlinear integral equations with new admissibility types in b-metric spaces. J. Fixed Point Theory Appl., 2016; 18: 397-416.

[65] Wardowski D., Fixed points of new type of contractive mappings in complete metric space. Fixed Point Theory Appl. 2012; doi:10.1186/1687-1812-2012-94.

[66] Wardowski D., Dung N.V., Fixed points of F-weak contractions on complete metric space. Demonstr. Math. 2014; 47:146-155.

[67] Yamaod O., Sintunavarat V, Some fixed point results for generalized contraction mappings with cyclic (α, β)-admissible mapping in multiplicative metric spaces. J. Inequal. Appl. 2014; 2014:488.

[68] Yamaod O., Sintunavarat V, Fixed point theorems for (α, β)-(ψ, φ)-contractive mapping in b-metric spaces with some numerical results and applications. J. Nonlinear Sci. Appl. 2016; 9:22-34.

Chapter 4

Fixed Point Theory in Fuzzy Metric Spaces

Dhananjay Gopal

4.1 Introduction

In 1965, Zadeh [33] introduced the concept of fuzzy set which transformed and stimulated most branches of science and engineering including mathematics.

A fuzzy set on a set can be defined by assigning to each element of a set a value in $[0, 1]$ representing its grade of membership in the fuzzy set. Mathematically, a fuzzy set A of X is a mapping $A : X \to [0, 1]$.

The concept of fuzziness found its place in probabilistic metric spaces due to Menger[16]. The main reason was that, in some cases, uncertainty in the distance between two points was due to fuzziness rather than randomness. With this idea, in 1975, Kramosil and Michalek [15] extended the concept of probabilistic metric spaces to the fuzzy situation. Since the definition of fuzzy metric space due to Kramozil and Michalek is closely related to the definition of Menger spaces; therefore we recall some definitions from probabilistic metric space theory.

Definition 4.1 *[22]* A mapping $F : \mathbb{R} \to \mathbb{R}^+$ is called a distribution function if it is nondecreasing and left continuous with $\inf_{t\in\mathbb{R}} F(t) = 0$ and $\sup_{t\in\mathbb{R}} F(t) = 1$.

We shall denote by D^+ the set of all distribution functions, while H will always denote the specific distribution function defined by

$$H(t) = \begin{cases} 0 & \text{if } t \leq 0, \\ 1 & \text{if } t > 0. \end{cases}$$

Definition 4.2 *[22]* A binary operation $* : [0,1] \times [0,1] \to [0,1]$ is a continuous t-norm if the following conditions hold:

(a) $*$ is commutative and associative.

(b) $*$ is continuous.

(c) $*(a,1) = a$ for all $a \in [0,1]$.

(d) $*(a,b) \leq *(c,d)$ whenever $a \leq c$ and $b \leq d$, for $a,b,c,d \in [0,1]$.

The following are three basic continuous t-norms:

(a) The minimum t-norm, say $*_M$, defined by $*_M(a,b) = \min\{a,b\}$.

(b) The product t-norm, say $*_p$, defined by $*_p(a,b) = a \cdot b$.

(c) The Lukasiewicz t-norm, say $*_L$, defined by $*_L(a,b) = \max\{a+b-1,0\}$.

These t-norms are related in the following way: $*_L \leq *_p \leq *_M$.

Definition 4.3 *[22]* A Menger probabilistic metric space (briefly, Menger PM-space) is a triple $(X,F,*)$ where X is a nonempty set, $*$ is a continuous t-norm, and F is a mapping from $X \times X$ into D^+ such that if $F_{x,y}$ denotes the value of F at the pair (x,y), the following conditions hold:

(PM1) $F_{x,y}(t) = H(t)$ if and only if $x = y$.

(PM2) $F_{x,y}(t) = F_{y,x}(t)$.

(PM3) $F_{x,y}(t+s) \geq *\,(F_{x,z}(t), F_{z,y}(s))$ for all $x,y,z \in X$ and $s,t \geq 0$.

4.2 Kramosil and Michalek fuzzy metric spaces and Grabiec's fixed point theorems

Definition 4.4 *[15]* The triple $(X,M,*)$ is a fuzzy metric space if X is a nonempty set, $*$ is a continuous t-norm and M is a fuzzy set on $X^2 \times [0,\infty)$ satisfying the following axioms:

(KM1) $M(x, y, 0) = 0$.

(KM2) $M(x, y, t) = 1$, for all $t > 0$ if and only if $x = y$.

(KM3) $M(x, y, t) = M(y, x, t)$.

(KM4) $M(x, y, t) * M(y, z, s) \leq M(x, z, t + s)$.

(KM5) The function $M(x, y, \cdot) : [0, \infty) \to [0, 1]$ is left continuous, for all $x, y, z \in X$ and $t, s > 0$.

In what follows, fuzzy metric spaces in the sense of Kramosil and Michalek [15] will be referred as KM-fuzzy metric spaces.

In 1988, M. Grabiec [2] initiated the study of fixed point theory in fuzzy metric space and established the fuzzy Banach contraction and fuzzy Edelstein contraction theorem. In order to obtain his theorems, Grabiec introduced the following notions.

Definition 4.5 *[2]* Let $(X, M, *)$ be a fuzzy metric space. Then

(a) A sequence $\{x_n\}_{n\in\mathbb{N}}$ in X is said to be convergent to $x \in X$, if $\lim_{n\to\infty} M(x_n, x, t) = 1$ for all $t > 0$.

(b) A sequence $\{x_n\}_{n\in\mathbb{N}}$ in X is said to be Cauchy (or G-Cauchy) if $\lim_{n\to\infty} M(x_n, x_{n+p}, t) = 1$ for each $p \in \mathbb{N}$ and $t > 0$.

A fuzzy metric space in which every Cauchy (or G-Cauchy) sequence is convergent is called complete (or G-complete). It is called compact if every sequence contains a convergent subsequence.

Lemma 4.1 *$M(x, y, \cdot)$ is nondecreasing for all $x, y \in X$.*

Proof. Suppose $M(x, y, t) > M(x, y, s)$ for some $0 < t < s$. Then $M(x, y, t) * M(y, y, s - t) \leq M(x, y, s) < M(x, y, t)$. By (KM2) $M(y, y, s - t) = 1$ and thus $M(x, y, t) < M(x, y, t)$, a contradiction.

Lemma 4.2 *Let $\lim_{n\to\infty} x_n = x$ and $\lim_{n\to\infty} y_n = y$. Then*

(a) $\lim_{n\to\infty} M(x_n, y_n, t) \geq M(x, y, t)$ for all $t > 0$;

(b) If $M(x, y, \cdot)$ is continuous, then $\lim_{n\to\infty} M(x_n, y_n, t) = M(x, y, t)$ for all $t > 0$.

Theorem 4.1 (Fuzzy Banach Contraction Theorem) *Let $(X, M, *)$ be a complete fuzzy metric space such that*

$$\lim_{t\to\infty} M(x, y, t) = 1 \text{ for all } x, y \in X. \tag{4.1}$$

Let $f : X \to X$ be a mapping satisfying

$$M(fx, fy, kt) \geq M(x, y, t) \tag{4.2}$$

for all $x, y \in X$ where $0 < k < 1$. Then f has a unique fixed point.

Proof. Let $x \in X$ and $x_n = f^n x$ $(n \in \mathbb{N})$. By a simple induction we get

$$M(x_n, x_{n+1}, kt) \geq M\left(x, x_1, \frac{t}{k^{n-1}}\right) \tag{4.3}$$

for all n and $t > 0$. Thus for any positive integer p we have

$$\begin{aligned} M(x_n, x_{n+p}, t) &\geq \overbrace{M\left(x_n, x_{n+1}, \frac{t}{p}\right) * \cdots M\left(x_{n+p-1}, x_{n+p}, \frac{t}{p}\right)}^{p} \\ &\geq M\left(x, x_1, \frac{t}{pk^n}\right) * \cdots * M\left(x, x_1, \frac{t}{pk^n}\right) \quad \text{by (4.3).} \end{aligned}$$

According to (4.1) we now have

$$\lim_{n\to\infty} M(x_{n+p}, x_n, t) \geq \overbrace{1 * 1 * 1 \cdots * 1}^{p} = 1,$$

i.e., $\{x_n\}$ is Cauchy and hence convergent. Call the limit y. Thus, we have

$$\begin{aligned} M(fy, y, t) &\geq M\left(fy, fx_n, \frac{t}{2}\right) * M\left(x_{n+1}, y, \frac{t}{2}\right) \\ &\geq M\left(y, x_n, \frac{t}{2k}\right) * M\left(x_{n+1}, y, \frac{t}{2}\right) \to 1 * 1 = 1 \end{aligned}$$

as $n \to \infty$ and by (KM4). By (KM2) we get $fy = y$, a fixed point.

To show uniqueness, assume $fz = z$ for some $z \in X$. Then

$$\begin{aligned} 1 \geq M(z, y, t) &= M(fz, fy, t) \geq M\left(z, y, \frac{t}{k}\right) = M\left(fz, fy, \frac{t}{k}\right) \\ &\geq M\left(z, y, \frac{t}{k^2}\right) \geq \ldots \geq M\left(z, y, \frac{t}{k^n}\right) \to 1 \end{aligned}$$

as $n \to \infty$. By (KM2), $z = y$.

Theorem 4.2 (Fuzzy Edelstein Contraction Theorem) *Let $(X, M, *)$ be a compact fuzzy metric space with $M(x, y, \cdot)$ continuous for all $x, y \in X$. Let $f : X \to X$ be a mapping satisfying*

$$M(fx, fy, t) > M(x, y, t) \tag{4.4}$$

for all $x \neq y$ and $t > 0$. Then f has a unique fixed point.

Proof. Let $x \in X$ and $x_n = f^n x (n \in \mathbb{N})$. Assume $x_n \neq x_{n+1}$ for each n (if not, $fx_n = x_n$). Now assume $x_n \neq x_m (n \neq m)$. For otherwise we get $M(x_n, x_{n+1}, t) = M(x_m, x_{m+1}, t) > M(x_{m-1}, x_m, t) > \cdots > M(x_n, x_{n+1}, t)$ where $m > n$, a contradiction. Since X is compact, $\{x_n\}$ has a convergent subsequence $\{x_{n_i}\}$. Let $y = \lim_{i\to\infty} x_{n_i}$. We also assume that $y, fy \notin \{x_n : n \in \mathbb{N}\}$ (if not, choose a subsequence with such a property). According to the above assumption we may now write

$$M(fx_{n_i}, fy, t) > M(x_{n_i}, y, t)$$

for all $i \in \mathbb{N}$ and $t > 0$. Since $M(x, y, \cdot)$ is continuous for all x, y in X, by (4.4) we obtain

$$\lim_{i\to\infty} M(fx_{n_i}, fy, t) \geq \lim_{i\to\infty} M(x_{n_i}, y, t) = 1$$

for each $t > 0$, and hence

$$\lim_{i\to\infty} fx_{n_i} = fy. \tag{4.5}$$

Similarly, we obtain

$$\lim_{i\to\infty} f^2x_{n_i} = f^2y \tag{4.6}$$

(recall that $fy \neq fx_{n_i}$ for all i). Now, observe that

$$M(x_{n_1}, fx_{n_1}, t) < M(fx_{n_1}, f^2x_{n_1}, t) < \ldots < M(x_{n_i}, fx_{n_i}, t)$$

$$\begin{aligned} < M(fx_{n_i}, f^2x_{n_i}, t) < \ldots < M(x_{n_{i+1}}, fx_{n_{i+1}}, t) \\ < M(fx_{n_{i+1}}, f^2x_{n_{i+1}}, t) < \ldots < 1 \end{aligned}$$

for all $t > 0$,. Thus $\{M(x_{n_i}, fx_{n_i}, t)\}$ and $\{M(fx_{n_i}, f^2x_{n_i}, t)\}$ $(t > 0)$ are convergent to a common limit. So by (4.5), (4.6) and ((b) of Lemma 4.1.2) we get

$$\begin{aligned} M(y, fy, t) &= M(\lim_{i\to\infty} x_{n_i}, f(\lim_{i\to\infty} x_{n_i}), t) \\ &= \lim_{i\to\infty} M(x_{n_i}, fx_{n_i}, t) \\ &= \lim_{i\to\infty} M(fx_{n_i}, f^2x_{n_i}, t) \\ &= M(fy, f^2y, t) \end{aligned}$$

for all $t > 0$.

Suppose $y \neq fy$. Then, by (4.4), $M(y, fy, t) < M(fy, f^2y, t)$ for $t > 0$, a contradiction. Hence $y = fy$, a fixed point. Uniqueness follows from (4.4).

These results of Grabiec [2] have further been generalized and extended by Subrahmanyam [26] and Vasuki [28] as follows.

Theorem 4.3 *[26] Let $(X, M, *)$ be a complete fuzzy metric space and let $f, g : X \to X$ be maps that satisfy the following conditions:*

(a) $g(X) \subseteq f(X)$;

(b) f is continuous;

(c) $M(g(x), g(y), \alpha t) \geq M(f(x), f(y), t)$ *for all* x, y *in* X *and* $0 < \alpha < 1$.

Then, f *and* g *have a unique common fixed point provided* f *and* g *commute.*

Theorem 4.4 *[28] Let* $\{f_n\}$ *be a sequence of mappings of a complete fuzzy metric space* $(X, M, *)$ *into itself such that for any two mappings* f_i *and* f_j *we have*

$$M(f_i^m x, f_j^m y, \alpha_{ij} t) \geq M(x, y, t)$$

for some $m \in \mathbb{N}$ *and* $0 < \alpha_{ij} < k < 1, i, j = 1, 2, \ldots, x, y \in X$. *The sequence* $\{f_n\}$ *has a unique common fixed point.*

In [3, 29] it has been observed that the notion of G-completeness has a disadvantage in that it is a very strong notion of completeness, in fact, if d is the Euclidean metric in $\mathbb{R}$, the induced fuzzy metric $(M_d, *)$ of Example 2.1 given in [29] is not G-complete.

In order to strengthen and obtain a Housedroff topology (the so-called M-topology), George and Veeramani [3, 4] imposed some stronger conditions on fuzzy metrics and modified the concept due to Kranmosil and Michalek.

4.3 George and Veeramani's fuzzy metric space and fuzzy contractive mappings

Definition 4.6 (George and Veeramani [3]) The triple $(X, M, *)$ is a fuzzy metric space if X is a nonempty set, $*$ is a continuous t-norm and M is a fuzzy set on $X^2 \times (0, \infty)$ satisfying the following axioms:

(GV1) $M(x, y, t) > 0$.

(GV2) $M(x, y, t) = 1$ if and only if $x = y$.

(GV3) $M(x, y, t) = M(y, x, t)$.

(GV4) $M(x, y, t) * M(y, z, s) \leq M(x, z, t + s)$.

(GV5) $M(x, y, .) : (0, \infty[\to [0, 1]$ is continuous, $x, y, z \in X$ and $t, s > 0$.

The axiom (GV1) is justified because in the same way that a classical metric space does not take the value ∞ then M cannot take the value 0. The axiom (GV2) gives the idea that only when $x = y$ the degree of nearness of x and y is perfect, or simply 1, and then $M(x, y, t) = 1$ for each $x \in X$ and for each $t > 0$. In this manner the values 0 and ∞ in the classical theory of metric space are identified with 1 and 0, respectively, in this fuzzy theory. Finally, in (GV5) the authors assume only that the variable t behaves nicely, i.e., they assume that for fixed x and y, the function $t \to M(x, y, t)$ is continuous without any imposition for M as $t \to \infty$. In what follows, fuzzy metric spaces in the sense of George and Veeramani [3] will be referred to as GV-fuzzy metric spaces.

Example 4.1 Let $X = \mathbb{R}$. Define $a * b = ab$ for all $a, b \in [0, 1]$ and

$$M(x, y, t) = \left[exp\left(\frac{|x-y|}{t}\right)\right]^{-1}$$

for all $x, y \in X$ and $t \in (0, \infty)$. Then $(X, M, *)$ is a fuzzy metric space.

The next example shows that every metric space induces a fuzzy metric space.

Example 4.2 Let (X, d) be a metric space. Define $a*b = ab$ for all $a, b \in [0, 1]$ and

$$M(x, y, t) = \frac{kt^n}{kt^n + md(x, y)}, \quad k, m, n \in N.$$

Then $(X, M, *)$ is a fuzzy metric space. In particular, taking $k = m = n = 1$, we get

$$M(x, y, t) = \frac{t}{t + d(x, y)},$$

that is called a standard fuzzy metric.

George and Veeramani proved in [3, 4] that every fuzzy metric M on X generates a topology τ_M on X which has as a base the family of open sets of the form $\{B_M(x, r, t) : x \in X, 0 < r < 1, t > 0\}$, where $B_M(x, r, t) = \{y \in X : M(x, y, t) > 1 - r\}$ for all $x \in X, r \in (0, 1)$ and $t > 0$. If (X, d) is a metric space, the topology generated by d coincides with the topology τ_{M_d} generated by the fuzzy metric M_d.

Remark 4.1 (George and Veeramani [3]) The metric space (X, d) is complete if and only if the standard fuzzy metric space $(X, M_d, *)$ is complete.

Definition 4.7 (George and Veeramani [3]) Let $(X, M, *)$ be a fuzzy metric space. Then a sequence $\{x_n\}_{n\in\mathbb{N}}$ in X is said to be Cauchy (or M-Cauchy) if for each $\epsilon \in (0, 1)$ and each $t > 0$ there is $n_0 \in \mathbb{N}$ such that $M(x_n, x_m, t) > 1 - \epsilon$ for all $n, m \geq n_0$.

A fuzzy metric space in which every Cauchy (or M-Cauchy) sequence is convergent is called complete (or M-complete).

Definition 4.8 *[14, 9]* Let $(X, M, *)$ be a fuzzy metric space. The fuzzy metric $(M, *)$ (or the fuzzy metric space $(X, M, *)$) is said to be non-Archimedean or strong if it satisfies for each $x, y, z \in X$ and each $t > 0$

$$M(x, y, t) \geq M(x, y, t) * M(y, z, t).$$

In order to obtain fuzzy version of the classical Banach contraction theorem, Gregori and Sapena [7] introduced the following concepts.

Definition 4.9 Let $(X, M, *)$ be a fuzzy metric space. A mapping $f : X \to X$ is said to be fuzzy contractive if $k \in (0, 1)$ such that

$$\frac{1}{M(f(x), f(y), t)} - 1 \leq k \left(\frac{1}{M(x, y, t)} - 1 \right)$$

for each $x, y \in X$ and $t > 0$.

Definition 4.10 Let $(X, M, *)$ be a fuzzy metric space. A sequence $\{x_n\}$ in X is said to be fuzzy contractive if $k \in (0, 1)$ such that

$$\frac{1}{M(x_{n+1}, x_{n+2}, t)} - 1 \leq k \left(\frac{1}{M(x_n, x_{n+1}, t)} - 1 \right)$$

for all $t > 0, n \in \mathbb{N}$.

Recall that a sequence $\{x_n\}$ in a metric space (X, d) is said to be contractive if $k \in (0, 1)$ such that $d(x_{n+1}, x_{n+2}) \leq kd(x_n, x_{n+1})$, for all $n \in \mathbb{N}$.

Proposition 4.1 *Let $(X, M_d, *)$ be the standard fuzzy metric space induced by the metric d on X. The sequence $\{x_n\}$ in X is contractive in (X, d) iff $\{x_n\}$ is fuzzy contactive in $(X, M_d, *)$.*

Theorem 4.5 *[3] A sequence $\{x_n\}$ in a fuzzy metric space $(X, M, *)$ converges to x iff $M(x_n, x, t) \to 1$ as $n \to \infty$.*

Theorem 4.6 (Fuzzy Banach contraction theorem) *Let $(X, M, *)$ be a complete fuzzy metric space (in the sense of George and Veeramani) in which fuzzy contractive sequences are Cauchy. Let $f : X \to X$ be a contractive mapping as k, the contractive constant. Then, f has a unique fixed point.*

Proof. Fix $x \in X$. Let $x_n = f^n(x), n \in \mathbb{N}$. We have for $t > 0$

$$\frac{1}{M(f(x), f^2(x), t)} - 1 \leq k \left(\frac{1}{M(x, x_1, t)} - 1 \right)$$

and by induction

$$\frac{1}{M(x_{n+1}, x_{n+2}, t)} - 1 \leq k \left(\frac{1}{M(x_n, x_{n+1}, t)} - 1 \right)$$

for all $n \in \mathbb{N}$. Then $\{x_n\}$ is a fuzzy contractive sequence and thus a Cauchy sequence; hence x_n converges to y, for some $y \in X$. We will see y is a fixed point for f. By Theorem 4.5, we have

$$\frac{1}{M(f(y), f(x_n), t)} - 1 \leq k \left(\frac{1}{M(y, x_n, t)} - 1 \right) \to 0$$

as $n \to \infty$, then $\lim_{n\to\infty} M(f(y), f(x_n), t) = 1$ for each $t > 0$. Therefore, $\lim_{n\to\infty} f(x_n) = f(y)$, i.e., $\lim_{n\to\infty} x_{n+1} = f(y)$ and then $f(y) = y$.

To show uniqueness, assume $f(z) = z$ for some $z \in X$. Then for $t > 0$ we have

$$\begin{aligned}\frac{1}{M(y,z,t)} - 1 &= \frac{1}{M(f(y),f(z),t)} - 1 \\ &\leq k\left(\frac{1}{M(y,z,t)} - 1\right) \\ &\leq \ldots \leq k^n\left(\frac{1}{M(y,z,t)} - 1\right) \to 0\end{aligned}$$

as $n \to \infty$. Hence, $M(y,z,t) = 1$ and then $y = z$.

Now suppose $(X, M_d, *)$ is a complete standard fuzzy metric space induced by the metric d on X. From Remark 4.1 (X,d) is complete; then if $\{x_n\}$ is a fuzzy contractive sequence, by Proposition 4.1 it is contractive in (X,d), hence convergent. From Theorem 4.6, we have the following corollary, which can be considered the fuzzy version of the classic Banach contraction theorem on complete metric space.

Corollary 4.1 *Let $(X, M_d, *)$ be a complete standard fuzzy metric space and let $f : X \to X$ be a fuzzy contractive mapping. Then, f has a unique fixed point.*

Theorem 4.7 (Fuzzy Banach contraction theorem) *Let $(X, M, *)$ be a G-complete fuzzy metric space (in the sense of Kramosil and Michalek) and let $f : X \to X$ be a fuzzy contractive mapping. Then, f has a unique fixed point.*

Proof. Let $k \in (0,1)$ and suppose f satisfies

$$\frac{1}{M(f(x),f(y),t)} - 1 \leq k\left(\frac{1}{M(x,y,t)} - 1\right);$$

$t > 0$. Fix $x \in X$. Let $x_n = f^n(x), n \in \mathbb{N}$. We have seen in the proof of Theorem 4.6 that $\{x_n\}$ is a fuzzy contractive sequence satisfying

$$\frac{1}{M(x_{n+1},x_{n+2},t)} - 1 \leq k\left(\frac{1}{M(x_n,x_{n+1},t)} - 1\right);$$

$n \in \mathbb{N}$. Thus,

$$\begin{aligned}\frac{1}{M(x_{n+1},x_{n+2},t)} - 1 &\leq k^2\left(\frac{1}{M(x_{n-1},x_n,t)} - 1\right) \\ &\leq \ldots \leq k^n\left(\frac{1}{M(x_1,x_2,t)} - 1\right) \\ &\to 0 \text{ as } n \to \infty\end{aligned}$$

and, therefore, $\lim_{n\to\infty} M(x_n, x_{n+1}, t) = 1, \forall t > 0$. For a fixed $p \in \mathbb{N}$ we have

$$M(x_n, x_{n+p}, t) \geq M\left(x_n, x_{n+1}, \frac{t}{p}\right) \tag{4.7}$$

$$* \ldots * M\left(x_{n+p-1}, x_{n+p}, \frac{t}{p}\right) \to \overbrace{1 * \ldots * 1}^{p} = 1 \tag{4.8}$$

and thus, $\{x_n\}$ is a G-Cauchy sequence. Therefore, $\{x_n\}$ converges to y for some $y \in X$. Now, imitating the proof of Theorem 4.6 one can prove that y is the unique fixed point for f.

Remark 4.2 Theorem 4.7 proved that each fuzzy contractive sequence is G-Cauchy whereas Theorem 4.6 assumed that fuzzy contractive sequences are M-Cauchy. This leads to the following question.

Question (Gregori and Sapena [7]). Is a fuzzy contractive sequence a Cauchy sequence in George and Veeramani's sense?

The above problem generated much interest by fuzzy fixed point theorists working on aspects of fuzzy contractive mapping and associated fixed points. Mihet [18] introduced point convergent and improved the result of Gregori and Sapena [7].

Definition 4.11 Let $(X, M, *)$ be a fuzzy metric space. A sequence (x_n) in X is said to be pointwise convergent to $x \in X$ (we write $x_n \to_p x$) if $t > 0$ such that

$$\lim_{n\to\infty} M(x_n, x, t) = 1.$$

It is easy to see that a GV fuzzy metric space $(X, M, *)$ endowed with the point convergence is a space with convergence in the sense of Fréchet, that is, one of the following holds:

(1) Every sequence in X has at most one limit point.

(2) Every constant sequence, $x_n = x, \forall n \in \mathbb{N}$, is convergent and $\lim_{n\to\infty} x_n = x$.

(3) Every subsequence of a convergent sequence is also convergent and has the same limit as the whole sequence.

Remark 4.3 It is worth noting that if the point convergence in a fuzzy metric space $(X, M, *)$ is Fréchet, then (GV2) holds. Thus, the uniqueness of the limit in the point convergence characterizes a fuzzy metric space in the sense of George and Veeramani. Indeed, let $x, y \in X, x \neq y$. If $M(x, y, t) = 1$ for some $t > 0$; then the sequence $(x_n)_{n\in\mathbb{N}} \subset X$ defined as $x, y, x, y, \cdots$ has two distinct limits; for the equality $M(x, x, t) = M(y, x, t) = 1$ implies $x_n \to_p x$, while $M(x, y, t) = M(y, y, t) = 1$ implies $x_n \to_p y$.

In the next example we will see p-convergent (but not convergent) sequences.

Example 4.3 Let $(x_n)_{n\in\mathbb{N}} \subset (0,\infty)$, $x_n \nearrow 1$ and $X = (x_n) \cup \{1\}$. Define $M(x_n, x_n, t) = 1 \forall n \in \mathbb{N},\ \forall t > 0,\ M(1,1,t) = 1 \forall t > 0,\ M(x_n, x_m, t) = \min\{x_n, x_m\}\ \forall n, m \in \mathbb{N},\ \forall t > 0$ and

$$M(x_n, 1, t) = \begin{cases} \min\{x_n, t\} & \text{if } 0 < t < 1, \\ x_n & \text{if } t > 1 \end{cases}$$

for all $n \in \mathbb{N}$. Then (X, M, T_M), where $T_M(a,b) = \min\{a,b\}$, is a fuzzy metric space (see [8], Example 2]). Since $\lim_{n\to\infty} M(x_n, 1, \frac{1}{2}) = \frac{1}{2}$, (x_n) is not convergent. Nevertheless it is p-convergent to 1, for $\lim_{n\to\infty} M(x_n, 1, 2) = 1$.

Theorem 4.8 *Let $(X, M, *)$ be a GV-fuzzy metric space and $f : X \to X$ be a fuzzy contractive mapping. Suppose that, for some $x \in X$, the sequence $(x_n)_{n\in\mathbb{N}}, x_n = f^n(x)$ of its iterates has a p-convergent subsequence. Then f has a unique fixed point.*

It should be noted that a similar theorem does not hold in KM-fuzzy metric spaces. This is illustrated in the following.

Example 4.4 Let X be the set $\mathbb{N} = \{1, 2, \cdots\}$. We define (for $p \neq q$) the fuzzy mapping M by

$$M(p, q, t) = \begin{cases} 0 & \text{if } t = 0, \\ 1 - 2^{-\min\{p,q\}} & \text{if } 0 < t \le 1, \\ 1 & \text{if } t > 1. \end{cases}$$

As $1 - 1/2^{-\min(p,r)} \ge \min\{1 - 1/2^{-\min(r,q)}, 1 - 1/2^{-\min(p,q)}\} \forall p, q, r \in N^*$, (X, M, T_M) is a KM-fuzzy metric space satisfying $M(x, y, t) \neq 0$ for all $x, y \in X$ and $t > 0$. The mapping $f : \mathbb{N} \to \mathbb{N}$, $f(x) = x + 1$ is fuzzy contractive. Indeed, if $t > 1$ then

$$\frac{1}{M(f(p), f(q), t)} - 1 = 0 < \frac{1}{2}\left(\frac{1}{M(p,q,t)} - 1\right)$$

for every $p, q \in N^*$, while if $0 < t \le 1$ and $p < q$, then

$$\begin{aligned} \frac{1}{M(f(p), f(q), t)} - 1 &= \frac{1}{2^{p+1} - 1} \\ &\le \frac{1}{2^{p+1} - 2} = \frac{1}{2}\left(\frac{1}{M(p,q,t)} - 1\right). \end{aligned}$$

As $\lim_{n\to\infty} M(f^n(x), 1, x) = 1$ for every $x \in X$ and $s > 1$, it follows that $x_n \to_p 1$. Nevertheless, 1 is not a fixed point of f.

Remark 4.4 (1) We note that in Examples 4.3 and 4.4, there are essentially no nonconstant convergent sequences. (2) It will be natural to continue the study of convergence spaces by finding more examples and introducing a similar concept for Cauchy sequence, p-completeness. Also, it would be interesting to compare different types of contraction maps in fuzzy metric spaces.

Yun *et al.* [32] introduced the notion of minimal slope of a map between fuzzy metric spaces and various properties of fuzzy contractive mapping which support the question proposed by Gregori and Sapena [7].

In 2008, Mihet [19], provided a partial affirmative answer to the above question proposed by Gregori and Sapena by introducing fuzzy Ψ-contractive mapping as follows:

Definition 4.12 *[19]* Let Ψ be the class of all mapping $\psi : [0,1] \to [0,1]$ such that ψ is continuous, nondecreasing and $\psi(t) > t, \ \forall\, t \in (0,1)$. Let $(X, M, *)$ be a fuzzy metric space and $\psi \in \Psi$. A mapping $f : X \to X$ is called a fuzzy ψ-contractive mapping if the following implication takes place:

$$M(x,y,t) > 0 \Longrightarrow M(f(x), f(y), t) \geq \psi(M(x,y,t)).$$

A fuzzy ψ-contractive sequence in a fuzzy metric space $(X, M, *)$ is any sequence $\{x_n\}_{n \in \mathbb{N}}$ in X such that

$$M(x_{n+2}, x_{n+1}, t) \geq \psi(M(x_{n+1}, x_n, t))$$

for all $n \in \mathbb{N}$ and $t > 0$.

Example 4.5 Let $X = [0, \infty), a * b = \min\{a, b\} \ \forall a, b \in [0,1]$ and

$$M(x,y,t) = \begin{cases} 0 & \text{if } t \leq |x-y|, \\ 1 & \text{if } t > |x-y|. \end{cases}$$

It is well known that $(X, M, *)$ is a KM-fuzzy metric space. Let ψ be a mapping in Ψ. Since $\psi(1) = 1$ and

$$\begin{aligned} M(x,y,t) > 0 &\Longrightarrow M(x,y,t) = 1 \\ &\Longrightarrow \psi(M(x,y,t)) = 1. \end{aligned}$$

It follows that any fuzzy contractive mapping on $(X, M, *)$ satisfying

$$|x - y| < t \Longrightarrow |f(x) - f(y)| < t,$$

that is,

$$|f(x) - f(y)| \leq |x - y| \ \ \forall x, y \in X.$$

Conversely, if $f : X \to X$ is such that $|f(x) - f(y)| \leq |x - y| \ \ \forall x, y \in X$, then f is a fuzzy ψ contractive mapping for every ψ in Ψ such that $\psi(0) = 0$. Thus, the mapping $f : X \to X, f(x) = x + 1, g(x) = x$ is fuzzy ψ_k-contractive on $(X, M, *)$.

Remark 4.5 (Mihet [17], Example 3.4) The sequence $\{x_n\}_{n\in\mathbb{N}}, x_n = n+1$ in the fuzzy metric space considered in Example 4.5, although fuzzy ψ_k-contractive, is not M-Cauchy.

We note that, for every $k \in (0,1)$, the mapping $\psi_k : [0,1] \to [0,1], \psi_k(t) = \dfrac{t}{t+k(1-t)}$ is in Ψ and a ψ_k-fuzzy contractive mapping is a fuzzy contractive mapping in the sense of Geogori and Sepena [7].

Theorem 4.9 *Let $(X, M, *)$ be an M-complete non-Archimedean fuzzy metric space and $f : X \to X$ be a fuzzy ψ- contractive mapping. If $x \in X$ such that $M(x, f(x), t) > 0$ for every $t > 0$, then f has a fixed point.*

Proof. Let $x \in X$ be such that $M(x, f(x), t) > 0$, $\forall\ t > 0$ and $x_n = f^n(x), n \in \mathbb{N}$. We have

$$\begin{aligned} M(x_1, x_2, t) &\geq \psi(M(x_0, x_1, t)) \\ &\geq M(x_0, x_1, t) > 0 \ \ \forall t > 0. \end{aligned}$$

Hence

$$\begin{aligned} M(x_2, x_3, t) &\geq \psi(M(x_1, x_2, t)) \\ &\geq M(x_1, x_2, t) > 0 \ \ \forall t > 0. \end{aligned}$$

By induction, $M(x_{n+1}, x_{n+2}, t) \geq M(x_n, x_{n+1}, t) > 0$ for all $t > 0$. Therefore, for every $t > 0$, $M(x_n, x_{n+1}, t)_{n\in\mathbb{N}}$ is a nondecreasing sequence of numbers in $(0, 1]$. Fix a $t > 0$ and denote $\lim\limits_{n\to\infty} M(x_n, x_{n+1}, t)$ by l. We have $l \in (0, 1]$ (for $M(x_0, x_1, t) > 0$) and since $M(x_n, x_{n+1}, t) \geq \psi(M(x_{n-1}, x_n, t))$ and ψ is continuous, $l \geq \psi(l)$. This implies $l = 1$, and therefore

$$\lim_{n\to\infty} M(x_n, x_{n+1}, t) = 1 \ \forall t > 0.$$

If $\{x_n\}$ is not M-Cauchy, $\epsilon \in (0,1)$ and $t > 0$ such that for each $k \in \mathbb{N}$ there exist $m(k), n(k) \in \mathbb{N}$ with $m(k) > n(k) \geq k$ and

$$M(x_{m(k)}, x_{n(k)}, t) \leq 1 - \epsilon.$$

For each k, let $m(k)$ be the least integer exceeding $n(k)$ satisfying the above property, that is, $M(x_{m(k)-1}, x_{n(k)-1}, t) > 1-\epsilon$ and $M(x_{m(k)}, x_{n(k)}, t) \leq 1-\epsilon$. Then, for each positive integer k,

$$\begin{aligned} 1 - \epsilon &\geq M(x_{m(k)}, x_{n(k)}, t) \\ &\geq *(M(x_{m(k)-1}, x_{n(k)}, t), M(x_{m(k)-1}, x_{m(k)}, t)) \\ &\geq *(1-\epsilon, M(x_{m(k)-1}, x_{m(k)}, t)). \end{aligned}$$

Since $\lim\limits_{k\to\infty} *(1-\epsilon, M(x_{m(k)-1}, x_{m(k)}, t)) = *(1-\epsilon, 1) = 1-\epsilon$, it follows that

$$\lim_{k\to\infty} M(x_{m(k)}, x_{n(k)}, t) = 1 - \epsilon.$$

Let us denote $M(x_{n(k)}, x_{n(k)+1}, t)$ by z_n. Then

$$\begin{aligned} M(x_{m(k)}, x_{n(k)}, t) &\geq *^2(z_n, M(x_{m(k)+1}, x_{n(k)+1}, t), z_m) \\ &\geq *^2(z_n, M(x_{m(k)}, x_{n(k)}, t), z_m). \end{aligned}$$

Letting $k \to \infty$ we obtain

$$\begin{aligned} 1 - \epsilon &\geq *^2(1, \psi(1-\epsilon), 1) \\ &= \psi(1-\epsilon) > 1 - \epsilon, \end{aligned}$$

which is a contradiction. Thus $\{x_n\}$ is a Cauchy sequence.
If $\lim_{n\to\infty} x_n = y$, then from $M(f(y), f(x_n), t) \geq \psi(M(y, x_n, t))$ it follows that $x_{n+1} \to f(y)$. From here we deduce that

$$M(y, f(y), t) \geq *^2(M(y, x_n, t), M(x_n, x_{n+1}, t), M(x_{n+1}, f(y), t)) \xrightarrow[n\to\infty]{} 1$$

for all $t > 0$, hence $f(y) = y$.

Theorem 4.10 *Let $(X, M, *)$ be an M-complete non-Archimedean fuzzy metric space satisfying the condition $M(x, y, t) > 0, \forall\, t > 0$ and $f : X \to X$ be a fuzzy ψ-contractive mapping, Then f has a unique fixed point.*

Example 4.6 Let $X = (0, \infty), a * b = ab, \forall\, a, b \in [0, 1]$ and

$$M(x, y, t) = \frac{\min(x, y)}{\max(x, y)}\ \forall\, t \in (0, \infty),\ \forall\, x, y > 0.$$

$(X,\ M,\ ^*)$ is an M-complete non-Archimedean fuzzy metric space. Since $\sqrt{t} > t\ \forall\, t \in (0, 1)$, the mapping $f : X \to X, f(x) = \sqrt{x}$ is a fuzzy ψ-contractive mapping with $\psi(t) = \sqrt{t}$. Thus, all the conditions of Theorem 4.10 are satisfied and the fixed point of f is $x = 1$.

More generalizations of results of Geogori and Sepena [7] and Mihet [19] can be found in [17, 18, 1, 30, 5, 27, 6]. Recently, Wardowski [36] introduced $\mathcal{H}$-contractive mappings, as generalizations of fuzzy contractive mappings, and established the conditions guaranteeing the existence and uniqueness of fixed points for such contractions in M-complete fuzzy metric spaces in the sense of George and Veeramani.

Definition 4.13 Let $\mathcal{H}$ be the family of the mappings η: $(0, 1] \to [0, \infty)$ satisfying the following conditions:

(H1) η transforms $(0, 1]$ onto $[0, \infty)$.

(H2) η is strictly decreasing.

Then the mapping $f : X \to X$ is called fuzzy $\mathcal{H}$-contractive (see, Wardowski [36]) with respect to $\eta \in \mathcal{H}$ if $k \in (0, 1)$ satisfies the following condition:

$$\eta(M(fx, fy, t)) \le k\eta(M(x, y, t))$$

for all $x, y \in X$ and $t > 0$.

Theorem 4.11 *[36] Let $(X, M, *)$ be an M-complete fuzzy metric space and $f\colon X \to X$ be a fuzzy $\mathcal{H}$-contractive mapping with respect to $\eta \in \mathcal{H}$ such that:*

(a) $\prod_{i=1}^{k} M(x, fx, t_i) \neq 0$ for all $x \in X$, $k \in \mathbb{N}$ and a sequence $(t_i)_{i \in \mathbb{N}} \subset (0, \infty), t_i \searrow 0$;

(b) $r * s > 0 \implies \eta(r * s) \le \eta(r) + \eta(s)$ for all $r, s \in \{M(x, fx, t)\colon x \in X, t > 0\}$;

(c) $\{\eta(M(x, fx, t_i))\colon i \in \mathbb{N}\}$ is bounded for all $x \in X$ and any sequence $(t_i)_{i\in\mathbb{N}} \subset (0.\infty), t_i \searrow 0$.

Then f has a unique fixed point $x^* \in X$ and for each $x_0 \in X$ the sequence $(f^n x_0)_{n\in\mathbb{N}}$ converges to x^*.

Gregori and Minana [11] observed that the main idea of Wardowski [36] is correct and different from the known concepts in the literature but they also showed that fuzzy $\mathcal{H}$-contractive mappings are included in the class of fuzzy Ψ-contractive mappings. They noted some drawbacks of the conditions used in Theorem 4.11.

In this direction of research work, a recent paper of Mihet [20] provides a larger perspective to study fixed points of fuzzy $\mathcal{H}$-contractive mappings.

4.4 Fuzzy metric-like spaces and fixed point results

Shukla and Abbas [23] considered the fuzzy metric-like, which is a combination of fuzzy metric and metric-like [13] and proved some basic versions of fixed point theorems in spaces endowed with fuzzy metric-like. In fuzzy metric-like spaces, the degree of nearness of points x and y of space, when $x = y$ is not perfect indicate the self fuzzy distance may not be unity.

First we recall some basic definitions and results from [23].

Definition 4.14 *[23]* The triplet $(X, \mathcal{F}, \star)$ is a fuzzy metric-like space if X is an arbitrary set, $\star$ is a continuous t-norm, and $\mathcal{F}$ is a fuzzy set in $X^2 \times (0, \infty)$ satisfying the following conditions:

(FML1) $\mathcal{F}(x, y, t) > 0$.

(FML2) if $\digamma(x,y,t)=1$ then $x=y$.

(FML3) $\digamma(x,y,t)=\digamma(y,x,t)$.

(FML4) $\digamma(x,y,t)\star\digamma(y,z,s)\leq\digamma(x,z,t+s)$.

(FML5) $\digamma(x,y,\cdot)\colon(0,\infty)\to[0,1]$ is a continuous mapping for all $x,y,z\in X$ and $s,t>0$.

$\digamma$ (endowed with $\star$) is called a fuzzy metric-like on X.

Remark 4.6 A fuzzy metric-like space satisfies all the conditions of a fuzzy metric space except that $\digamma(x,x,t)$ may be less than 1 for all $t>0$ and for some (or may be for all) $x\in X$. Also, every fuzzy metric space is fuzzy metric-like with self fuzzy distance, that is, with $\digamma(x,x,t)=1$ for all $t>0$ and for all $x\in X$.

Note that axiom (FM2) in Definition 4.14 indicates when $x=y$, the degree of nearness of x and y is perfect or simply 1, and $M(x,x,t)=1$ for each $x\in X$ and for each $t>0$. While in fuzzy metric-like space, $\digamma(x,x,t)$ may be less than 1, that is, the concept of fuzzy metric-like is applicable when the degree of nearness of x and y is not perfect for $x=y$.

Example 4.7 Let $X=[0,1]$. Then, the triplet $(X,\digamma,\star_l)$ is a fuzzy metric-like space, where the fuzzy set $\digamma$ is defined by

$$\digamma(x,y,t)=\begin{cases}1 & \text{if } x=y=0,\\ \dfrac{x+y}{2} & \text{otherwise}\end{cases}$$

for all $t>0$.

By using the following propositions, several examples of fuzzy metric-like spaces can be obtained.

Proposition 4.2 *[23] Let (X,σ) be any metric-like space (for the related definitions we refer to [13]). Then the triplet $(X,\digamma,\star_p)$ is a fuzzy metric-like space, where the fuzzy set $\digamma$ is given by*

$$\digamma(x,y,t)=\frac{kt^n}{kt^n+m\sigma(x,y)}\quad \textit{for all } x,y\in X,t>0,$$

where $k\in\mathbb{R}^+,m>0$ and $n\geq 1$.

Remark 4.7 *[23]* Proposition 4.2 shows that every metric-like space induces a fuzzy metric-like space. For $k=n=m=1$ the induced fuzzy metric-like space $(X,\digamma_\sigma,\star_p)$ is called the standard fuzzy metric-like space, where

$$\digamma_\sigma(x,y,t)=\frac{t}{t+\sigma(x,y)}\quad \text{for all } x,y\in X,t>0.$$

Proposition 4.3 *[23] Let (X, σ) be any metric-like space. Then triplet $(X, F, \star_p)$ is a fuzzy metric-like space, where the fuzzy set F is defined by*

$$F(x, y, t) = e^{-\sigma(x,y)/t^n} \quad \textit{for all } x, y \in X, t > 0,$$

where $n \geq 1$.

We point out that Propositions 4.2 and 4.3 hold even if we employ the minimum t-norm $\star_m$ rather than the product t-norm $\star_p$ (see [23]).

Now we define convergent and Cauchy sequences in fuzzy metric-like spaces and the completeness of fuzzy metric-like spaces.

Definition 4.15 *[23]* Let $(X, F, \star)$ be a fuzzy metric-like space and $\{x_n\}$ be a sequence in X. Then

(1) $\{x_n\}$ is said to be convergent to $x \in X$ and x is called a limit of $\{x_n\}$ if

$$\lim_{n\to\infty} F(x_n, x, t) = F(x, x, t) \quad \text{for all } t > 0;$$

(2) $\{x_n\}$ is said to be Cauchy if, for all $t > 0$ and each $p \geq 1$, the limit $\lim_{n\to\infty} F(x_{n+p}, x_n, t)$ exists;

(3) $(X, F, \star)$ is said to be complete if every Cauchy sequence $\{x_n\}$ in X converges to some $x \in X$ such that $\lim_{n\to\infty} F(x_n, x, t) = F(x, x, t) = \lim_{n\to\infty} F(x_{n+p}, x_n, t)$ for all $t > 0$ and each $p \geq 1$.

Remark 4.8 *[23]* In a fuzzy metric-like space, the limit of a convergent sequence need not be unique and a convergent sequence need not be Cauchy.

Definition 4.16 *[23]* Let $(X, F, \star)$ be a fuzzy metric-like space. We say that a mapping $f\colon X \to X$ is fuzzy contractive if $k \in (0, 1)$ such that

$$\frac{1}{F(fx, fy, t)} - 1 \leq k \left[\frac{1}{F(x, y, t)} - 1\right] \tag{4.9}$$

for all $x, y \in X$ and $t > 0$. Here k is called a fuzzy contractive constant of f.

Definition 4.17 Let $(X, F, \star)$ be a fuzzy metric-like space. We say that a mapping $f : X \to X$ is Tirado-fuzzy contractive (see [27]) if $k \in (0, 1)$ such that

$$1 - F(fx, fy, t) \leq k\,[1 - F(x, y, t)] \tag{4.10}$$

for all $x, y \in X$ and $t > 0$. Here k is called a Tirado-fuzzy contractive constant of f.

There are two notions of the Cauchy sequences and completeness in fuzzy metric spaces, namely, G-Cauchy sequence and G-completeness (due to [2]) and M-Cauchy sequence and M-completeness (due to [3]). It is well-known that the definition of completeness due to George and Veeramani is more general than that of Grabiec. The Cauchy sequences and completeness of fuzzy metric-like spaces defined by [23] are extensions of Grabiec's definition in fuzzy metric-like spaces. Shukla *et al.* [24] modified the previous notions of Cauchy sequence and complete fuzzy metric-like space by introducing the more general concepts of 1-Cauchy sequence and 1-complete fuzzy metric-like space. Shukla *et al.* [24] established some new fixed point theorems that extend and unify previous results.

Definition 4.18 Let $(X, \mathcal{F}, \star)$ be a fuzzy metric-like space and let $\{x_n\}$ be a sequence in X. The $\{x_n\}$ is called a 1-G-Cauchy sequence if $\lim_{n\to\infty} \mathcal{F}(x_{n+p}, x_n, t) = 1$ for all $t > 0$ and all $p \geq 1$. The space $(X, \mathcal{F}, \star)$ is called 1-G-complete if every 1-G-Cauchy sequence in X converges to some $x \in X$ such that $\mathcal{F}(x, x, t) = 1$ for all $t > 0$.

It is obvious that every 1-G-Cauchy sequence is a Cauchy sequence (in the sense of Shukla and Abbas [23]) in $(X, \mathcal{F}, \star)$, and every complete fuzzy metric-like space (in the sense of Shukla and Abbas) is 1-G-complete. On the other hand, the following examples show that a Cauchy sequence (in the sense of Shukla and Abbas) need not be a 1-G-Cauchy sequence, and there exists a fuzzy metric-like space which is 1-G-complete, but not complete (in the sense of Shukla and Abbas).

Example 4.8 Let $X = \mathbb{N} = \{1, 2, 3, \ldots\}$. Define the fuzzy set $\mathcal{F}$ in $X^2 \times (0, \infty)$ by

$$\mathcal{F}(x, y, t) = \frac{\min(x, y)}{\max(x, y)^3} = \begin{cases} \dfrac{x}{y^3} & \text{if } x \leq y, \\ \dfrac{y}{x^3} & \text{if } y \leq x. \end{cases}$$

Then $(X, \mathcal{F}, \star_p)$ is a fuzzy metric-like space. Consider the sequence $\{x_n\}$ in X. For each $p \geq 1$, $\lim_{n\to\infty} \mathcal{F}(x_n, x_{n+p}, t) = \lim_{n\to\infty} \dfrac{n}{(n+p)^3} = 0$. If this limit exists, $\{x_n\}$ is a Cauchy sequence (in the sense of [23]) in $(X, \mathcal{F}, \star_p)$. However, it is not a 1-G-Cauchy sequence.

Example 4.9 Let $X = [0, \infty) \cap \mathbb{Q}$ and define $\sigma\colon X \times X \to \mathbb{R}^+$ by $\sigma(x, y) = \max\{x, y\}$. Then (X, σ) is a metric-like space. It is easy to see that the standard fuzzy metric-like space $(X, \mathcal{F}_\sigma, \star_p)$ is 1-G-complete fuzzy metric-like. Indeed, if $\{x_n\}$ is a 1-G-Cauchy sequence in X, we must have $x_n \to 0$ as $n \to \infty$, and so, $M(x_n, 0, t) \to 1$ as $n \to \infty$ for all $t > 0$. This shows that every 1-G-Cauchy sequence in X converges to $0 \in X$. On the other hand, it is easy to see that (X, σ) is not a complete fuzzy metric-like space (in the sense of [23]).

Definition 4.19 Let $(X, F, \star)$ be a fuzzy metric-like space and let $\{x_n\}$ be a sequence in X. The $\{x_n\}$ is called a 1-M-Cauchy sequence if $\lim_{n,m\to\infty} F(x_n, x_m, t) = 1$ for all $t > 0$. The space $(X, F, \star)$ is called 1-M-complete if every 1-M-Cauchy sequence in X converges to some $x \in X$ such that $F(x, x, t) = 1$ for all $t > 0$.

Remark 4.9 In view of Remark 4.8, every complete fuzzy metric space in the sense of [2] is 1-G-complete fuzzy metric-like and every complete fuzzy metric space in the sense of [3] is 1-M-complete fuzzy metric-like.

It is obvious that the 1-G-completeness implies 1-M-completeness. It is well known that every fuzzy metric space which is complete in the sense of [2] is also complete in the sense of [3]. In view of Remarks 4.8 and 4.9, one can conclude that 1-M-completeness is more general than 1-G-completeness.

Next, we introduce a kind of control function which will be used to define the notion of *$\mathcal{B}$-contraction*. This class of mappings will be useful to unify the notions of fuzzy contractive and Tirado-fuzzy contractive mappings.

Definition 4.20 Let $\mathcal{B}$ denote the family of functions $\beta\colon [0,1] \to [0,1]$ satisfying the following conditions:

($\mathcal{B}$1) β is nondecreasing.

($\mathcal{B}$2) $\lim_{n\to\infty} \beta^n(t) = 1$ for each $t \in (0,1)$.

Remark 4.10 If $\beta \in \mathcal{B}$, then

(1) $t < \beta(t)$ for all $t \in (0,1)$;

(2) $\beta(1) = \lim_{t\to 1^-} \beta(t) = 1$.

Example 4.10 The following functions $\beta : [0,1] \to [0,1]$, belong to class $\mathcal{B}$.

(1) $\beta(t) = 1$;

(2) $\beta(t) = t^k$, where $k \in (0,1)$;

(3) $\beta(t) = \dfrac{t}{k + (1-k)t}$, where $k \in (0,1)$;

(4) $\beta(t) = \dfrac{2t}{1+t}$;

(5) $\beta(t) = 1 - k(1-t)$, where $k \in (0,1)$.

Definition 4.21 Let $(X, F, \star)$ be a fuzzy metric-like space. A mapping $f : X \to X$ is said to be a $\mathcal{B}$-contraction if $\beta \in \mathcal{B}$ such that

$$\beta(F(x,y,t)) \leq F(fx, fy, t) \text{ for all } x, y \in X \text{ and all } t > 0. \tag{4.11}$$

In such a case, we will say that f is a $\mathcal{B}$-contraction with respect to β.

Remark 4.11 Every fuzzy contractive mapping with contractive constant k is a $\mathcal{B}$-contraction with respect to $\beta \in \mathcal{B}$, where $\beta(t) = \dfrac{t}{k + (1-k)t}$ for all $t \in [0,1]$.

Remark 4.12 Every Tirado-fuzzy contractive mapping with Tirado-fuzzy contractive constant k is a $\mathcal{B}$-contraction with respect to $\beta \in \mathcal{B}$, where $\beta(t) = 1 - k(1-t)$ for all $t \in [0,1]$.

Remark 4.13 A mapping $f : X \to X$ is called a strict contraction if the following condition holds:

$$F(x, y, t) < F(fx, fy, t) \text{ for all } x, y \in X \text{ with } x \neq y \text{ and all } t > 0.$$

It is clear from Remark 4.10 and Definition 4.21 that $\mathcal{B}$-contractions form a subclass of strict contractions.

The following theorem ensures the existence and uniqueness of the fixed point of a $\mathcal{B}$-contraction in 1-complete fuzzy metric spaces and generalizes Theorem 3.1 given in [23].

Theorem 4.12 *Let $(X, F, *)$ be a 1-complete fuzzy metric-like space and let $f \colon X \to X$ be a $\mathcal{B}$-contraction. Then f has a unique fixed point $u \in X$ and $F(u, u, t) = 1$ for all $t > 0$.*

Proof. Assume that f is a $\mathcal{B}$-contraction with respect to $\beta \in \mathcal{B}$. Choose an arbitrary x_0 in X, and define the iterative Picard sequence $\{x_n\}$ given by $x_n = fx_{n-1}$ for all $n \in \mathbb{N}$. If there exists some $n_0 \in \mathbb{N}$ such that $x_{n_0} = x_{n_0-1}$, then x_{n_0-1} is a fixed point of f. In this case, the existence part is finished. Conversely, assume that $x_n \neq x_{n-1}$ for all $n \in \mathbb{N}$. We shall show that $\{x_n\}$ is a 1-M-Cauchy sequence. For a fixed $t > 0$, let

$$I_n = \inf_{m \geq n} F(x_n, x_m, t), \; n \in \mathbb{N}.$$

Now, since f is a $\mathcal{B}$-contraction with respect to $\beta \in \mathcal{B}$, by Remark 4.10 for all $n, m \in \mathbb{N}, m \geq n$, we have

$$\begin{aligned} F(x_n, x_{m-1}, t) &\leq \beta(F(x_n, x_{m-1}, t)) \\ &\leq F(fx_n, fx_{m-1}, t) \\ &= F(x_{n+1}, x_m, t). \end{aligned}$$

Taking infimum over m ($\geq n+1$) we obtain

$$I_n \leq I_{n+1} \quad \text{for all} \quad n \in \mathbb{N}.$$

Therefore, there exists I such that $\lim_{n \to \infty} I_n = I$.

Again, since f is a $\mathcal{B}$-contraction with respect to $\beta \in \mathcal{B}$, for all $n, m \in \mathbb{N}, m > n$, we have

$$\beta(F(x_{n-1}, x_{m-1}, t)) \leq F(fx_{n-1}, fx_{m-1}, t) = F(x_n, x_m, t).$$

From the above inequality, if $n \geq 2$, $m > n$ then

$$\beta(F(x_{n-2}, x_{m-2}, t)) \leq F(x_{n-1}, x_{m-1}, t).$$

Taking into account that β is nondecreasing, it follows that

$$\beta^2(F(x_{n-2}, x_{m-2}, t)) \leq \beta(F(x_{n-1}, x_{m-1}, t)) \leq F(x_n, x_m, t).$$

By induction, we obtain that

$$\beta^n(F(x_0, x_{m-n}, t)) \leq F(x_n, x_m, t) \quad \text{for all } n \in \mathbb{N}.$$

Taking infimum over m ($> n$) we obtain

$$\inf\left\{\beta^n(F(x_0, x_1, t)), \beta^n(F(x_0, x_2, t)), \dots\right\} \leq I_n \quad \text{for all } n \in \mathbb{N}.$$

Using ($\mathcal{B}2$), it follows from the above inequality that

$$\lim_{n\to\infty} I_n = I = 1. \tag{4.12}$$

By definition of I_n, for all $n, m \in \mathbb{N}, m > n$ we obtain $I_n \leq F(x_n, x_m, t)$, and so, by (4.12)

$$\lim_{n,m\to\infty} F(x_n, x_m, t) = 1.$$

Therefore, $\{x_n\}$ is a 1-M-Cauchy sequence in X. By 1-M-completeness of $(X, F, *)$, there exists $u \in X$ such that

$$\lim_{n\to\infty} F(x_n, u, t) = \lim_{n,m\to\infty} F(x_n, x_m, t) = F(u, u, t) = 1 \text{ for all } t > 0. \tag{4.13}$$

We shall show that u is the unique fixed point of f. Indeed, by using (4.11),

$$\beta(F(x_n, u, t)) \leq F(fx_n, fu, t) = F(x_{n+1}, fu, t)$$

for all $n \in \mathbb{N}$ and all $t > 0$. Taking into account (4.13) and Remark 4.10, the above inequality implies that

$$\lim_{n\to\infty} F(x_{n+1}, fu, t) = 1 \text{ for all } t > 0. \tag{4.14}$$

By (FML4) we have

$$F(u, fu, t) \geq F(u, x_{n+1}, t/2) * F(x_{n+1}, fu, t/2).$$

Using (4.13), (4.14) and the above inequality, we deduce that

$$F(u, fu, t) = 1 \text{ for all } t > 0,$$

which means that $fu = u$. Next, we claim that u is unique. Conversely, suppose that $v \in X$ is another fixed point of f, that is, $fv = v$ and $s > 0$ such that $\digamma(u, v, s) < 1$. Then, from (4.11) and the fact that $t < \beta(t)$ for all $t \in (0, 1)$, we have

$$\digamma(u, v, s) < \beta(\digamma(u, v, s)) \leq \digamma(fu, fv, s) = \digamma(u, v, s),$$

which is a contradiction. As a consequence, u is the unique fixed point of f, and it verifies $\digamma(u, u, t) = 1$ for all $t > 0$.

The following corollary generalizes Theorem 3.1 of Shukla and Abbas [23] in 1-complete fuzzy metric-like spaces.

Corollary 4.2 *Let $(X, \digamma, \star)$ be a 1-M-complete fuzzy metric-like space and let $f\colon X \to X$ be a fuzzy contractive mapping with fuzzy contractive constant $k \in (0, 1)$. Then f has a unique fixed point $u \in X$ and $\digamma(u, u, t) = 1$ for all $t > 0$.*

Proof. It follows from Theorem 4.12 that $\beta(t) = \dfrac{t}{k + (1-k)t}$ for all $t \in [0, 1]$ where $k \in (0, 1)$, taking into account Remark 4.11.

Corollary 4.3 *Let $(X, \digamma, \star)$ be a 1-M-complete fuzzy metric-like space and let $f : X \to X$ be a Tirado-fuzzy contractive mapping with Tirado-fuzzy contractive constant $k \in (0, 1)$. Then f has a unique fixed point $u \in X$ and $\digamma(u, u, t) = 1$ for all $t > 0$.*

Proof. It follows from Theorem 4.12 that the function $\beta(t) = 1 - k(1 - t)$ for all $t \in [0, 1]$ where $k \in (0, 1)$, taking into account Remark 4.12.

Remark 4.14 We point out that Corollary 4.2 implies Corollary 4.3. To prove it, given $k \in (0, 1)$, let $\beta_{GS}(t) = \dfrac{t}{k + (1-k)t}$ and $\beta_T(t) = 1 - k(1 - t)$ for all $t \in [0, 1]$. It is not difficult to check that $\beta_{GS}(t) \leq \beta_T(t)$ for all $t \in [0, 1]$. Thus, in view of Remarks 4.11 and 4.12, we deduce that every Tirado-fuzzy contractive mapping is a fuzzy contractive mapping with same contractive constant k. However, the converse is not true, as illustrated in the following example.

Example 4.11 Let $X = \{2^{3^n} : n \in \mathbb{N}\} \cup \{2\}$. Define the fuzzy set $\digamma$ in $X^2 \times (0, \infty)$, for all $x, y \in X$ and all $t > 0$, by

$$\digamma(x, y, t) = \begin{cases} 1 & \text{if } x = y = 2, \\ \dfrac{\min(x, y)}{\max(x, y)^3} & \text{otherwise.} \end{cases}$$

Then $(X, M, \star_p)$ is a fuzzy metric-like space. Define a mapping $f : X \to X$ by

$$fx = \begin{cases} 2 & \text{if } x = 2, \\ \sqrt[3]{x} & \text{if } x \in X \setminus \{2\}. \end{cases}$$

Then, the mapping f is a fuzzy contractive mapping with $k \in \left[\frac{3}{255}, 1\right)$. On the other hand, f is not a Tirado-fuzzy contractive mapping in the fuzzy metric-like space $(X, \mathcal{F}, \star_p)$. Indeed, if $x, y \in X \setminus \{2\}$ with $x < y$, then for all $t > 0$, we have $\mathcal{F}(fx, fy, t) = \frac{\sqrt[3]{x}}{y}$ and $\mathcal{F}(x, y, t) = \frac{x}{y^3}$. Therefore, as $y \to \infty$, it is not possible to choose $k \in [0, 1)$ such that (4.10) is satisfied.

Example 4.12 Let $\{x_n\} \subset (0, 1)$ be a strictly increasing sequence convergent to 1 with respect to the usual metric in $\mathbb{R}$ and $X = \{x_n\} \cup \{1\}$. Define a fuzzy set on $X^2 \times (0, \infty)$ by

$$\mathcal{F}(x, y, t) = \min\{x, y\} \quad \text{for all } x, y \in X.$$

Then $(X, \mathcal{F}, *)$ is a 1-complete metric-like space, where $a * b = \min\{a, b\}$. Define $f : X \to X$ by

$$fx_n = x_{n+2} \text{ for all } n \in \mathbb{N} \quad \text{and} \quad f(1) = 1.$$

Now, define $\beta\colon [0, 1] \to [0, 1]$ by

$$\beta(t) = \begin{cases} 0 & \text{if } t = 0, \\ x_2 & \text{if } 0 < t < x_2, \\ x_{n+1} & \text{if } x_n \leq t < x_{n+1}, n \geq 2, \\ 1 & \text{if } t = 1. \end{cases}$$

It is easy to see that $\beta \in \mathcal{B}$ and f is a $\mathcal{B}$-contraction. All the conditions of Theorem 4.12 are satisfied and f has a unique fixed point of 1, and this point verifies $\mathcal{F}(1, 1, t) = 1$ for all $t > 0$.

Remark 4.15 Take $x_n = 1 - \frac{1}{n+1}$ for all $n \in \mathbb{N}$ in the above example. Then f is not a fuzzy contractive mapping. To see this, note that for all $n \in \mathbb{N}$,

$$\begin{aligned} \frac{1}{\mathcal{F}(fx_n, fx_{n+1}, t)} - 1 \leq k\left[\frac{1}{\mathcal{F}(x_n, x_{n+1}, t)} - 1\right] &\iff \frac{1}{x_{n+2}} - 1 \leq k\left[\frac{1}{x_n} - 1\right] \\ &\iff \frac{n}{n+2} \leq k. \end{aligned}$$

Therefore, when $n \to \infty$, the above inequality contradicts the condition $k \in (0, 1)$. Thus, the mapping f in the above example is a $\mathcal{B}$-contraction in each case, but it is not a fuzzy contractive mapping nor a Tirado-fuzzy contractive mapping. As a consequence, the main result obtained by [23] (that is, Corollary 4.2) and Corollary 4.3 are not applicable to f.

Let Ψ be the class of all functions $\psi\colon [0, 1] \to [0, 1]$ such that ψ is continuous, nondecreasing and $\psi(t) > t$ for all $t \in (0, 1)$. Mihet [19] introduced and studied the so-called fuzzy ψ-contractive mappings in fuzzy metric spaces and also showed that ψ-contractive mappings generalize fuzzy contractive mappings introduced in [7], and those defined in [27].

Remark 4.16 The class Ψ is contained properly in the class $\mathcal{B}$.

Proof. Let $\psi \in \Psi$; then by definition, $\psi^n(t) \leq \psi^{n+1}(t)$ for all $t \in (0,1)$ and $n \geq 1$. Therefore, $\{\psi^n(t)\}$ is a nondecreasing bounded sequence in $[0,1]$ for all $t \in (0,1)$, and so must converge to some $a_t \in [0,1]$ and $\psi^n(t) \leq a_t$ for all $n \geq 1$. Without loss of generality we can assume that $a_t > 0$. Suppose, $a_t < 1$; then by definition of Ψ, we have $a_t < \psi^n(a_t) \leq a_t$ for all $n \in \mathbb{N}$. This contradiction shows that $a_t = 1$, that is,

$$\lim_{n\to\infty} \psi^n(t) = 1 \; \forall \, t \in (0,1].$$

Thus, $\Psi \subseteq \mathcal{B}$. Conversely, there exist functions $\beta \in \mathcal{B}$ such that $\beta \notin \Psi$. For instance, let a and b are two positive reals such that $a > 1, b < 1, ab < 1$, and let $\beta \colon [0,1] \to [0,1]$ be defined by

$$\beta(t) = \begin{cases} at & \text{if } t \in [0,b], \\ 1 & \text{if } t \in (b,1]. \end{cases}$$

Then $\beta \in \mathcal{B}$ but $\beta \notin \Psi$ (since β is not continuous), which completes the proof.

As an immediate consequence of the above remark, we obtain the following version of Mihet's result in 1-M-complete fuzzy metric-like spaces.

Corollary 4.4 *Let $(X, F, \star)$ be a 1-M-complete fuzzy metric-like space and $f : X \to X$ be a ψ-contraction, that is,*

$$\psi(F(x,y,t)) \leq F(fx, fy, t) \;\; \textit{for all } x, y \in X \textit{ and all } t > 0,$$

where $\psi \in \Psi$. Then f has a unique fixed point $u \in X$ and $F(u,u,t) = 1$ for all $t > 0$.

The following theorem generalizes and extends Grabiec's result in [2] and Theorem 3.7 by [23] in 1-M-complete fuzzy metric-like space.

Theorem 4.13 *Let $(X, F, \star)$ be a 1-M-complete fuzzy metric-like space and $f : X \to X$ be a mapping satisfying*

$$F(fx, fy, t) \geq F(x, y, g(t)) \tag{4.15}$$

for all $x, y \in X$ and all $t > 0$, where $g \colon (0,\infty) \to (0,\infty)$ is such that $F(x,y,g(t)) \geq F(x,y,t)$ and $\lim_{n\to\infty} F(x,y,g^n(t)) = 1$ for all $x, y \in X$ and $t > 0$. Then f has a unique fixed point $u \in X$ and $F(u,u,t) = 1$ for all $t > 0$.

Proof. Choose an arbitrary x_0 in X and define the Picard sequence $\{x_n\}$ given by $x_n = fx_{n-1}$ for all $n \in \mathbb{N}$. Without loss of generality, we can assume that $x_n \neq x_{n-1}$ for all $n \in \mathbb{N}$. We claim that the sequence $\{x_n\}$ is a 1-M-Cauchy sequence. For a fixed $t > 0$, let

$$I_n = \inf_{m \geq n} F(x_n, x_m, t), \; n \in \mathbb{N}.$$

Now, for all $n, m \in \mathbb{N}, m \geq n$, we obtain from (4.15) that

$$F(x_{n+1}, x_m, t) = F(fx_n, fx_{m-1}, t) \geq F(x_n, x_{m-1}, g(t)) \geq F(x_n, x_{m-1}, t).$$

Taking infimum over $m(\geq n+1)$ we obtain

$$I_n \leq I_{n+1} \text{ for all } n \in \mathbb{N}.$$

Therefore, there exists I such that $\lim_{n\to\infty} I_n = I$. Again, for any $p \in \mathbb{N}$ from (4.15), we have

$$\begin{aligned} F(x_{n+p}, x_{m+p}, t) &= F(fx_{n+p-1}, fx_{m+p-1}, t) \geq F(x_{n+p-1}, x_{m+p-1}, g(t)) \\ &\geq F(x_{n+p-1}, x_{m+p-1}, t). \end{aligned}$$

It follows from the above inequality that

$$F(x_{n+p-1}, x_{m+p-1}, g(t)) \geq F(x_{n+p-2}, x_{m+p-2}, g^2(t)) \geq F(x_{n+p-2}, x_{m+p-2}, t).$$

Repeating this process and using the inequalities thus obtained, we have

$$F(x_{n+p}, x_{m+p}, t) \quad \geq \quad F(x_n, x_m, g^p(t)) \geq F(x_n, x_m, t).$$

Taking infimum over $m(\geq n)$ we obtain

$$I_n \leq \inf\left\{F(x_n, x_{n+1}, g^p(t)), F(x_n, x_{n+2}, g^p(t)), \dots\right\} \leq I_{n+p}.$$

Since $\lim_{p\to\infty} F(x, y, g^p(t)) = 1$ for all $x, y \in X, t > 0$, we obtain from the above inequality, that

$$\lim_{n\to\infty} I_n = I = 1 \tag{4.16}$$

By definition of I_n, for all $n, m \in \mathbb{N}, m > n$ we obtain $I_n \leq F(x_n, x_m, t)$, and so, by (4.16)

$$\lim_{n,m\to\infty} F(x_n, x_m, t) = 1.$$

Therefore, $\{x_n\}$ is a 1-M-Cauchy sequence in X. By 1-M-completeness of $(X, F, \star)$, there exists $u \in X$ such that

$$\lim_{n\to\infty} F(x_n, u, t) = \lim_{n,m\to\infty} F(x_n, x_m, t) = F(u, u, t) = 1 \text{ for all } t > 0. \tag{4.17}$$

We shall show that u is the unique fixed point of f. Indeed, it follows from (4.15) that

$$\begin{aligned} F(u, fu, t) &\geq F(u, x_{n+1}, t/2) \star F(x_{n+1}, fu, t/2) \\ &= F(u, x_{n+1}, t/2) \star F(fx_n, fu, t/2) \\ &\geq F(u, x_{n+1}, t/2) \star F(x_n, u, g(t/2)). \end{aligned}$$

Using (4.17), $F(u, fu, t) = 1$ for all $t > 0$, that is, $fu = u$. Thus, u is a fixed point of f.

For the uniqueness of the fixed point, let $v \in X$ be another fixed point of f, that is, $fv = v \neq u = fu$. From (4.15), we have

$$\mathcal{F}(u, v, s) = \mathcal{F}(fu, fv, s) \geq \mathcal{F}(u, v, g(s)).$$

Repeating this process we obtain

$$\mathcal{F}(u, v, s) \geq \mathcal{F}(u, v, g^n(s)).$$

Letting $n \to \infty$ as and using the properties of the function g, we obtain $\mathcal{F}(u, v, s) = 1$, that is, $u = v$. Thus, u is the unique fixed point of f, which verifies $\mathcal{F}(u, u, t) = 1$ for all $t > 0$.

Example 4.13 Let α, β and δ be rationals such that $0 < \alpha < \beta < \delta$ and $X = [0, \delta] \cap \mathbb{Q}$. Define the fuzzy set $\mathcal{F}$ in $X^2 \times (0, \infty)$ by

$$\mathcal{F}(x, y, t) = 1 - \frac{\max\{x, y\}}{\delta + t} \text{ for all } x, y \in X, t > 0.$$

Then $(X, M, \star_l)$ is a 1-M-complete fuzzy metric-like space, but it is not a fuzzy metric space, as $\mathcal{F}(x, x, t) = 1 - \dfrac{x}{\delta + t} \neq 1$ for all $x \in X \setminus \{0\}$. Define the mapping $f : X \to X$ by

$$fx = \begin{cases} 0 & \text{if } x \in [0, \beta] \cap \mathbb{Q}, \\ \alpha & \text{if } x \in (\beta, \delta] \cap \mathbb{Q}. \end{cases}$$

Define the mapping $g : (0, \infty) \to (0, \infty)$ by

$$g(t) = \left(\frac{\beta - \alpha}{\alpha}\right)\delta + \left(\frac{\beta}{\alpha}\right)t$$

for all $t > 0$. Then it is easy to see that $\mathcal{F}(x, y, g(t)) \geq \mathcal{F}(x, y, t)$ and $\lim_{n\to\infty} \mathcal{F}(x, y, g^n(t)) = 1$ for all $x, y \in X$ and $t > 0$. Then, f and g satisfy all the conditions of Theorem 4.13 and f has a unique fixed point $0 \in X$ with $\mathcal{F}(0, 0, t) = 1$ for all $t > 0$.

4.5 Fuzzy Prešić-Ćirić operators and unified fixed point theorems

In 2016, Shukla *et al.* [25] introduced the notion of Fuzzy Prešić-Ćirić operators and established unified fixed point theorems for such operators and also discussed some new properties of fuzzy contractive sequences in the framework of George and Veeramani fuzzy metric space.

Definition 4.22 Let $(X, M, *)$ be a fuzzy metric space, k a positive integer and $f\colon X^k \to X$ a mapping. Then f is called a fuzzy Prešić-Ćirić operator if

$$\frac{1}{M(f(x_1, \ldots, x_k), f(x_2, \ldots, x_{k+1}), t)} - 1 \leq \lambda \cdot \max_{1 \leq i \leq k} \left\{ \frac{1}{M(x_i, x_{i+1}, t)} - 1 \right\}, \tag{4.18}$$

for all $x_1, \ldots, x_k, x_{k+1} \in X$ and $t > 0$, where $\lambda \in (0, 1)$. Alternatively, the above condition may be written as

$$M(f(x_1, \ldots, x_k), f(x_2, \ldots, x_{k+1}), t) \geq \left[\lambda \cdot \max_{1 \leq i \leq k} \left\{ \frac{1}{M(x_i, x_{i+1}, t)} - 1 \right\} + 1 \right]^{-1}, \tag{4.19}$$

for all $x_1, \ldots, x_k, x_{k+1} \in X$ and $t > 0$, where $\lambda \in (0, 1)$.

Remark 4.17 Taking M as M_d in condition (4.18) (or (4.19)), we get

$$d(f(x_1, \ldots, x_k), f(x_2, \ldots, x_{k+1})) \leq \lambda \max\{d(x_i, x_{i+1})\colon 1 \leq i \leq k\},$$

for all $x_1, \ldots, x_k, x_{k+1} \in X$, where $\lambda \in (0, 1)$.

The following definition will be needed to prove a fixed point theorem for the fuzzy Prešić-Ćirić operators in M-complete fuzzy metric spaces.

Let $*$ be a given t-norm. For $a_1, a_2, \ldots, a_n \in [0, 1]$, we use the notation

$$\mathop{*}_{i=1}^{n} a_i = a_1 * a_2 * \cdots * a_n.$$

Let $a \in [0, 1]$. Then we can define the sequence $\{*^n a\}_{n \in \mathbb{N}}$ by $*^1 a = a$ and $*^{n+1} a = (*^n a) * a$, for $n \geq 1$.

Definition 4.23 (Hadžić and Pap [12]) A t-norm $*$ is said to be of H-type if the sequence $\{*^n a\}_{n \in \mathbb{N}}$ is equicontinuous at 1, that is, for all $\varepsilon \in (0, 1)$, there exists $\delta \in (0, 1)$ such that, $a \in (1 - \delta, 1]$ implies $*^n a > 1 - \varepsilon$ for all $n \in \mathbb{N}$.

An important H-type t-norm is $*_m$. Some other examples of H-type t-norms can be found in [12]. We denote the class of all Hadžić-type t-norms by $\mathcal{H}$.

Theorem 4.14 *Let $(X, M, *)$ be an M-complete fuzzy metric space, k a positive integer and $f\colon X^k \to X$ a fuzzy Prešić-Ćirić operator. Suppose that one of the following conditions holds:*

(H1) *$* \in \mathcal{H}$ and there exist $x_1, x_2 \ldots, x_k \in X$ such that*

$$\inf_{t>0} M(x_i, x_{i+1}, t) > 0, i = 1, 2, \ldots, k-1, \quad \inf_{t>0} M(x_k, f(x_1, \ldots, x_k), t) > 0.$$

(H2) *$x_1, x_2 \ldots, x_k \in X$ such that the following property holds: for each $\varepsilon \in (0, 1)$ and $t > 0$, there exists $n_0 \in \mathbb{N}$ such that, for $m, n > n_0$, with*

$m > n$, we get $*_{j=n}^{m-1}\left[1+\mu(ts_j^{(n)})\lambda^{\frac{j}{k}}\right]^{-1} > 1-\varepsilon$, *where* $s_j^{(n)} = \frac{1}{2^{j-n+1}}$, $j = n, \ldots, m-2$, $s_{m-1}^{(n)} = s_{m-2}^{(n)}$ *and*

$$\mu(z) := \max\left\{\max_{1\le i\le k-1} \frac{1}{\lambda^{\frac{i}{k}}}\left[\frac{1}{M(x_i, x_{i+1}, z)} - 1\right], \right.$$
$$\left. \frac{1}{\lambda}\left[\frac{1}{M(x_k, f(x_1, \ldots, x_k), z)} - 1\right]\right\}.$$

Then f has a fixed point in X. If, in addition, we suppose that on the diagonal $\Delta \subset X^k$,

$$M(f(u, \ldots, u), f(v, \ldots, v), t) > M(u, v, t), \quad \forall\, t > 0 \tag{4.20}$$

holds for $u, v \in X$ with $u \neq v$, then f has a unique fixed point.

Proof. Let $x_1, x_2, \ldots, x_k$ be the points in X given by hypothesis. Define a sequence $\{x_n\}$ by $x_{n+k} = f(x_n, x_{n+1}, \ldots, x_{n+k-1})$, $\forall\, n \in \mathbb{N}$. For notational convenience, set $M(x_n, x_{n+1}, t) = M_n(t)$, for all $n \in \mathbb{N}$ and $t > 0$ and consider $\mu(t) := \max\left\{\frac{1}{\theta^i}\left[\frac{1}{M_i(t)} - 1\right] : 1 \le i \le k\right\}$, where $\theta = \lambda^{\frac{1}{k}}$. By mathematical induction, we show that

$$\frac{1}{M_n(t)} - 1 \le \mu(t)\theta^n, \qquad \forall\, n \in \mathbb{N},\ \forall\, t > 0. \tag{4.21}$$

By the definition of $\mu(t)$, it is obvious that (4.21) is true for $n = 1, 2, \ldots, k$. Let the following k inequalities, for $t > 0$, $\frac{1}{M_n(t)} - 1 \le \mu(t)\theta^n$, $\frac{1}{M_{n+1}(t)} - 1 \le \mu(t)\theta^{n+1}, \ldots, \frac{1}{M_{n+k-1}(t)} - 1 \le \mu(t)\theta^{n+k-1}$ be the induction hypothesis.
Since $\theta = \lambda^{\frac{1}{k}} < 1$, from (4.18) we have

$$\begin{aligned}
\frac{1}{M_{n+k}(t)} - 1 &= \frac{1}{M(f(x_n, x_{n+1}, \ldots, x_{n+k-1}), f(x_{n+1}, x_{n+2}, \ldots, x_{n+k}), t)} - 1 \\
&\le \lambda \max\left\{\frac{1}{M(x_{n+i-1}, x_{n+i}, t)} - 1 \colon 1 \le i \le k\right\} \\
&= \lambda \max\left\{\frac{1}{M_{n+i-1}(t)} - 1 \colon 1 \le i \le k\right\} \\
&\le \lambda \max\left\{\mu(t)\,\theta^{n+i-1} \colon 1 \le i \le k\right\} \le \lambda\mu(t)\theta^n = \mu(t)\theta^{n+k},\ t > 0.
\end{aligned}$$

Hence, by induction, (4.21) is true for all $n \in \mathbb{N}$.

Next, we show that $\{x_n\}$ is an M-Cauchy sequence. Consider $\varepsilon \in (0, 1)$

and $t > 0$ fixed. For $n, m \in \mathbb{N}$ with $m > n$, using (4.21),

$$\begin{aligned}
M(x_n, x_m, t) \geq & M(x_n, x_{n+1}, t/2) * M(x_{n+1}, x_m, t/2) \\
\geq & M(x_n, x_{n+1}, t/2) * M(x_{n+1}, x_{n+2}, t/2^2) \\
& * \cdots * M(x_{m-2}, x_{m-1}, t/2^{m-n-1}) * M(x_{m-1}, x_m, t/2^{m-n-1}) \\
= & \left(*_{j=n}^{m-2} M_j(t/2^{j-(n-1)}) \right) * M_{m-1}(t/2^{m-n-1}) \\
\geq & \left(*_{j=n}^{m-2} \left[1 + \mu(t/2^{j-(n-1)})\theta^j\right]^{-1} \right) * \left[1 + \mu(t/2^{m-n-1})\theta^{m-1}\right]^{-1} \\
\geq & \left(*_{j=n}^{m-2} \left[1 + \mu(t/2^{j-(n-1)})\theta^n\right]^{-1} \right) * \left[1 + \mu(t/2^{m-n-1})\theta^{n}\right]^{-1}.
\end{aligned}$$

Under condition (H1), it is easy to check that $\mu := \sup_{t>0} \mu(t) \in [0, \infty)$, since

$$\sup_{t>0} \mu(t) \leq \max_{1 \leq i \leq k} \left\{ \frac{1}{\theta^i} \sup_{t>0} \left[\frac{1}{M_i(t)} - 1 \right] \right\} = \max_{1 \leq i \leq k} \left\{ \frac{1}{\theta^i} \left[\frac{1}{\inf_{t>0} M_i(t)} - 1 \right] \right\}$$

$$\text{and} \quad \inf_{t>0} M_i(t) = \inf_{t>0} M(x_i, x_{i+1}, t) > 0, \; i = 1, 2, \ldots, k-1,$$

$$\inf_{t>0} M_k(t) = \inf_{t>0} M(x_k, x_{k+1}, t) = \inf_{t>0} M(x_k, f(x_1, \ldots, x_k), t) > 0.$$

In these conditions, for $n, m \in \mathbb{N}$ with $m > n$,

$$M(x_n, x_m, t) \geq [1 + \mu\theta^n]^{-1} * \cdots * [1 + \mu\theta^n]^{-1} = *^{m-n} [1 + \mu\theta^n]^{-1}. \quad (4.22)$$

Since $* \in \mathcal{H}$, there exists $\delta \in (0,1)$ such that if $[1 + \mu\theta^n]^{-1} \in (1 - \delta, 1]$, then $*^{m-n} [1 + \mu\theta^n]^{-1} > 1 - \varepsilon$, for $m > n$. As $0 < \theta < 1$, given $\delta > 0$, there exists $n_0 \in \mathbb{N}$ such that $[1 + \mu\theta^n]^{-1} \in (1 - \delta, 1]$, for all $n > n_0$ (it suffices to take $n_0 \in \mathbb{N}$ such that $1 + \mu\theta^{n_0} < \frac{1}{1-\delta}$). With this choice, we obtain $*^{m-n} [1 + \mu\theta^n]^{-1} > 1 - \varepsilon$, $\forall\, n > n_0$, $m > n$. The above inequality with (4.22) and the properties of M give $M(x_n, x_m, t) > 1 - \varepsilon$, $\forall\, n, m > n_0$, $\forall\, t > 0$.

On the other hand, if (H2) holds, $n_0 \in \mathbb{N}$ such that for $m, n > n_0$, with $m > n$, we get

$$\begin{aligned}
M(x_n, x_m, t) \geq & \left(*_{j=n}^{m-2} \left[1 + \mu(t/2^{j-(n-1)})\theta^j\right]^{-1} \right) * \left[1 + \mu(t/2^{m-n-1})\theta^{m-1}\right]^{-1} \\
= & \mathop{*}_{j=n}^{m-1} \left[1 + \mu(ts_j^{(n)})\theta^j\right]^{-1} > 1 - \varepsilon.
\end{aligned}$$

Thus, in both situations, $\{x_n\}$ is an M-Cauchy sequence. By M-completeness of X, there exists $u \in X$ such that

$$\lim_{n \to \infty} M(x_n, u, t) = 1, \; \forall\, t > 0. \quad (4.23)$$

Now we show that u is a fixed point of f. Indeed, for any $n \in \mathbb{N}$ and $t > 0$, we have

$$\begin{aligned}M(x_{n+k}, f(u,\ldots,u),t) &= M(f(x_n,\ldots,x_{n+k-1}), f(u,\ldots,u),t)\\ &\geq M(f(x_n,\ldots,x_{n+k-1}), f(x_{n+1},\ldots,x_{n+k-1},u),t/2)\\ &* M(f(x_{n+1},\ldots,x_{n+k-1},u), f(x_{n+2},\ldots,x_{n+k-1},u,u),t/2^2)\\ &* \cdots * M(f(x_{n+k-2},x_{n+k-1},u,\ldots,u), f(x_{n+k-1},u,\ldots,u),t/2^{k-1})\\ &* M(f(x_{n+k-1},u,\ldots,u), f(u,\ldots,u),t/2^{k-1}). \qquad (4.24)\end{aligned}$$

Using (4.19), (4.21) and $\theta \in (0,1)$, we have

$$\begin{aligned}&M(f(x_n,\ldots,x_{n+k-1}), f(x_{n+1},\ldots,x_{n+k-1},u),t)\\ &\geq \left[\lambda \cdot \max\left\{\max_{1\leq i\leq k-1}\left\{\frac{1}{M(x_{n+i-1},x_{n+i},t)}-1\right\}, \frac{1}{M(x_{n+k-1},u,t)}-1\right\}+1\right]^{-1}\\ &= \left[\lambda \cdot \max\left\{\max_{1\leq i\leq k-1}\left\{\frac{1}{M_{n+i-1}(t)}-1\right\}, \frac{1}{M(x_{n+k-1},u,t)}-1\right\}+1\right]^{-1}\\ &\geq \left[\lambda \cdot \max\left\{\max_{1\leq i\leq k-1}\{\mu(t)\theta^{n+i-1}\}, \frac{1}{M(x_{n+k-1},u,t)}-1\right\}+1\right]^{-1}\\ &\geq \left[\lambda \cdot \max\left\{\mu(t)\theta^{n}, \frac{1}{M(x_{n+k-1},u,t)}-1\right\}+1\right]^{-1}.\end{aligned}$$

Using (4.23) and $0 < \theta < 1$, it follows from the above inequality that $\lim_{n\to\infty} M(f(x_n,\ldots,x_{n+k-1}), f(x_{n+1},\ldots,x_{n+k-1},u),t) = 1, \ \forall\, t > 0$. Similarly,

$$\lim_{n\to\infty} M(f(x_{n+1},\ldots,x_{n+k-1},u), f(x_{n+2},\ldots,x_{n+k-1},u,u),t) = 1, \ \forall\, t > 0, \ldots,$$

$$\lim_{n\to\infty} M(f(x_{n+k-1},u,\ldots,u), f(u,\ldots,u),t) = 1, \ \forall\, t > 0.$$

The above properties with (4.24) imply that

$$\lim_{n\to\infty} M(x_{n+k}, f(u,\ldots,u),t) = 1, \ \forall\, t > 0. \qquad (4.25)$$

Therefore, for any $n \in \mathbb{N}$ and $t > 0$, we have

$$M(u, f(u,u,\ldots,u),t) \ \geq \ M(u, x_{n+k}, t/2) * M(x_{n+k}, f(u,\ldots,u),t/2),$$

which, together with (4.23) and (4.25), gives $M(u, f(u,u,\ldots,u),t) = 1, \forall\, t > 0$. Thus, $f(u,u,\ldots,u) = u$, that is, u is a fixed point of f.

Finally, for uniqueness, suppose that $v \in X$ is another fixed point of f with $u \neq v$. Then, from (4.20), we have $M(u,v,t) = M(f(u,\ldots,u), f(v,\ldots,v),t) > M(u,v,t)$, for every $t > 0$. This contradiction shows that $u = v$. Thus, under condition (4.20), the fixed point of f is unique.

Remark 4.18 The uniqueness condition (4.20) in Theorem 4.14 is reduced, for $M = M_d$, to $d(T(u,\ldots,u),T(v,\ldots,v)) < d(u,v)$, $\forall\, u,v \in X$ with $u \neq v$. From the proof of Theorem 4.14, it is obvious that uniqueness is also derived by considering the weaker hypothesis: for each $u, v \in X$ fixed with $u \neq v$, there exists $t > 0$ such that

$$M(f(u,\ldots,u), f(v,\ldots,v), t) > M(u,v,t). \tag{4.26}$$

Corollary 4.5 *Let $(X, M, *)$ be an M-complete fuzzy metric space and $f\colon X \to X$ be a fuzzy contractive mapping (see Gregori and Sapena [7]), that is,*

$$\frac{1}{M(fx,fy,t)} - 1 \leq \lambda\left(\frac{1}{M(x,y,t)} - 1\right),\ \forall x,y \in X,\ \forall t > 0,$$

where $\lambda \in (0,1)$. Suppose that one of the following conditions holds:

(h1) *$* \in \mathcal{H}$ and there exists $x_1 \in X$ such that $\inf_{t>0} M(x_1, f(x_1), t) > 0$.*

(h2) *$x_1 \in X$ such that the following property holds: for each $\varepsilon \in (0,1)$ and $t > 0$, there exists $n_0 \in \mathbb{N}$ such that, for m, $n > n_0$, with $m > n$, we get $*_{j=n}^{m-1}\left[1 + \mu(ts_j^{(n)})\lambda^j\right]^{-1} > 1 - \varepsilon$, where $s_j^{(n)} = \frac{1}{2^{j-n+1}}$, $j = n,\ldots,m-2$, $s_{m-1}^{(n)} = s_{m-2}^{(n)}$ and $\mu(z) := \frac{1}{\lambda}\left[\frac{1}{M(x_1,f(x_1),z)} - 1\right]$.*

Then f has a unique fixed point in X.

Proof. Consider $k = 1$ in Theorem 4.14; then the existence of a fixed point $u \in X$ follows. Further, for x, $y \in X$ fixed with $x \neq y$, we get $\frac{1}{M(x,y,t)} - 1 > 0$ for some $t > 0$. Thus, since f is a fuzzy contractive mapping, we have $\dfrac{1}{M(fx,fy,t)} - 1 < \dfrac{1}{M(x,y,t)} - 1$, for some $t > 0$, that is, $M(fx,fy,t) > M(x,y,t)$, for some $t > 0$, where $x, y \in X$ are fixed with $x \neq y$. Hence, the uniqueness condition (4.26) of Remark 4.18 is satisfied. Therefore, the fixed point of f is unique.

Remark 4.19 Note that if we select $M = M_d$, conditions (H1) in Theorem 4.14 and (h1) in Corollary 4.5 are valid only if the mapping f has fixed points. Indeed, for $k = 1$, $\inf_{t>0} M_d(x_1, f(x_1), t) = \inf_{t>0} \dfrac{t}{t + d(x_1, f(x_1))} = 1$, if $f(x_1) = x_1$ and $\inf_{t>0} M_d(x_1, f(x_1), t) = 0$, if $f(x_1) \neq x_1$. Similarly, in the general case $k \in \mathbb{N}$, for the validity of (H1), $x_1 = x_2 = \cdots = x_k = f(x_1, x_2, \ldots, x_k)$. Thus, for the standard fuzzy metric induced by the metric d, the restriction (H1) is not useful, since it requires starting the process with a fixed point.

However, condition (H2) is interesting in the general case specifically for the standard fuzzy metric induced by the metric d.

Indeed, for $k = 1$, $M = M_d$ and any $x_1 \in X$, we have

$$\mu(z) := \frac{1}{\lambda}\left[\frac{1}{\dfrac{z}{z + d(x_1, f(x_1))}} - 1\right] = \frac{d(x_1, f(x_1))}{\lambda z}$$

and, thus, for $t > 0$ and $m, n \in \mathbb{N}$ with $m > n$,

$$\mathop{*}_{j=n}^{m-1}\left[1 + \mu(ts_j^{(n)})\lambda^j\right]^{-1} = \mathop{*}_{j=n}^{m-1}\left[1 + \frac{d(x_1, f(x_1))}{t}\frac{\lambda^{j-1}}{s_j^{(n)}}\right]^{-1}.$$

Note that if $f(x_1) = x_1$, then μ is null and for every $\varepsilon > 0$, $\mathop{*}_{j=n}^{m-1}\left[1 + \mu(ts_j^{(n)})\lambda^j\right]^{-1} = 1 > 1 - \varepsilon$, for all $m, n \in \mathbb{N}$ with $m > n$ is satisfied. Consider the general case $f(x_1) \neq x_1$ or $f(x_1) = x_1$. If we take $* = *_m$, for $t > 0$ and $m, n \in \mathbb{N}$ with $m > n$,

$$\begin{aligned}
\mathop{*}_{j=n}^{m-1}\left[1 + \mu(ts_j^{(n)})\lambda^j\right]^{-1} &= \min_{n \le j \le m-1}\left[1 + \frac{d(x_1, f(x_1))}{t}\frac{\lambda^{j-1}}{s_j^{(n)}}\right]^{-1} \\
&= \left[1 + \frac{d(x_1, f(x_1))}{t}\max_{n \le j \le m-1}\frac{\lambda^{j-1}}{s_j^{(n)}}\right]^{-1}.
\end{aligned}$$

Replacing $s_j^{(n)} = \frac{1}{2^{j-n+1}}$, $j = n, \ldots, m-2$ and $s_{m-1}^{(n)} = s_{m-2}^{(n)}$, for $t > 0$ and $m, n \in \mathbb{N}$ with $m > n$,

$$\begin{aligned}
&\max_{n \le j \le m-1}\frac{\lambda^{j-1}}{s_j^{(n)}} = \max\left\{2\lambda^{n-1}, 4\lambda^n, \ldots, 2^{m-n-1}\lambda^{m-3}, 2^{m-n-1}\lambda^{m-2}\right\} \\
&= \max\left\{(2\lambda)^{n-1}2^{-n+2}, (2\lambda)^n 2^{-n+2}, \ldots, (2\lambda)^{m-3}2^{-n+2}, (2\lambda)^{m-2}2^{-n+1}\right\} \\
&\le \max\left\{(2\lambda)^{n-1}2^{-n+2}, (2\lambda)^n 2^{-n+2}, \ldots, (2\lambda)^{m-3}2^{-n+2}, (2\lambda)^{m-2}2^{-n+2}\right\} \\
&\le \max\left\{(2\lambda)^{n-1}, (2\lambda)^n, \ldots, (2\lambda)^{m-3}, (2\lambda)^{m-2}\right\}2^{-n+2}.
\end{aligned}$$

If $\lambda \in (0, \frac{1}{2}]$, since $m > n$, we get $0 \le \max_{n \le j \le m-1}\frac{\lambda^{j-1}}{s_j^{(n)}} \le (2\lambda)^{n-1}2^{-n+2} = 2\lambda^{n-1} \to 0$, as $n \to \infty$, so that for $\varepsilon \in (0, 1)$ and $t > 0$ fixed, $n_0 \in \mathbb{N}$ such that for $m, n > n_0$, with $m > n$, $\max_{n \le j \le m-1}\frac{\lambda^{j-1}}{s_j^{(n)}} < \frac{t}{d(x_1, f(x_1))}\frac{\varepsilon}{1 - \varepsilon}$ and hence, $*_{j=n}^{m-1}\left[1 + \mu(ts_j^{(n)})\lambda^j\right]^{-1} = \left[1 + \frac{d(x_1, f(x_1))}{t}\max_{n \le j \le m-1}\frac{\lambda^{j-1}}{s_j^{(n)}}\right]^{-1} > 1 - \varepsilon$. This proves that condition (h2) holds for $M = M_d$ and $* = *_m$ if $\lambda \in (0, \frac{1}{2}]$, independently of the choice of $x_1 \in X$.

Moreover, for a general $k \in \mathbb{N}$, using $M = M_d$ and any $x_1, \ldots, x_k \in X$, we get

$$\begin{aligned}\mu(z) :=& \max\left\{\max_{1\le i\le k-1} \frac{1}{\lambda^{\frac{i}{k}}}\left[\frac{1}{\frac{z}{z+d(x_i,x_{i+1})}} - 1\right], \frac{1}{\lambda}\left[\frac{1}{\frac{z}{z+d(x_k,f(x_1,\ldots,x_k))}} - 1\right]\right\}\\ =& \max\left\{\max_{1\le i\le k-1} \frac{1}{\lambda^{\frac{i}{k}} z} d(x_i, x_{i+1}), \frac{1}{\lambda z} d(x_k, f(x_1, \ldots, x_k))\right\}\\ =& \frac{1}{z}\max\left\{\max_{1\le i\le k-1} \frac{d(x_i, x_{i+1})}{\lambda^{\frac{i}{k}}}, \frac{d(x_k, f(x_1, \ldots, x_k))}{\lambda}\right\}.\end{aligned}$$

Hence, for $t > 0$ and $m, n \in \mathbb{N}$ with $m > n$, $*_{j=n}^{m-1}\left[1 + \mu(ts_j^{(n)})\lambda^{\frac{j}{k}}\right]^{-1}$ is equal to

$$\mathop{*}_{j=n}^{m-1}\left[1 + \frac{1}{ts_j^{(n)}}\max\left\{\max_{1\le i\le k-1}\frac{d(x_i, x_{i+1})}{\lambda^{\frac{i}{k}}}, \frac{d(x_k, f(x_1, \ldots, x_k))}{\lambda}\right\}\lambda^{\frac{j}{k}}\right]^{-1}.$$

Using $* = *_m$, for $t > 0$ and $m, n \in \mathbb{N}$ with $m > n$, the previous expression is

$$\left[1 + \frac{1}{t}\max_{n\le j\le m-1}\frac{1}{s_j^{(n)}}\max\left\{\max_{1\le i\le k-1}\frac{d(x_i, x_{i+1})}{\lambda^{\frac{i}{k}}}, \frac{d(x_k, f(x_1, \ldots, x_k))}{\lambda}\right\}\lambda^{\frac{j}{k}}\right]^{-1}.$$

Now, since $\lambda > 0$ and $s_j^{(n)} > 0$, for every $j = n, \ldots, m-1$, we get

$$\begin{aligned}&\max_{n\le j\le m-1}\frac{1}{s_j^{(n)}}\max\left\{\max_{1\le i\le k-1}\frac{d(x_i, x_{i+1})}{\lambda^{\frac{i}{k}}}, \frac{d(x_k, f(x_1, \ldots, x_k))}{\lambda}\right\}\lambda^{\frac{j}{k}}\\ &= \max_{n\le j\le m-1}\max\left\{\max_{1\le i\le k-1}\frac{d(x_i, x_{i+1})\lambda^{\frac{j}{k}}}{s_j^{(n)}\lambda^{\frac{i}{k}}}, \frac{d(x_k, f(x_1, \ldots, x_k))\lambda^{\frac{j}{k}}}{s_j^{(n)}\lambda}\right\}\\ &\le \max\left\{\max_{1\le i\le k-1}\max_{n\le j\le m-1}\frac{d(x_i, x_{i+1})\lambda^{\frac{j}{k}}}{s_j^{(n)}\lambda^{\frac{i}{k}}}, \max_{n\le j\le m-1}\frac{d(x_k, f(x_1, \ldots, x_k))\lambda^{\frac{j}{k}}}{s_j^{(n)}\lambda}\right\}\\ &= \max\left\{\max_{1\le i\le k-1}\frac{d(x_i, x_{i+1})}{\lambda^{\frac{i}{k}}}\max_{n\le j\le m-1}\frac{\lambda^{\frac{j}{k}}}{s_j^{(n)}}, \frac{d(x_k, f(x_1, \ldots, x_k))}{\lambda}\max_{n\le j\le m-1}\frac{\lambda^{\frac{j}{k}}}{s_j^{(n)}}\right\}\\ &= \max\left\{\max_{1\le i\le k-1}\frac{d(x_i, x_{i+1})}{\lambda^{\frac{i}{k}}}, \frac{d(x_k, f(x_1, \ldots, x_k))}{\lambda}\right\}\max_{n\le j\le m-1}\frac{\lambda^{\frac{j}{k}}}{s_j^{(n)}}.\end{aligned}$$

For $s_j^{(n)} = \frac{1}{2^{j-n+1}}$, $j = n, \ldots, m-2$ and $s_{m-1}^{(n)} = s_{m-2}^{(n)}$, for $t > 0$ and $m, n \in \mathbb{N}$ with $m > n$,

$$\begin{aligned}&\max_{n\le j\le m-1}\frac{\lambda^{\frac{j}{k}}}{s_j^{(n)}} = \max\left\{2\lambda^{\frac{n}{k}}, 4\lambda^{\frac{n+1}{k}}, \ldots, 2^{m-n-1}\lambda^{\frac{m-2}{k}}, 2^{m-n-1}\lambda^{\frac{m-1}{k}}\right\}\\ &= \max\left\{(2\lambda^{\frac{1}{k}})^n 2^{-n+1}, (2\lambda^{\frac{1}{k}})^{n+1}2^{-n+1}, \ldots, (2\lambda^{\frac{1}{k}})^{m-2}2^{-n+1}, (2\lambda^{\frac{1}{k}})^{m-1}2^{-n}\right\}\\ &\le \max\left\{(2\lambda^{\frac{1}{k}})^n, (2\lambda^{\frac{1}{k}})^{n+1}, \ldots, (2\lambda^{\frac{1}{k}})^{m-2}, (2\lambda^{\frac{1}{k}})^{m-1}\right\}2^{-n+1}.\end{aligned}$$

If $\lambda \in (0, \frac{1}{2^k}]$, then $2\lambda^{\frac{1}{k}} \leq 1$ and since $m > n$,

$$0 \leq \max_{n \leq j \leq m-1} \frac{\lambda^{\frac{j}{k}}}{s_j^{(n)}} \leq (2\lambda^{\frac{1}{k}})^n 2^{-n+1} = 2\lambda^{\frac{n}{k}} \to 0, \text{ as } n \to \infty,$$

so that for $\varepsilon \in (0,1)$ and $t > 0$ fixed, $n_0 \in \mathbb{N}$ such that for $m, n > n_0$, with $m > n$,

$$\max_{n \leq j \leq m-1} \frac{1}{s_j^{(n)}} \max\left\{\max_{1 \leq i \leq k-1} \frac{d(x_i, x_{i+1})}{\lambda^{\frac{i}{k}}}, \frac{d(x_k, f(x_1, \ldots, x_k))}{\lambda}\right\} \lambda^{\frac{j}{k}} < t\frac{\varepsilon}{1-\varepsilon}$$

and $*_{j=n}^{m-1}\left[1 + \mu(ts_j^{(n)})\lambda^{\frac{j}{k}}\right]^{-1} > 1 - \varepsilon$. Therefore, (H2) holds for $M = M_d$ and $* = *_m$ if $\lambda \in (0, \frac{1}{2^k}]$, independently of the choice of $x_1, \ldots, x_k \in X$.

Next, we give an example which illustrates Theorem 4.14.

Example 4.14 Let $0 < \alpha < \dfrac{1}{2}$, $x_n = \dfrac{2\alpha + 2^{n-1} - 1}{2^n}$, $n \in \mathbb{N}$, and consider the set $X = \{1\} \cup \{x_n : n \in \mathbb{N}\}$. Define the fuzzy set $M\colon X^2 \times (0, \infty) \to [0,1]$ by

$$M(x,y,t) = \begin{cases} x *_m y & \text{if } x \neq y, \\ 1 & \text{if } x = y, \end{cases} \quad \forall x, y \in X, \ \forall t > 0.$$

Then $(X, M, *_m)$ is an M-complete fuzzy metric space and $*_m \in \mathcal{H}$.

For $k = 2$, let $f\colon X^2 \to X$ be defined by

$$f(x,y) = \begin{cases} 2(x_i *_m x_j) & \text{if } x = x_i,\ y = x_j,\ i < j, \\ 1 & \text{otherwise.} \end{cases}$$

Now, by some routine calculations, f satisfies

$$\frac{1}{M(f(y_1,y_2), f(y_2,y_3), t)} - 1 \leq \lambda \cdot \max\left\{\frac{1}{M(y_1,y_2,t)} - 1, \frac{1}{M(y_2,y_3,t)} - 1\right\},$$

for all $y_1, y_2, y_3 \in X$ and $t > 0$, where $\lambda = \dfrac{1}{2}$. Therefore, f is a fuzzy Prešić-Ćirić operator with $\lambda = \dfrac{1}{2}$.

We prove that condition (H1) is satisfied. Indeed, $*_m \in \mathcal{H}$ and starting with the points $x_1 = \alpha < x_2 = \frac{1}{4} + \frac{\alpha}{2}$ in X, we get $\inf_{t>0} M(x_1, x_2, t) = \min\{x_1, x_2\} = x_1 = \alpha > 0$ and

$$\begin{aligned}\inf_{t>0} M(x_2, f(x_1,x_2), t) &= \inf_{t>0} M(x_2, 2\min\{x_1,x_2\}, t) = \inf_{t>0} M\left(\frac{1}{4} + \frac{\alpha}{2}, 2\alpha, t\right) \\ &= \begin{cases} 1 & \text{if } \alpha = \frac{1}{6}, \\ \min\{\frac{1}{4} + \frac{\alpha}{2}, 2\alpha\} & \text{if } \alpha \neq \frac{1}{6} \end{cases} > 0.\end{aligned}$$

Also, by definition of f, for $x, y \in X$ with $x \neq y$, we have $M(f(x,x), f(y,y), t) = M(1,1,t) = 1 > M(x,y,t)$, $\forall t > 0$. Hence, all the conditions of Theorem 4.14 are satisfied and we can conclude the existence of a unique fixed point of f. In fact, 1 is the unique fixed point.

Next, we give a sufficient condition for the validity of condition (4.20) under hypothesis (4.18) (or, equivalently, (4.19)) provided that $k \geq 2$. This condition is related to M and the t-norm $*$ and allows us to establish the following corollary of Theorem 4.14.

Corollary 4.6 *Let $(X, M, *)$ be an M-complete fuzzy metric space, k an integer with $k \geq 2$ and $f\colon X^k \to X$ a fuzzy Prešić-Ćirić operator. Suppose that one of the conditions (H1) or (H2) holds. Then f has a fixed point in X. If, in addition, we suppose that for each $u, v \in X$ fixed with $u \neq v$, there exists $t > 0$ such that*

$$\mathop{*}_{i=1}^{k} \left[\lambda \cdot \left(\frac{1}{z_i} - 1\right) + 1\right]^{-1} > M(u,v,t), \tag{4.27}$$

where $z_i = M(u, v, t/2^i)$, for $i = 1, \ldots, k-1$, and $z_k = z_{k-1}$, then f has a unique fixed point.

Proof. The first part of the corollary follows from the proof of Theorem 4.14. To find the uniqueness of fixed point, suppose that u, $v \in X$ are fixed points of f with $u \neq v$. Then, for any $t > 0$, we have

$$\begin{aligned}
&M(u,v,t) = M(f(u,\ldots,u), f(v,\ldots,v), t)\\
&\geq M(f(u,\ldots,u), f(u,\ldots,u,v), t/2) * M(f(u,\ldots,u,v), f(v,\ldots,v), t/2)\\
&\geq M(f(u,\ldots,u), f(u,\ldots,u,v), t/2) * M(f(u,\ldots,u,v), f(u,\ldots,u,v,v), t/2^2)\\
&* \cdots * M(f(u,u,v,\ldots,v), f(u,v,\ldots,v), t/2^{k-1})\\
&* M(f(u,v,\ldots,v), f(v,\ldots,v), t/2^{k-1}).
\end{aligned} \tag{4.28}$$

Using (4.19), for every $t > 0$, we get,

$$M(f(u,\ldots,u), f(u,\ldots,u,v), t/2) \geq \left[\lambda \cdot \left(\frac{1}{M(u,v,t/2)} - 1\right) + 1\right]^{-1}$$

and similarly,

$$M(f(u,\ldots,u,v), f(u,\ldots,u,v,v), t/2^2) \geq \left[\lambda \cdot \left(\frac{1}{M(u,v,t/2^2)} - 1\right) + 1\right]^{-1}, \ldots,$$

$$\cdots M(f(u,u,v,\ldots,v), f(u,v,\ldots,v), t/2^{k-1}) \geq \left[\lambda \cdot \left(\frac{1}{M(u,v,t/2^{k-1})} - 1\right) + 1\right]^{-1}$$

$$M(f(u,v,\ldots,v), f(v,\ldots,v), t/2^{k-1}) \geq \left[\lambda \cdot \left(\frac{1}{M(u,v,t/2^{k-1})} - 1\right) + 1\right]^{-1}.$$

In consequence, by (4.28), for every $t > 0$, the following inequality holds

$$M(u,v,t) \geq \left[\lambda \cdot \left(\frac{1}{M(u,v,t/2)} - 1\right) + 1\right]^{-1} * \left[\lambda \cdot \left(\frac{1}{M(u,v,t/2^2)} - 1\right) + 1\right]^{-1}$$
$$* \cdots * \left[\lambda \cdot \left(\frac{1}{M(u,v,t/2^{k-1})} - 1\right) + 1\right]^{-1} * \left[\lambda \cdot \left(\frac{1}{M(u,v,t/2^{k-1})} - 1\right) + 1\right]^{-1}.$$

From the previous inequality and (4.27), we can affirm that $t > 0$ such that $M(u,v,t) > M(u,v,t)$, which is a contradiction, so that $u = v$ and the fixed point of f is unique.

Next, we present the following extension of Theorem 4.14.

Theorem 4.15 *Let $(X, M, *)$ be an M-complete fuzzy metric space, k a positive integer and $f\colon X^k \to X$ a fuzzy Prešić-Ćirić operator. Suppose that one of the following conditions holds:*

(H1$*$) *See condition (H1).*

(H2$*$) *There exist $x_1, x_2 \ldots, x_k \in X$ such that the following property holds: for each $\varepsilon \in (0,1)$ and $t > 0$, there exists $n_0 \in \mathbb{N}$ such that for m, $n > n_0$, with $m > n$, we get $*_{j=n}^{m-1}\left[1 + \mu(ts_j^{(n,m)})\lambda^{\frac{j}{k}}\right]^{-1} > 1 - \varepsilon$, for some collection of values $s_j^{(n,m)} > 0$, $j = n, \ldots, m-1$, with $\sum_{j=n}^{m-1} s_j^{(n,m)} \leq 1$, where $\mu(z)$ is given in Theorem 4.14.*

Then f has a fixed point in X. If, in addition, on the diagonal $\Delta \subset X^k$, condition (4.20) holds for $u, v \in X$ with $u \neq v$, then f has a unique fixed point.

Proof. It is similar to the proof of Theorem 4.14. To prove that $\{x_n\}$ is an M-Cauchy sequence, we consider the choice for $s_j^{(n,m)}$ given in the statement. Using the nondecreasing character of $M(x, y, \cdot)$ for every x, $y \in X$ and following the proof of Theorem 4.8 [7], for $t > 0$ and $n, m \in \mathbb{N}$ with $m > n$,

$$M(x_n, x_m, t) \geq M(x_n, x_{n+1}, ts_n^{(n,m)}) * M(x_{n+1}, x_{n+2}, ts_{n+1}^{(n,m)})$$
$$* \cdots * M(x_{m-2}, x_{m-1}, ts_{m-2}^{(n,m)}) * M(x_{m-1}, x_m, ts_{m-1}^{(n,m)})$$
$$= \mathop{*}_{j=n}^{m-1} M_j(ts_j^{(n,m)}) \geq \mathop{*}_{j=n}^{m-1}\left[1 + \mu(ts_j^{(n,m)})\theta^j\right]^{-1}.$$

The case (H1$*$) is analogous to the proof of Theorem 4.14. Under condition (H2$*$), given $\varepsilon > 0$, there exists $n_0 \in \mathbb{N}$ such that, for m, $n > n_0$ with $m > n$, we get $M(x_n, x_m, t) \geq *_{j=n}^{m-1}\left[1 + \mu(ts_j^{(n,m)})\theta^j\right]^{-1} > 1 - \varepsilon$. The proof is completed similarly to that of Theorem 4.14.

Corollary 4.7 *Let $(X, M, *)$ be an M-complete fuzzy metric space and $f\colon X \to X$ be a fuzzy contractive mapping, that is, $\dfrac{1}{M(fx, fy, t)} - 1 \leq \lambda\left(\dfrac{1}{M(x,y,t)} - 1\right)$, $\forall x, y \in X$, $\forall t > 0$, where $\lambda \in (0,1)$. Suppose that one of the following conditions holds:*

(h1$*$) *Condition (h1) is satisfied.*

(h2$*$) *There exists $x_1 \in X$ such that the following property holds: for each $\varepsilon \in (0,1)$ and $t > 0$, there exists $n_0 \in \mathbb{N}$ such that for $m, n > n_0$, with $m > n$, we get $*_{j=n}^{m-1}\left[1 + \mu(ts_j^{(n,m)})\lambda^j\right]^{-1} > 1 - \varepsilon$, for some collection of values $s_j^{(n,m)} > 0$, $j = n, \ldots, m-1$, with $\sum_{j=n}^{m-1} s_j^{(n,m)} \leq 1$, where μ is given in Corollary 4.5.*

Then f has a unique fixed point in X.

Remark 4.20 Theorem 4.15 and Corollary 4.7 are more general than Theorem 4.14 and Corollary 4.5, respectively. Conditions (H2*) and (h2*) show that the relevant point in the choice of the values $s_j^{(n,m)}$ is that $s_j^{(n,m)} > 0$ for every $j = n, \ldots, m-1$, and $\sum\limits_{j=n}^{m-1} s_j^{(n,m)} \leq 1$. Hence, we can select different values of $s_j^{(n,m)}$ as long as these requirements are fulfilled. Note also that it is possible to use the expressions of $s_j^{(n,m)}$ independently of m, that is, in the form $s_j^{(n)}$, if we select them as positive numbers such that $\sum\limits_{j=n}^{\infty} s_j^{(n)} \leq 1$. Moreover, we can take the expressions of $s_j^{(n,m)}$ to be independent of n and m, in the form s_j, if we select them as positive numbers such that $\sum\limits_{j=1}^{\infty} s_j \leq 1$.

Remark 4.21 It is important to note that where $M = M_d$ and $* = *_m$, condition (H2*) (resp. (h2*)) is trivially valid for arbitrary choices of $x_1, \ldots, x_k$ (resp., x_1) and for any value of $\lambda \in (0,1)$, since we can choose $s_j^{(n,m)} = \frac{1}{j(j+1)}$ for $j = n, \ldots, m-1$, which are positive and such that $\sum\limits_{j=1}^{\infty} \frac{1}{j(j+1)} = 1$.

We start with $k = 1$. For $M = M_d$, $* = *_m$, any $x_1 \in X$ and taking $s_j^{(n,m)} = \frac{1}{j(j+1)}$, $j = n, \ldots, m-1$, we have, for $t > 0$ and $m, n \in \mathbb{N}$ with $m > n$, $\max\limits_{n \leq j \leq m-1} \dfrac{\lambda^{j-1}}{s_j^{(n,m)}} = \max\limits_{n \leq j \leq m-1} j(j+1)\lambda^{j-1}$. We study the function $\varphi(x) := x(x+1)\lambda^{x-1}$, whose derivative is $\varphi'(x) = (2x+1)\lambda^{x-1} + x(x+1)\log(\lambda)\lambda^{x-1} = \lambda^{x-1}(2x + 1 + x(x+1)\log(\lambda))$. Since $\lambda \in (0,1)$, the quadratic function $\psi(x) := \log(\lambda)\, x^2 + (\log \lambda + 2)x + 1$ is concave and has its vertex at $x = -\frac{1}{2} - \frac{1}{\log(\lambda)}$, which is arbitrarily large if λ is arbitrarily close to zero. However, there exists $n_1 \in \mathbb{N}$ large enough (depending just on λ) such that for every $x > n_1$, $\psi(x) < 0$. Therefore, for every $x > n_1$, $\varphi'(x) < 0$ and φ is decreasing on $(n_1, +\infty)$. In consequence, if we take $t > 0$ and $m, n \in \mathbb{N}$ with $m > n > n_1$, then $\max\limits_{n \le j \le m-1} \dfrac{\lambda^{j-1}}{s_j^{(n,m)}} = n(n+1)\lambda^{n-1} \to 0$, as $n \to \infty$. Hence, for $\varepsilon \in (0,1)$ and $t > 0$ fixed, there exists $n_0 \in \mathbb{N}$ with $n_0 \ge n_1$ such that for $m, n > n_0$, with $m > n$, $\max\limits_{n \le j \le m-1} \dfrac{\lambda^{j-1}}{s_j^{(n,m)}} < \dfrac{t}{d(x_1, f(x_1))} \dfrac{\varepsilon}{1-\varepsilon}$ and, in consequence,

$$*_{j=n}^{m-1} \left[1 + \mu(ts_j^{(n,m)})\lambda^j\right]^{-1} = *_{j=n}^{m-1} \left[1 + \frac{d(x_1, f(x_1))}{t} j(j+1)\lambda^{j-1}\right]^{-1} > 1 - \varepsilon.$$

We considered that $d(x_1, f(x_1)) > 0$ since the condition $f(x_1) = x_1$ leads to a trivial case. Therefore, since $\lambda \in (0,1)$, condition (h2*) holds for $M = M_d$ and $* = *_m$, independently of the choice of x_1.

Now, we consider the general case $k \in \mathbb{N}$. For $M = M_d$, $* = *_m$ and any $x_1, \ldots, x_k \in X$, for $t > 0$ and $m, n \in \mathbb{N}$ with $m > n$, taking $s_j^{(n,m)} = \frac{1}{j(j+1)}$, $j = n, \ldots, m-1$,

$$\begin{aligned} &\max_{n \le j \le m-1} \frac{\lambda^{\frac{j}{k}}}{s_j^{(n,m)}} \\ &= \max\left\{n(n+1)\lambda^{\frac{n}{k}}, \ldots, (m-2)(m-1)\lambda^{\frac{m-2}{k}}, (m-1)m\lambda^{\frac{m-1}{k}}\right\}. \end{aligned}$$

We consider the function $\widetilde{\varphi}(x) := x(x+1)\nu^x$, being $\nu = \lambda^{\frac{1}{k}}$ where $\widetilde{\varphi}'(x) = \nu^x\left(2x + 1 + x(x+1)\log(\nu)\right)$. The sign of $\widetilde{\varphi}'$ coincides with the sign of the function $\widetilde{\psi}$, given by $\widetilde{\psi}(x) := \log(\nu)\, x^2 + (\log \nu + 2)x + 1$. Since $\lambda \in (0,1)$, $\nu \in (0,1)$ and the graph of $\widetilde{\psi}$ is a concave parabola with vertex at $x = -\frac{1}{2} - \frac{1}{\log(\nu)}$. Similarly to $k = 1$, there exists $\widetilde{n_1} \in \mathbb{N}$ large enough (depending on λ) such that for every $x > \widetilde{n_1}$, $\widetilde{\psi}(x) < 0$; hence, for every $x > \widetilde{n_1}$, $\widetilde{\varphi}'(x) < 0$ and $\widetilde{\varphi}$ is decreasing on $(\widetilde{n_1}, +\infty)$. Therefore, for fixed $t > 0$ and taking $m, n \in \mathbb{N}$ with $m > n > \widetilde{n_1}$, we get $\max\limits_{n \le j \le m-1} \dfrac{\lambda^{j-1}}{s_j^{(n,m)}} = n(n+1)\lambda^{\frac{n}{k}} \to 0$, as $n \to \infty$. This proves that, for $\varepsilon \in (0,1)$ and $t > 0$ fixed, there exists $n_0 \in \mathbb{N}$ with $n_0 \ge \widetilde{n_1}$ such that for $m, n > n_0$, with $m > n$,

$$\max_{n \le j \le m-1} \frac{\lambda^{\frac{j}{k}}}{s_j^{(n,m)}} < \frac{t}{\max\left\{\max\limits_{1 \le i \le k-1} \dfrac{d(x_i, x_{i+1})}{\lambda^{\frac{i}{k}}}, \dfrac{d(x_k, f(x_1, \ldots, x_k))}{\lambda}\right\}} \frac{\varepsilon}{1-\varepsilon}.$$

Since $\lambda > 0$ and $s_j^{(n,m)} > 0$, for every $j = n, \ldots, m-1$, following the calculations in Remark 4.19, for $t > 0$ and $m, n \in \mathbb{N}$ with $m > n$,

$$\mathop{*}_{j=n}^{m-1}\left[1+\mu(ts_j^{(n,m)})\lambda^{\frac{j}{k}}\right]^{-1}$$

$$\geq \left[1+\frac{1}{t}\max\left\{\max_{1\leq i\leq k-1}\frac{d(x_i,x_{i+1})}{\lambda^{\frac{i}{k}}},\frac{d(x_k,f(x_1,\ldots,x_k))}{\lambda}\right\}\max_{n\leq j\leq m-1}\frac{\lambda^{\frac{j}{k}}}{s_j^{(n,m)}}\right]^{-1}.$$

Hence, we proved that for $\varepsilon \in (0,1)$ and $t > 0$ fixed, there exists $n_0 \in \mathbb{N}$ such that for $m, n > n_0$ with $m > n$, $*_{j=n}^{m-1}\left[1+\mu(ts_j^{(n,m)})\lambda^{\frac{j}{k}}\right]^{-1} > 1-\varepsilon$. In the previous inequalities, we assumed that

$$\max\left\{\max_{1\leq i\leq k-1}\frac{d(x_i,x_{i+1})}{\lambda^{\frac{i}{k}}},\frac{d(x_k,f(x_1,\ldots,x_k))}{\lambda}\right\}>0,$$

since the opposite situation leads to a trivial case.

Again, since $\lambda \in (0,1)$, condition (H2*) holds for $M = M_d$ and $* = *_m$, independently of the choice of $x_1, \ldots, x_k$.

Corollary 4.8 *Let $(X, M, *)$ be an M-complete fuzzy metric space, k an integer with $k \geq 2$ and $f\colon X^k \to X$ a fuzzy Prešić-Ćirić operator. Suppose that one of the conditions (H1*) or (H2*) holds. Then f has a fixed point in X. If, in addition, we suppose that for each $u, v \in X$ fixed with $u \neq v$, there exists $t > 0$ such that*

$$\mathop{*}_{i=1}^{k}\left[\lambda\cdot\left(\frac{1}{z_i}-1\right)+1\right]^{-1} > M(u,v,t), \tag{4.29}$$

where $z_i = M(u, v, t\,r_i)$, for $i = 1, \ldots, k$, for some sequence of values $r_i > 0$, $i = 1, \ldots, k$, $\sum_{i=1}^{k} r_i \leq 1$, f has a unique fixed point.

Proof. The existence of fixed points follows from the proof of Theorem 4.15. For the uniqueness of the fixed point, we suppose that u, and $v \in X$ are fixed points of f with $u \neq v$. Then, using the nondecreasing character of $M(x, y, \cdot)$ for all x, $y \in X$, for any $t > 0$,

$$\begin{aligned}
M(u,v,t) &= M(f(u,\ldots,u),f(v,\ldots,v),t)\\
&\geq M(f(u,\ldots,u),f(u,\ldots,u,v),t\,r_1) * M(f(u,\ldots,u,v),f(u,\ldots,u,v,v),t\,r_2)\\
&\quad *\cdots* M(f(u,u,v,\ldots,v),f(u,v,\ldots,v),t\,r_{k-1})\\
&\quad\quad * M(f(u,v,\ldots,v),f(v,\ldots,v),t\,r_k).
\end{aligned} \tag{4.30}$$

Similarly to the proof of Corollary 4.6, using (4.19), for every $t > 0$, $M(f(u,\ldots,u),f(u,\ldots,u,v),t\,r_1) \geq \left[\lambda\cdot\left(\frac{1}{M(u,v,t\,r_1)}-1\right)+1\right]^{-1}$ and, similarly,

$$M(f(u,\ldots,u,v),f(u,\ldots,u,v,v),tr_2) \geq \left[\lambda\cdot\left(\frac{1}{M(u,v,tr_2)}-1\right)+1\right]^{-1},\ldots,$$

$$M(f(u,u,v,\ldots,v),f(u,v,\ldots,v),tr_{k-1}) \geq \left[\lambda\cdot\left(\frac{1}{M(u,v,tr_{k-1})}-1\right)+1\right]^{-1}$$

$$M(f(u,v,\ldots,v),f(v,\ldots,v),tr_k) \geq \left[\lambda\cdot\left(\frac{1}{M(u,v,tr_k)}-1\right)+1\right]^{-1}.$$

Therefore, the previous inequalities and (4.30) imply for every $t>0$ that $M(u,v,t) \geq *_{i=1}^{k}\left[\lambda\cdot\left(\frac{1}{M(u,v,tr_i)}-1\right)+1\right]^{-1}$. Hence, from (4.29), $t>0$ such that $M(u,v,t)>M(u,v,t)$ and we obtain a contradiction again the fixed point of f is unique.

Remark 4.22 Concerning condition (4.29), the expression of the values r_i, $i=1,\ldots,k$, can be of similar type to $s_j^{(n,m)}$ in (H2*) or different, provided that the requirements $r_i>0$, $i=1,\ldots,k$, and $\sum_{i=1}^{k} r_i \leq 1$ are fulfilled.

Remark 4.23 Instead of $(X,M,*)$ an M-complete fuzzy metric space, if we consider the hypothesis that $(X,M,*)$ is a G-complete fuzzy metric space, we can remove the restrictions (H1), (H2) in Theorem 4.14 and Corollary 4.6 and (h1), (h2) in Corollary 4.5. This comes from the proof of Theorem 4.14. We can start with arbitrary points $x_1,\ldots,x_k$ in X and to prove that the sequence defined is G-Cauchy, we note that for $t>0$ and $p\in\mathbb{N}$ fixed,

$$\begin{aligned} M(x_n,x_{n+p},t) &\geq \mathop{*}_{j=n}^{n+p-1} M(x_j,x_{j+1},ts_j^{(n)}) \\ &= \mathop{*}_{j=n}^{n+p-1} M_j(ts_j^{(n)}) \geq \mathop{*}_{j=n}^{n+p-1}\left[1+\mu(ts_j^{(n)})\theta^n\right]^{-1}, \end{aligned}$$

where $s_j^{(n)}=\dfrac{1}{2^{j-n+1}}$, $j=n,\ldots,n+p-1$.

Note that the last term in the previous inequality consists of a fixed number of terms (for every n), that is, p terms that tend to 1 as $n\to\infty$ due to $\theta\in(0,1)$ and the fact that $s_j^{(n)}$ represents a constant sequence for each j fixed, in the sense that $s_n^{(n)}=\frac{1}{2}$ for every n, $s_{n+1}^{(n)}=\frac{1}{2^2}$, for every n, $\ldots$, $s_{n+p-1}^{(n)}=\frac{1}{2^p}$, for every n. Hence $\lim_{n\to\infty} M(x_n,x_{n+p},t)=1*\overset{p)}{\cdots}*1=1$ for each $t>0$ and $p>0$, and $\{x_n\}$ is G-Cauchy.

We include some conclusions on fuzzy contractive sequences that are derived from the proofs of the main results in the previous section.

Definition 4.24 Let $(X,M,*)$ be a fuzzy metric space and k a positive integer. We say that $\{x_n\}\subset X$ is a fuzzy contractive sequence if $\lambda\in(0,1)$ such that

$$\frac{1}{M(x_{n+k},x_{n+k+1},t)}-1 \leq \lambda\cdot\max_{1\leq i\leq k}\left\{\frac{1}{M(x_{n+i-1},x_{n+i},t)}-1\right\}, \quad (4.31)$$

for all $t > 0$ and $n \in \mathbb{N}$.

Condition (4.31) can also be written as

$$M(x_{n+k}, x_{n+k+1}, t) \geq \left[\lambda \cdot \max_{1 \leq i \leq k} \left\{ \frac{1}{M(x_{n+i-1}, x_{n+i}, t)} - 1 \right\} + 1 \right]^{-1},$$

for all $t > 0$ and $n \in \mathbb{N}$, where $\lambda \in (0, 1)$.

This notion is a generalization of Definition 3.8 [7] since, for $k = 1$, it is reduced to $\dfrac{1}{M(x_{n+1}, x_{n+2}, t)} - 1 \leq \lambda \cdot \left(\dfrac{1}{M(x_n, x_{n+1}, t)} - 1 \right)$, for all $t > 0$ and $n \in \mathbb{N}$, where $\lambda \in (0, 1)$.

For $k = 1$, Gregori proposed [7] the following question: Is a fuzzy contractive sequence a Cauchy sequence in George and Veeramani's sense (that is, an M-Cauchy sequence)? We study this problem for an arbitrary $k \in \mathbb{N}$, by imposing sufficient conditions which guarantee the validity of this assertion.

For a given sequence $\{x_n\}$, consider $\mu(z) := \max_{1 \leq i \leq k} \dfrac{1}{\lambda^{\frac{i}{k}}} \left[\dfrac{1}{M(x_i, x_{i+1}, z)} - 1 \right]$, and the hypotheses:

(HS1) $* \in \mathcal{H}$ and $\inf_{t>0} M(x_i, x_{i+1}, t) > 0$, for all $i = 1, 2, \ldots, k$.

(HS2) For each $\varepsilon \in (0, 1)$ and $t > 0$, there exists $n_0 \in \mathbb{N}$ such that for $m, n > n_0$, with $m > n$, we get $*_{j=n}^{m-1} \left[1 + \mu(ts_j^{(n,m)}) \lambda^{\frac{j}{k}} \right]^{-1} > 1 - \varepsilon$, for some collection of values $s_j^{(n,m)} > 0$, $j = n, \ldots, m-1$, such that $\sum_{j=n}^{m-1} s_j^{(n,m)} \leq 1$.

Theorem 4.16 *Let $(X, M, *)$ be a fuzzy metric space and k a positive integer. Let $\{x_n\} \subset X$ be a fuzzy contractive sequence. Suppose that one of the conditions (HS1) or (HS2) holds. Then $\{x_n\}$ is an M-Cauchy sequence.*

Proof. As in the proof of Theorem 4.14, we denote $M_n(t) := M(x_n, x_{n+1}, t)$, for $n \in \mathbb{N}$ and $t > 0$ and $\mu(t) = \max_{1 \leq i \leq k} \left\{ \dfrac{1}{\theta^i} \left[\dfrac{1}{M_i(t)} - 1 \right] \right\}$, where $\theta = \lambda^{\frac{1}{k}}$. Similarly to the proof of Theorem 4.14, by induction, we prove that

$$\frac{1}{M_n(t)} - 1 \leq \mu(t)\theta^n, \qquad \forall\, n \in \mathbb{N},\ \forall\, t > 0. \tag{4.32}$$

Indeed, it is true for $n = 1, 2, \ldots, k$. Assuming that it is true for $n, n+1, \ldots, n+k-1$, we have, from (4.31),

$$\frac{1}{M_{n+k}(t)} - 1 = \frac{1}{M(x_{n+k}, x_{n+k+1}, t)} - 1 \leq \lambda \max_{1 \leq i \leq k} \left\{ \frac{1}{M(x_{n+i-1}, x_{n+i}, t)} - 1 \right\}$$
$$= \lambda \max_{1 \leq i \leq k} \left\{ \frac{1}{M_{n+i-1}(t)} - 1 \right\} \leq \lambda \max_{1 \leq i \leq k} \left\{ \mu(t)\, \theta^{n+i-1} \right\} \leq \lambda \mu(t) \theta^n = \mu(t) \theta^{n+k},$$

for $t > 0$, where we used $\theta = \lambda^{\frac{1}{k}} < 1$.

To check that $\{x_n\}$ is an M-Cauchy sequence, $\varepsilon \in (0,1)$ and $t > 0$ are fixed. Then, by (4.32), using the nondecreasing character of $M(x, y, \cdot)$ for every $x, y \in X$ and following the proof of Theorem 4.8 [7], for $n, m \in \mathbb{N}$ with $m > n$, that $M(x_n, x_m, t) \geq *_{j=n}^{m-1} M_j(ts_j^{(n,m)}) \geq *_{j=n}^{m-1} \left[1 + \mu(t\, s_j^{(n,m)})\theta^j\right]^{-1}$, for any collection of values $s_j^{(n,m)} > 0$, we have $j = n, \ldots, m-1$, with $\sum_{j=n}^{m-1} s_j^{(n,m)} \leq 1$.

If (HS1) holds, $\mu := \sup\limits_{t>0} \mu(t) \in [0, \infty)$; therefore, for $n, m \in \mathbb{N}$ with $m > n$, we get $M(x_n, x_m, t) \geq *^{m-n} \left[1 + \mu\theta^n\right]^{-1}$. Since $* \in \mathcal{H}$, the proof is complete similarly to the proof of Theorem 4.14.

On the other hand, (HS2) provides trivially the character of M-Cauchy sequence for $\{x_n\}$.

Remark 4.24 If $M = M_d$, condition (HS1) is satisfied only for constant sequences $\{x_n\}$.

Remark 4.25 The role of a specific class of t-norm, i.e., H-type t-norm, can easily be observed from the proofs of Theorem 4.14 and Theorem 4.15. Then the following question naturally arises.

Question. Do Theorems 4.14 and 4.15 remain true if the H-type t-norm is replaced by any arbitrary t-norm ?

References

[1] Abbas M., Imdad M., Gopal D., ϕ-weak contractions in fuzzy metric spaces. Iranian J. Fuzzy Syst. 2011; 8(5):141-148.

[2] Grabiec M., Fixed points in fuzzy metric spaces. Fuzzy Sets Syst. 1988; 27:385-389.

[3] George A., Veeramani P., On some results in fuzzy metric spaces. Fuzzy Sets Syst. 1994; 64:395-399.

[4] George A., Veeramani P., On some results of analysis for fuzzy metric spaces. Fuzzy Sets Syst. 1997; 90(3):365-368.

[5] Gopal D., Imdad M., Vetro C., Hasan M., Fixed point theory for cyclic weak ϕ-contraction in fuzzy metric spaces. J. Nonlinear Anal. Appl. 2012; Art. jnaa-00110.

[6] Gopal D., Vetro C., Some new fixed point theorems in fuzzy metric spaces. Iranian J. Fuzzy Syst. 2014; 11(3):95-107.

[7] Gregori V., Sapena A., On fixed point theorems in fuzzy metric spaces. Fuzzy Sets Syst. 2002; 125:245-252.

[8] Gregori V., Romaguera S., Characterizing completable fuzzy metric spaces. Fuzzy Sets Syst. 2004; 144:411-420.

[9] Gregori V., Morillas S., Sapena A., On a class of completable fuzzy metric spaces. Fuzzy Sets Syst. 2010; 161:2193-2205.

[10] Gregori V., Morillas S., Sapena A., Examples of fuzzy metrics and applications. Fuzzy Sets Syst. 2011; 170:95-111.

[11] Gregori V., Minana J.J., Remarks on fuzzy contractive mappings. Fuzzy Sets Syst. 2014; 251:101-103.

[12] Hadžić O., Pap E., Fixed Point Theory in Probabilistic Metric Spaces. Dordrecht: Kluwer, 2001.

[13] Harandi A.A., Metric-like spaces, partial metric spaces and fixed points. Fixed Point Theory Appl. 2012; 2012(1): Art. 204.

[14] Istratescu V., An introduction to theory in probabilistic metric spaces, with applications. Ed tehica Bucuresti. 1974 (in Romanian).

[15] Kramosil I., Michalek J., Fuzzy metric and statistical metric spaces. Kybernetica. 1975; 15:326-334.

[16] Menger K., Statistical metric. Proc. Natl. Acad. Sci. USA. 1942; 28:535–537.

[17] Mihet D., A Banach contraction theorem in fuzzy metric spaces. Fuzzy Sets Syst. 2004; 144:431-439.

[18] Mihet D., On fuzzy contractive mappings in fuzzy metric spaces. Fuzzy Sets Syst. 2007; 158:915-921.

[19] Mihet D., Fuzzy ψ-contractive mappings in non-Archimedean fuzzy metric spaces. Fuzzy Sets Syst. 2008; 159:739-744.

[20] Mihet D., A note on fuzzy contractive mappings in fuzzy metric. Fuzzy Sets Syst. 2014; 251:83-91.

[21] Romagucra S., Thc complction of fuzzy metric spaces and of other related structure. in P.G. Chareton (Ed.), Computational Mathematics: Theory, Method and Apllications, New York: Nova Science, 2011.

[22] Schweizer B., Sklar A., Probabilistic Metric Spaces. New York: Elsevier, 1983.

[23] Shukla S., Abbas M., Fixed point results in fuzzy metric-like spaces. Iranian J. Fuzzy Syst. 2015; 11(5):81-92.

[24] Shukla S., Gopal D., de-Hierro A.-F. R.-L., Some fixed point theorems in 1-M complete fuzzy metric-like spaces. Intl. J. Genl. Syst. 2016; 3:1-15.

[25] Shukla S., Gopal D., de-Hierro A.-F. R.-L., Fuzzy Prešić-Ćirić operators and applications to certain nonlinear differential equations. Math. Model. Anal. 2016; 21(6):811-835.

[26] Subrahmanyam P.V., A common fixed point theorem in a fuzzy metric space. Inform. Sci. 1995; 83:109-112.

[27] Tirado P., On compactness and G-completeness in fuzzy metric spaces. Iranian J. Fuzzy Syst. 2012; 9(4):151-158.

[28] Vasuki R., A common fixed point theorem in a fuzzy metric space. Fuzzy Sets Syst. 1998; 97(3):395-397.

[29] Vasuki R., Veeramani P., Fixed point theorems and Cauchy sequences in fuzzy metric spaces. Fuzzy Sets Syst. 2003; 135:415-417.

[30] Wang S., Answers to some open questions on fuzzy ψ-contractions in fuzzy metric spaces. Fuzzy Sets Syst. 2013; 222:115-119.

[31] Wardowski D., Fuzzy contractive mappings and fixed points in fuzzy metric space. Fuzzy Sets and Syst. 2013; 222:108-114.

[32] Yun G., Hwang S., Chang J., Fuzzy Lipschitz maps and fixed point theorems in fuzzy metric spaces. Fuzzy Sets Syst. 2010; 161(8):1117-1130.

[33] Zadeh L.A., Fuzzy sets. Inform. Control. 1965; 8:338-353.

Chapter 5

Fixed Point Theory in Soft Metric Spaces: Rise and Fall

Mujahid Abbas, Ghulam Murtaza and Salvador Romaguera

5.1 Introduction

Mathematical models have been used extensively in real world problems related to engineering, computer sciences, economics, social, natural and medical sciences. They are valuable mathematical tools to study the behavior and aspects of a system and its subsystems.

It is very natural to deal with uncertainties and imprecise data in various situations. Fuzzy set theory has evolved as an important tool (initiated by Zadeh [42]) to resolve the issues of uncertainty and ambiguity. The contribution made by probability theory, fuzzy set theory, vague sets, rough sets and interval mathematics to deal with uncertainty is of critical.

There are certain limitations and deficiencies pertaining to the parametrization in fuzzy set theory (see [31]). The problem of inadequacy of parameters has been successfully solved by soft set theory which provides enough tools to deal with uncertainty in a data and represent it in a useful way. The distinguishing attribute of soft set theory is that unlike probability theory and fuzzy set theory, it does not uphold a precise quantity. This attribute facilitated applications in decision making, demand analysis, forecasting, information sciences, mathematics and other disciplines (see [12, 14, 13, 15, 16, 23, 33, 35, 44, 45]).

A lot of activity has advanced soft set theory (see [3, 4, 10, 17, 20, 18, 19, 21, 25, 29, 26, 28, 34, 37, 38, 39, 40, 41, 43]) since Molodtsov ([31]) initiated the concept. Soft topology on a soft set was introduced by Cagman et. al ([5]) and basic properties of soft topological spaces were studied.

This chapter is a modified version of our papers [1] and [2]. We established the Banach contraction principle in the framework of soft metric spaces and other theorems of fixed point theory. This chapter has two parts: the first deals with the results of fixed point theory based on existing literature of soft set theory and the second deals with restrictions in soft metric spaces.

In the sequel, the letters U, E and $P(U)$ will denote the universal set, the set of parameters and the power set of U, respectively. Let A be a nonempty subset of E.

Definition 5.1 *[31]* If F is a set valued mapping on $A \subset E$ taking values in $P(U)$; a pair (F, A) is called a soft set over U. In other words, a soft set over U is a parameterized family of subsets of the universe U. For $\varepsilon \in A$, $F(\varepsilon)$ is called the set of ε-approximate elements of the soft set (F, A).

A soft set (F, A) could be identified with the set $\{(e, F(e)) : e \in A\}$. A soft set (F, A) is said to be (a) an absolute soft set denoted by $\tilde{U}$ if for each $\varepsilon \in A$, $F(\varepsilon) = U$ or (b) a null soft set denoted by Φ if for each $\varepsilon \in A$, $F(\varepsilon) = \phi$

Example 5.1 A soft set (F, A) describes the attractiveness of the houses which Mr. X is going to buy. U is the set of houses under consideration, e.g. $\{h_1, h_2, h_3, h_4, h_5\}$. E is the set of parameters {expansive, beautiful, wooden, cheap, in green surroundings, modern, in good repair, in bad repair} $= \{p_1, p_2, p_3, p_4, p_5, p_6, p_7, p_8\}$. Then

$$(F, A) = \{(p_1, \{h_1, h_2\}), (p_2, \{h_2, h_3\}), (p_3, \{h_4, h_5\})\}.$$

Other basic notions and properties related to soft set theory may be found in [11, 27, 31].

We denote the collection of soft sets over a common universe U by $S(U)$.

Das and Samanta introduced [9] soft real sets and soft real numbers and discussed their properties. They also gave applications of these concepts to real life problems. They introduced in [8] the concept of a soft metric and showed that soft metric space is also a soft topological space. Wardowski ([36]) introduced soft mapping and obtained its fixed point in the setup of soft topological spaces. Now we will discuss definitions from Das and Samanta.

Definition 5.2 *[9]* If f is a single valued mapping on $A \subset E$ taking values in U, the pair (f, A), or simply f, is called a soft element of U. Let $(F, A) \in S(U)$. A soft element f of U is said to belongs to (F, A), denoted by $f\tilde{\in}(F, A)$, if $f(e) \in F(e)$, for each $e \in A$. Thus for a soft set (F, A) over U, we have $F(e) = \{f(e), f\tilde{\in}(F, A)\}$, $e \in A$.

Let $\mathbb{R}$ be the set of real numbers. We denote the collection of all nonempty bounded subsets of $\mathbb{R}$ by $B(\mathbb{R})$.

Definition 5.3 *[9]* A soft real set denoted by $(\widehat{f}, A)$ or simply by $\widehat{f}$ is a mapping $\widehat{f} : A \to B(\mathbb{R})$. If $\widehat{f}$ is a single valued mapping on $A \subset E$ taking values in $\mathbb{R}$, the pair $(\widehat{f}, A)$ or simply $\widehat{f}$, is called a soft element of $\mathbb{R}$ or a soft real number. If $\widehat{f}$ is a single valued mapping on $A \subset E$ taking values in the set $\mathbb{R}^+$ of non-negative real numbers, a pair $(\widehat{f}, A)$ or simply $\widehat{f}$ is called a non-negative soft real number. We shall denote the set of non-negative soft real numbers (corresponding to A) by $\mathbb{R}(A)^*$. A null soft number $\overline{0}$ is a soft real number defined by $\overline{0}(e) = 0$ for all $e \in A$. A unit soft number $\overline{1}$ is a soft real number defined by $\overline{1}(e) = 1$ for all $e \in A$. A constant soft real number $\overline{c}$ is a soft real number such that for each $e \in A$, we have $\overline{c}(e) = c$, where c is some real number.

Definition 5.4 *[8]* A soft set (F, A) over U is said to be a soft point if there is exactly one $\lambda \in A$ such that $F(\lambda) = \{x\}$ for some $x \in U$ and $F(e) = \emptyset$ for all $e \in A \backslash \{\lambda\}$. We shall denote such a soft point by (F_λ^x, A) or simply by F_λ^x.

Definition 5.5 *[8]* Let (F, A) be a soft set over U. A soft point F_λ^x is said to belong to (F, A), denoted by $F_\lambda^x \tilde{\in} (F, A)$, if $F_\lambda^x(\lambda) = \{x\} \subset F(\lambda)$.

Definition 5.6 *[8]* Two soft points F_λ^x, F_μ^y are said to be equal if $\lambda = \mu$ and $F(\lambda) = F(\mu)$, that is, $x = y$. Thus $F_\lambda^x \neq F_\mu^y$ if and only if $x \neq y$ or $\lambda \neq \mu$.

Proposition 5.1 *[8] Let (F, A) be a soft set over U. Then*

$$(F, A) = \cup \{F_\lambda^x : F_\lambda^x \tilde{\in} (F, A)\}.$$

Remark 5.1 *[8]* The collection of all soft points of (F, A) is denoted $SP(F, A)$.

Definition 5.7 *[8]* For two soft real numbers $\widehat{f}, \widehat{g}$, we say that

(i) $\widehat{f} \tilde{\leq} \widehat{g}$ if $\widehat{f}(e) \leq \widehat{g}(e)$ for all $e \in A$,

(ii) $\widehat{f} \tilde{\geq} \widehat{g}$ if $\widehat{f}(e) \geq \widehat{g}(e)$ for all $e \in A$,

(iii) $\widehat{f} \tilde{<} \widehat{g}$ if $\widehat{f}(e) < \widehat{g}(e)$ for all $e \in A$,

(iv) $\widehat{f} \tilde{>} \widehat{g}$ if $\widehat{f}(e) > \widehat{g}(e)$ for all $e \in A$.

Remark 5.2 The notion of a soft mapping may be found in [1, 36]. Recall that if f is a soft mapping from a soft set (F, A) to a soft set (G, B) (denoted by $f : (F, A) \tilde{\to} (G, B)$), for each soft point $F_\lambda^x \tilde{\in} (F, A)$ there exists only one soft point $G_\mu^y \tilde{\in} (G, B)$ such that $f(F_\lambda^x) = G_\mu^y$.

The definition of a soft metric introduced in [8] is given below:

Definition 5.8 *[8]* Let U be a universe, A be a nonempty subset of parameters and $\tilde{U}$ be the absolute soft set, i.e., $F(\lambda) = U$ for all $\lambda \in A$, where $(F, A) = \tilde{U}$. A mapping $d : SP(\tilde{U}) \times SP(\tilde{U}) \to \mathbb{R}(A)^*$ is said to be a soft metric on $\tilde{U}$ if for any $U_\lambda^x, U_\mu^y, U_\gamma^z \in SP(\tilde{U})$ (equivalently, $U_\lambda^x, U_\mu^y, U_\gamma^z \tilde{\in} \tilde{U}$), the following hold:

(M1) $d(U_\lambda^x, U_\mu^y)\tilde{\geq}\bar{0}$.

(M2) $d(U_\lambda^x, U_\mu^y) = \bar{0}$ if and only if $U_\lambda^x = U_\mu^y$.

(M3) $d(U_\lambda^x, U_\mu^y) = d(U_\mu^x, U_\lambda^y)$.

(M4) $d(U_\lambda^x, U_\gamma^z)\tilde{\leq}d(U_\lambda^x, U_\mu^y) + d(U_\mu^y, U_\gamma^z)$.

The soft set $\tilde{U}$ endowed with a soft metric d is called a soft metric space and is denoted by $(\tilde{U}, d, A)$ or simply by $(\tilde{U}, d)$ if no confusion will arise.

See [8] for several basic properties of the structure of soft metric spaces. In order to help the reader we recall the following notions which will be used later.

Given a soft metric space $(\tilde{U}, d)$, a net $\{U_{\lambda_\alpha}^{x_\alpha}\}_{\alpha\in\Lambda}$ of soft points in $\tilde{U}$ will be simply denoted by $\{U_{\lambda,\alpha}^x\}_{\alpha\in\Lambda}$. In particular, a sequence $\{U_{\lambda_n}^{x_n}\}_{n\in\mathbb{N}}$ of soft points in $\tilde{U}$ will be denoted by $\{U_{\lambda,n}^x\}_n$.

Definition 5.9 *[8]* Let $(\tilde{U}, d)$ be a soft metric space. A sequence $\{U_{\lambda,n}^x\}_n$ of soft points in $\tilde{U}$ is said to be convergent in $(\tilde{U}, d)$ if there is a soft point $U_\mu^y\tilde{\in}\tilde{U}$ such that $d(U_{\lambda,n}^x, U_\mu^y) \to \bar{0}$ as $n \to \infty$. This means for every $\widehat{\varepsilon}\tilde{>}\bar{0}$, chosen arbitrarily, $m \in \mathbb{N}$ such that $d(U_{\lambda,n}^x, U_\mu^y)\tilde{<}\widehat{\varepsilon}$, whenever $n \geq m$.

Proposition 5.2 *[8] Limit of a sequence $\{U_{\lambda,n}^x\}_n$ in a soft metric space $(\tilde{U}, d)$, is unique.*

Proposition 5.3 *[8] Let $(\tilde{U}, d)$ be a soft metric space and $(F, A)\tilde{\subset}\tilde{U}$. $U_\mu^y\tilde{\in}\tilde{U}$ is a soft limit point of (F, A) if and only if there is a sequence $\{U_{\lambda,n}^x\}_n$ in (F, A) other than $\{U_\mu^y\}_n$ which converges to U_μ^y.*

Proposition 5.4 *[8] Let (F, A) be a soft subset in a soft metric space $(\tilde{U}, d)$. (F, A) is soft closed if and only if $\{U_{\lambda,n}^x\}_n$ in (F, A) which converges in $\tilde{U}$ cannot converge to a soft point of $(F, A)^c$.*

Definition 5.10 *[8]* A sequence $\{U_{\lambda,n}^x\}_n$ of soft points in a soft metric space $(\tilde{U}, d)$ is said to be a Cauchy sequence in $(\tilde{U}, d)$ if for each $\widehat{\varepsilon}\tilde{>}\bar{0}$, there exists $m \in \mathbb{N}$ such that $d(U_{\lambda,i}^x, U_{\lambda,j}^x)\tilde{<}\widehat{\varepsilon}$, for all $i, j \geq m$. That is, $d(U_{\lambda,i}^x, U_{\lambda,j}^x) \to \bar{0}$ as $i, j \to \infty$.

Proposition 5.5 *[8] Every convergent sequence $\{U_{\lambda,n}^x\}_n$ in a soft metric space $(\tilde{U}, d)$ is a Cauchy sequence.*

Definition 5.11 *[8]* A soft metric space $(\tilde{U}, d)$ is called complete if every Cauchy sequence in $(\tilde{U}, d)$ converges to some soft point of $\tilde{U}$. In this case, we say that the soft metric d is complete.

5.2 Banach contraction theorem with restriction

In the begining of this section, we will discuss propositions and definitions. In the sequel, soft real numbers will be denoted with tildes instead of widehats, i.e., we will write $\tilde{\varepsilon}, \tilde{\delta}$, etc., instead of $\widehat{\varepsilon}, \widehat{\delta}$, etc.

Proposition 5.6 *A soft subset (F, A) of a complete soft metric space $(\tilde{U}, d)$ is complete if and only if (F, A) is soft closed in $(\tilde{U}, d)$.*

Proof. Suppose that (F, A) is complete. By Proposition 5.3, we know that every $F^y_\mu \tilde{\in} \overline{(F, A)}$ has a sequence $\{F^x_{\lambda,n}\}_n$ in (F, A) which converges to F^y_μ. As $\{F^x_{\lambda,n}\}_n$ is a Cauchy sequence (Proposition 5.5) and (F, A) is complete, $\{F^x_{\lambda,n}\}_n$ converges in (F, A). By uniqueness of the limit (Proposition 5.2) we see that $F^y_\mu \tilde{\in} (F, A)$. This proves that (F, A) is soft closed. Conversely, if (F, A) is soft closed and $\{F^x_{\lambda,n}\}_n$ is a Cauchy sequence in (F, A), by the completeness of $(\tilde{U}, d)$ we have $F^x_{\lambda,n} \to F^y_\mu \tilde{\in} \tilde{U}$, which by Proposition 5.4 further implies that $F^y_\mu \tilde{\in} (F, A)$. Hence (F, A) is complete.

The following definitions are similar to those of Wardowski [36].

Definition 5.12 Let (F, A), $(G, A) \tilde{\in} S(U)$. The soft Cartesian product of (F, A) and (G, A), denoted by $(F, A) \tilde{\times} (G, A)$, is defined as

$$(F, A) \tilde{\times} (G, A) = \{((p_1, p_2), F(p_1) \times G(p_2)) : p_1, p_2 \in A\}.$$

Example 5.2 Suppose that $U = \{h_1, h_2, h_3\}$ and $A = \{p_1, p_2, p_3\}$. Define soft sets (F, A) and (G, A) as follows:

$$(F, A) = \{(p_1, \{h_1, h_2\}), (p_2, \{h_2, h_3\}), (p_3, \{h_1\})\} \text{ and}$$
$$(G, A) = \{(p_1, \{h_1\}), (p_2, \{h_1, h_3\}), (p_3, \{h_1, h_2\})\}.$$

Then

$$\begin{aligned}(F, A) \tilde{\times} (G, A) = \; & \{((p_1, p_1), \{h_1, h_2\} \times \{h_1\}), ((p_1, p_2), \{h_1, h_2\} \times \{h_1, h_3\}), \\ & ((p_1, p_3), \{h_1, h_2\} \times \{h_1, h_2\}), ((p_2, p_1), \{h_2, h_3\} \times \{h_1\}), \\ & ((p_2, p_2), \{h_2, h_3\} \times \{h_1, h_3\}), ((p_2, p_3), \{h_2, h_3\} \times \{h_1, h_2\}), \\ & ((p_3, p_1), \{h_1\} \times \{h_1\}), ((p_3, p_2), \{h_1\} \times \{h_1, h_3\}), \\ & ((p_3, p_3), \{h_1\} \times \{h_1, h_2\})\}.\end{aligned}$$

Definition 5.13 Let $(F, A), (G, A)$ be two soft sets over a common universe U. A soft relation R is a soft set such that $(R, A \times A) \tilde{\subset} (F, A) \tilde{\times} (G, A)$. That is,

$$(R, A \times A) = \{((p, q), U_p \times U_q) : p, q \in A, U_p \subseteq F(p), U_q \subseteq G(q)\}.$$

We will denote $((p,q),U_p \times U_q) \in (R, A \times A)$ as $(p,U_p)R(q,U_q)$.

Example 5.3 Let $(F,A),(G,A)$ be as in Example 5.2. Then

$$R = \{((p_1,p_1),\{(h_1,h_1)\}),((p_2,p_1),\{(h_2,h_1)\}),\\((p_2,p_3),\{(h_2,h_1),((h_3,h_2))\})\}$$

and we can write

$$(p_1,\{h_1\})R(p_1,\{h_1\}),(p_2,\{h_2\})R(p_1,\{h_1\}),(p_2,\{h_2,h_3\})R(p_3,\{h_1,h_2\}).$$

Definition 5.14 Let (F,A) and (G,A) be two soft sets. A soft relation $(T,A\times A)\tilde{\subset}(F,A)\tilde{\times}(G,A)$ is called a soft mapping from (F,A) to (G,A) if for each soft point $F_\lambda^x\tilde{\in}(F,A)$ there exists only one soft point F_μ^y such that $F_\lambda^x T F_\mu^y$. We will denote $F_\lambda^x T F_\mu^y$ by $T(F_\lambda^x) = F_\mu^y$. If $(T, A \times A)\tilde{\subset}(F,A)\tilde{\times}(G,A)$ is soft mapping from (F,A) to (G,A), we write it as $T:(F,A)\tilde{\to}(G,A)$.

Example 5.4 Let $(F,A),(G,A)$ be as in Example 5.2. Suppose that $(T,A\times A)\tilde{\subset}(F,A)\tilde{\times}(G,A)$ is defined as:

$$T = \{((p_1,p_3),\{(h_1,h_2)\}),((p_1,p_2),\{(h_2,h_1)\}),((p_2,p_1),\{(h_3,h_1)\}),\\((p_2,p_2),\{(h_2,h_1)\}),((p_3,p_2),\{(h_1,h_3)\})\}.$$

Therefore we can write $T(F_{p_1}^{h_1}) = F_{p_3}^{h_2}$, $T(F_{p_1}^{h_2}) = F_{p_2}^{h_1}$, $T(F_{p_2}^{h_3}) = F_{p_1}^{h_1}$, $T(F_{p_2}^{h_2}) = F_{p_2}^{h_1}$ and $T(F_{p_3}^{h_1}) = F_{p_2}^{h_3}$.

Definition 5.15 Let (F,A) and (G,A) be two soft sets and $T : (F,A)\tilde{\to}(G,A)$ a soft mapping. The image of $(H,A)\tilde{\subset}(F,A)$ under the soft mapping T is the soft set denoted by $T((H,A))$ and defined as follows

$$T((H,A)) = \tilde{\cup}\{T\{F_\lambda^x\} : F_\lambda^x\tilde{\in}(H,A)\}.$$

Definition 5.16 Let (F,A) and (G,A) be two soft sets and $T : (F,A)\tilde{\to}(G,A)$ a soft mapping. The inverse of $(Y,A)\tilde{\subset}(G,A)$ under the soft mapping T is the soft set denoted by $T^{-1}((Y,A))$ and defined as:

$$T^{-1}((Y,A)) = \tilde{\cup}\{\{F_\lambda^x\} : F_\lambda^x\tilde{\in}F, T\{F_\lambda^x\}\tilde{\in}Y\}.$$

Definition 5.17 Let (F,A) be a soft set and $T:(F,A)\tilde{\to}(F,A)$ a soft mapping. A soft point $F_\lambda^x\tilde{\in}(F,A)$ is called a soft fixed point of T if $T(F_\lambda^x) = F_\lambda^x$.

Example 5.5 Let $U = \{h_1,h_2,h_3\}$, $A = \{p_1,p_2\}$. Define the soft set (F,A) as

$$(F,A) = \{(p_1,\{h_1,h_2\}),(p_2,\{h_2,h_3\})\}.$$

If $T:(F,A)\tilde{\to}(F,A)$ is defined as:

$$T(F_{p_1}^{h_1}) = F_{p_1}^{h_1},\ T(F_{p_1}^{h_2}) = F_{p_2}^{h_2}, T(F_{p_2}^{h_2}) = F_{p_2}^{h_3},\ \text{and}\ T(F_{p_2}^{h_3}) = F_{p_1}^{h_2},$$

then $F_{p_1}^{h_1}$ is the fixed point of T.

Definition 5.18 Let $(\tilde{U}_1, d_1)$ and $(\tilde{U}_2, d_2)$ be two soft metric spaces. A soft mapping $T : \tilde{U}_1 \tilde{\to} \tilde{U}_2$ is said to be soft continuous at a soft point $U_\lambda^x \tilde{\in} \tilde{U}$ if for every $\tilde{\varepsilon} \tilde{>} \bar{0}$, there is a $\tilde{\delta} \tilde{>} \bar{0}$ such that $d_2(T(U_\lambda^x), T(U_\mu^y)) \tilde{<} \tilde{\varepsilon}$ whenever $d_1(U_\lambda^x, U_\mu^y) \tilde{<} \tilde{\delta}$. If T is soft continuous at every soft point of $\tilde{U}$, we say that T is soft continuous on $\tilde{U}$.

Proposition 5.7 *Let $(\tilde{U}, d_1)$ and $(\check{Y}, d_2)$ be two soft metric spaces. For a soft mapping $T : \tilde{U} \tilde{\to} \check{Y}$ the following are equivalent:*

(i) T is soft continuous on $\tilde{U}$.

(ii) For any $U_\mu^y \tilde{\in} \tilde{U}$, if $U_{\lambda,n}^x \to U_\mu^y$ in $\tilde{U}$, then $T(U_{\lambda,n}^x) \to T(U_\mu^y)$ in $\check{Y}$.

(iii) If (W, A) is soft closed in $\check{Y}$, then $T^{-1}(W, A)$ is soft closed in $\tilde{U}$.

(iv) If (V, A) is soft open in $\check{Y}$, then $T^{-1}(V, A)$ is soft open in $\tilde{U}$.

Proof. (i)⇒(ii): Suppose that $U_{\lambda,n}^x \to U_\mu^y$ in $\tilde{U}$. Given $\tilde{\varepsilon} \tilde{>} \bar{0}$, let $\tilde{\delta} \tilde{>} \bar{0}$ such that $T(SS(B(U_\mu^y, \tilde{\delta}))) \tilde{\subset} SS(B(T(U_\mu^y), \tilde{\varepsilon}))$. Then, since $U_{\lambda,n}^x \to U_\mu^y$, we have $\{U_{\lambda,n}^x\}_n$ eventually in $SS(B(U_\mu^y, \tilde{\delta}))$. This implies that $\{T(U_{\lambda,n}^x)\}_n$ eventually in $SS(B(T(U_\mu^y), \tilde{\varepsilon}))$. Since $\tilde{\varepsilon}$ is arbitrary, $T(U_{\lambda,n}^x) \to T(U_\mu^y)$.

(ii)⇒(iii): Let (W, A) be soft closed in $\check{Y}$. Given $\{U_{\lambda,n}^x\}_n$ in $T^{-1}(W, A)$ such that $U_{\lambda,n}^x \to U_\mu^y$ in $\tilde{U}$, we show that $U_\mu^y \tilde{\in} T^{-1}(W, A)$. But $\{U_{\lambda,n}^x\}_n$ in $T^{-1}(W, A)$ implies that $\{T(U_{\lambda,n}^x)\}_n$ is in (W, A), while $U_{\lambda,n}^x \to U_\mu^y$ in $\tilde{U}$ tells us that $T(U_{\lambda,n}^x) \to T(U_\mu^y)$ in $\check{Y}$ from (ii). Thus, since (W, A) is soft closed, $T(U_\mu^y) \tilde{\in} (W, A)$ or $U_\mu^y \tilde{\in} T^{-1}(W, A)$.

(iii)⇔(iv): It is obvious, since $T^{-1}((V, A)^c) = (T^{-1}(V, A))^c$.

(iv)⇒(i): Given $U_\lambda^x \tilde{\in} \tilde{U}$ and $\tilde{\varepsilon} \tilde{>} \bar{0}$, the set $SS(B(T(U_\lambda^x), \tilde{\varepsilon}))$ is open in $\check{Y}$ and by (iv), the set $T^{-1}(SS(B(T(U_\lambda^x), \tilde{\varepsilon})))$ is open in $\tilde{U}$. Then

$$SS(B(U_\lambda^x, \tilde{\delta})) \tilde{\subset} T^{-1}(SS(B(T(U_\lambda^x), \tilde{\varepsilon})))$$

for some $\tilde{\delta} \tilde{>} \bar{0}$, because $U_\lambda^x \tilde{\in} SS(B(T(U_\lambda^x), \tilde{\varepsilon}))$.

Definition 5.19 Let $(\tilde{U}, d)$ be a soft metric space and $T : \tilde{U} \tilde{\to} \tilde{U}$ a soft mapping. T is said to be a soft contraction if

$$d(T(U_\lambda^x), T(U_\mu^y)) \tilde{\leq} \bar{c} d(U_\lambda^x, U_\mu^y)$$

for all U_λ^x, $U_\mu^y \in SU(\tilde{U})$, where $\bar{0} \leq \bar{c} < \bar{1}$. We will call $\bar{c}$ a soft contraction constant.

Remark 5.3 A soft contraction on a soft metric space is a soft continuous mapping.

Now we will discuss the soft contraction theorem:

Theorem 5.1 *Let $(\tilde{U}, d)$ be a complete soft metric space, where $d : SP(\tilde{U}) \times SP(\tilde{U}) \to \mathbb{R}(A)^*$ with A a (nonempty) finite set, and let T be a soft contraction. T has unique soft fixed point U_μ^y. Moreover, for any soft point U_λ^x, the sequence of iterates $\{T^n(U_\lambda^x)\}_n$ converges to U_μ^y, and the following hold:*

$$d(U_{\lambda,n}^x, U_{\lambda,m}^x) \tilde{\leq} \frac{\bar{c}^m}{1-\bar{c}} d(T(U_{\lambda,0}^x), U_{\lambda,0}^x),$$
$$d(U_{\lambda,m}^x, U_\mu^y) \tilde{\leq} \bar{c} d(U_{\lambda,m-1}^x, U_\mu^y),$$

and

$$d(U_{\lambda,m}^x, U_\mu^y) \tilde{\leq} \frac{\bar{c}}{1-\bar{c}} d(U_{\lambda,m-1}^x, U_{\lambda,m}^x),$$

where $U_{\lambda,0}^x = U_\lambda^x$ and $U_{\lambda,n+1}^x = T(U_{\lambda,n}^x)$ for all $n = 0, 1, 2...$

Proof. Choose a soft point U_λ^x. Put $U_{\lambda,0}^x = U_\lambda^x$ and note that

$$\begin{aligned} d(U_{\lambda,n+1}^x, U_{\lambda,n}^x) &= d(T(U_{\lambda,n}^x), T(U_{\lambda,n-1}^x)) \tilde{\leq} \bar{c} d(U_{\lambda,n}^x, U_{\lambda,n-1}^x) \\ &\tilde{\leq} \bar{c}^2 d(U_{\lambda,n-1}^x, U_{\lambda,n-2}^x) \\ &\tilde{\leq} \cdots \\ &\tilde{\leq} \bar{c}^n d(U_{\lambda,1}^x, U_{\lambda,0}^x). \end{aligned}$$

For $n > m$, we have

$$\begin{aligned} d(U_{\lambda,n}^x, U_{\lambda,m}^x) &\tilde{\leq} d(U_{\lambda,n}^x, U_{\lambda,n-1}^x) + d(U_{\lambda,n-1}^x, U_{\lambda,n-2}^x) + \cdots + d(U_{\lambda,m+1}^x, U_{\lambda,m}^x) \\ &\tilde{\leq} (\bar{c}^{n-1} + \bar{c}^{n-2} + \cdots + \bar{c}^m) d(U_{\lambda,1}^x, U_{\lambda,0}^x) \\ &\tilde{\leq} \frac{\bar{c}^m}{1-\bar{c}} d(U_{\lambda,1}^x, U_{\lambda,0}^x). \end{aligned} \tag{5.1}$$

Now we show that $\{U_{\lambda,n}^x\}_n$ is a Cauchy sequence. Indeed, choose an arbitrary soft real number $\tilde{\varepsilon} \tilde{>} \bar{0}$. Since A is finite, we can write $A = \{\lambda_1, ..., \lambda_k\}$. Then, for each $i \in \{1, ..., k\}$, $N_i \in \mathbb{N}$ such that

$$\left(\frac{\bar{c}^{N_i}}{1-\bar{c}} d(U_{\lambda,1}^x, U_{\lambda,0}^x)\right)(\lambda_i) < \tilde{\varepsilon}(\lambda_i).$$

Consider $N = \max\{N_1, ..., N_k\}$. Therefore, for any $n > m \geq N$ and any $i \in \{1, ..., k\}$, we have

$$\begin{aligned} d(U_{\lambda,n}^x, U_{\lambda,m}^x)(\lambda_i) &\leq \left(\frac{\bar{c}^m}{1-\bar{c}} d(U_{\lambda,1}^x, U_{\lambda,0}^x)\right)(\lambda_i) \\ &\leq \left(\frac{\bar{c}^N}{1-\bar{c}} d(U_{\lambda,1}^x, U_{\lambda,0}^x)\right)(\lambda_i) < \tilde{\varepsilon}(\lambda_i), \end{aligned}$$

i.e.,

$$d(U^x_{\lambda,n}, U^x_{\lambda,m}) \tilde{<} \tilde{\varepsilon},$$

whenever $n > m \geq N$. Hence $\{U^x_{\lambda,n}\}_n$ is a Cauchy sequence. By the completeness of $(\tilde{U}, d)$ there is a $U^y_\mu \tilde{\in} \tilde{U}$ such that $d(U^x_{\lambda,n}, U^y_\mu) \to \bar{0}$ as $n \to \infty$.
Since

$$\begin{aligned} d(U^y_\mu, T(U^y_\mu)) &\tilde{\leq} d(U^x_{\lambda,n}, U^y_\mu) + d(U^x_{\lambda,n}, T(U^y_\mu)) \\ &\tilde{\leq} d(U^x_{\lambda,n}, U^y_\mu) + \bar{c}d(U^x_{\lambda,n-1}, U^y_\mu), \end{aligned}$$

we can make the second term smaller than any $\tilde{\varepsilon}\tilde{>}\bar{0}$ as $U^x_{\lambda,n} \to U^y_\mu$. Hence $d(U^y_\mu, T(U^y_\mu)) = \bar{0}$. This implies $T(U^y_\mu) = U^y_\mu$. So U^y_μ is a fixed point of T.

If U^z_γ is another fixed point of T,

$$d(U^y_\mu, U^z_\gamma) = d(T(U^y_\mu), T(U^z_\gamma)) \tilde{\leq} \bar{c}d(U^y_\mu, U^z_\gamma)$$

implies that $d(U^y_\mu, U^z_\gamma) = \bar{0}$ as $\bar{c}\tilde{<}\bar{1}$. Hence $U^y_\mu = U^z_\gamma$. Therefore the fixed point of T is unique.
As for $n > m$, we have

$$\begin{aligned} d(U^x_{\lambda,n}, U^x_{\lambda,m}) &\tilde{\leq} \frac{\bar{c}^m}{1-\bar{c}} d(U^x_{\lambda,1}, U^x_{\lambda,0}) \\ &= \frac{\bar{c}^m}{1-\bar{c}} d(T(U^x_{\lambda,0}), U^x_{\lambda,0}). \end{aligned}$$

Taking the limit as $n \to \infty$, we obtain

$$\begin{aligned} d(U^x_{\lambda,m}, U^y_\mu) &= d(T(U^x_{\lambda,m-1}), T(U^y_\mu)) \\ &\tilde{\leq} cd(U^x_{\lambda,m-1}, U^y_\mu) \\ &\tilde{\leq} \bar{c}[d(U^x_{\lambda,m-1}, U^x_{\lambda,m}) + d(U^x_{\lambda,m}, U^y_\mu)]. \end{aligned}$$

This implies

$$d(U^x_{\lambda,m}, U^y_\mu) \tilde{\leq} \frac{\bar{c}}{1-\bar{c}} d(U^x_{\lambda,m-1}, U^x_{\lambda,m}).$$

Remark 5.4 When $T : \tilde{U} \tilde{\to} \tilde{U}$ is a soft contraction with constant c, any iterate T^n is a soft contraction with constant $\bar{c}^n$. The unique soft fixed point of T will also be the unique soft fixed point of T^n.

Corollary 5.1 *Let $(\tilde{U}, d)$ be a complete soft metric space, where $d : SP(\tilde{U}) \times SP(\tilde{U}) \to \mathbb{R}(A)^*$ with A a (nonempty) finite set, $T : \tilde{U} \tilde{\to} \tilde{U}$ a soft contraction and $Y \tilde{\subset} \tilde{U}$ a soft closed subset such that $T(Y) \tilde{\subset} Y$. Then the unique soft fixed point of T is a soft point of Y.*

Proof. Since Y is a soft closed subset of a complete soft metric space, it is complete. Then by applying soft contraction theorem to T on Y, we obtain a soft fixed point of T in Y. Since T has only one fixed point in $\tilde{U}$, it must lie in Y.

Theorem 5.2 *Let T be a soft mapping on a complete soft metric space $(\tilde{U}, d)$, where $d : SP(\tilde{U}) \times SP(\tilde{U}) \to \mathbb{R}(A)^*$ with A a (nonempty) finite set . Suppose T is a soft contraction on a soft closed ball $SS(B[U^{x}_{\lambda,0}, \tilde{r}])$ with contraction constant $\bar{c}$ and $d(T(U^{x}_{\lambda,0}), U^{x}_{\lambda,0}) \tilde{<} (1-\bar{c})\tilde{r}$. Then T has a unique soft fixed point in $SS(B[U^{x}_{\lambda,0}, \tilde{r}])$.*

Proof. Construct a sequence $\{U^{x}_{\lambda,m}\}_m$ as in the previous theorem starting from $U^{x}_{\lambda,0}$. Now changing $m = 0$ in (5.1) to n, we have

$$d(U^{x}_{\lambda,m}, U^{x}_{\lambda,0}) \tilde{\leq} \frac{1}{1-\bar{c}} d(U^{x}_{\lambda,1}, U^{x}_{\lambda,0})$$
$$= \frac{1}{1-\bar{c}} d(T(U^{x}_{\lambda,0}), U^{x}_{\lambda,0}) \tilde{<} \tilde{r}.$$

Hence all $U^{x}_{\lambda,m}$'s are in $SS(B[U^{x}_{\lambda,0}, \tilde{r}])$. Since $\{U^{x}_{\lambda,m}\}_m$ is a Cauchy sequence, by the completeness of $(\tilde{U}, d)$ we have $U^{x}_{\lambda,m} \to U^{y}_{\mu} \tilde{\in} \tilde{U}$. As $SS(B[U^{x}_{\lambda,0}, \tilde{r}])$ is soft closed, so $U^{y}_{\mu} \tilde{\in} SS(B[U^{x}_{\lambda,0}, \tilde{r}])$. Hence the result.

We conclude this section with two examples that illustrate our result. They are based on the example in [8, Example 4.3].

Example 5.6 Let U and A be nonempty subsets of $\mathbb{R}$. Define $d : SP(\tilde{U}) \times SP(\tilde{U}) \to \mathbb{R}(A)^*$ as

$$d(U^{x}_{\lambda}, U^{y}_{\mu}) = |\overline{x} - \overline{y}| + |\overline{\lambda} - \overline{\mu}|,$$

for all $U^{x}_{\lambda}, U^{y}_{\mu} \in \tilde{U}$, where $|.|$ denotes the modulus of soft real numbers (recall that for each $x \in U$, $\overline{x}$ is the constant soft real number defined by $\overline{x}(\lambda) = x$ for all $\lambda \in A$, and, similarly, for each $\lambda \in A$, $\overline{\lambda}$ is the constant soft real number defined by $\overline{\lambda}(\mu) = \lambda$ for all $\mu \in A$). Then, according to [8, Example 4.3], d is a soft metric on $\tilde{U}$.

Example 5.7 Let $U = A = \{1/n : n \in \mathbb{N}\}$. Let d be the soft metric on $\widetilde{U}$ as constructed in Example 5.6. We show that $(\widetilde{U}, d)$ is complete. Indeed, suppose that $\{U^{x}_{\lambda,n}\}_n$ is a Cauchy sequence in $(\widetilde{U}, d)$. Take the soft real number $\widetilde{\varepsilon}$ such that $\widetilde{\varepsilon}(\lambda) = \lambda$ for all $\lambda \in A$, i.e., $\widetilde{\varepsilon}(1/k) = 1/k$ for all $k \in \mathbb{N}$. Then $m \in \mathbb{N}$ such that

$$d(U^{x}_{\lambda,i}, U^{x}_{\lambda,j}) \tilde{<} \widetilde{\varepsilon},$$

for all $i, j \geq m$. This implies that $d(U^{x}_{\lambda,i}, U^{x}_{\lambda,j})(1/k) < \widetilde{\varepsilon}(1/k)$ for all $k \in \mathbb{N}$. Hence

$$(|\overline{x_i} - \overline{x_j}| + |\overline{\lambda_i} - \overline{\lambda_j}|)(\frac{1}{k}) < \frac{1}{k},$$

for all $i, j \geq m$ and for all $k \in \mathbb{N}$. Consequently

$$|x_i - x_j| + |\lambda_i - \lambda_j| < \frac{1}{k},$$

for all $i, j \geq m$ and for all $k \in \mathbb{N}$. In particular, for any $j \geq m$,

$$|x_j - x_{j+1}| + |\lambda_j - \lambda_{j+1}| < \frac{1}{k},$$

for all $k \in \mathbb{N}$.

Therefore $x_j = x_{j+1}$ and $\lambda_j = \lambda_{j+1}$ for all $j \geq m$. We deduce that $x_j = x_m$ and $\lambda_j = \lambda_m$ for all $j \geq m$. Thus the sequence $\{U^x_{\lambda,n}\}_n$ is eventually constant, and hence convergent. We conclude that $(\tilde{U}, d)$ is complete.

Now let $T : \tilde{U} \to \tilde{U}$ be defined as $T(U^x_\lambda) = U_1^{x/2}$ for all $x \in U$, $\lambda \in A$. Clearly T has no fixed point. However it is a soft contraction with soft constant contraction $\overline{c}$ defined as $\overline{c}(\lambda) = 1/2$ for all $\lambda \in A$. Indeed, we fix $x, y \in U$ and $\lambda, \mu \in A$, then for each $\eta \in A$ we have

$$\begin{aligned} d(T(U^x_\lambda), T(U^y_\mu))(\eta) &= d(U_1^{x/2}, U_1^{y/2})(\eta) = \left|\frac{x}{2} - \frac{y}{2}\right| \\ &\leq \frac{1}{2}(|x-y| + |\lambda - \mu|) = \overline{c}(d(T(U^x_\lambda), T(U^y_\mu)))(\eta). \end{aligned}$$

Hence $d(T(U^x_\lambda), T(U^y_\mu)) \tilde{\leq} \overline{c}d(T(U^x_\lambda), T(U^y_\mu))$.

Remark 5.5 We discovered that Murat I. Yazar, Cigdem Gunduz (Aras) and Sadi Bayramov established in Theorem 4.8 of "Fixed point theorems of soft contractive mappings" ([32]), a similar result to our theorem (Theorem 5.1) but without assuming that the set A is finite. Our Example 5.7 shows that their result is not correct (the error seems occur on line 3 of page 9). Furthermore, if in Example 5.7 we use $x/3$ instead of $x/2$, a counterexample to Theorem 4.9 of Yazar-Gunduz-Bayramov's paper is also obtained.

5.3 Role of restricted condition

In this section we see that if $(\tilde{U}, d, A)$ is a soft metric space with A a (nonempty) finite set, d induces in a natural way a compatible metric on $SP(\tilde{U})$. We will see the role of this fact in establishing our fixed point theorems.

Theorem 5.3 *Let $(\tilde{U}, d, A)$ be a soft metric space with A a finite set. Define a function $m_d : SP(\tilde{U}) \times SP(\tilde{U}) \to \mathbb{R}^+$ as*

$$m_d(U^x_\lambda, U^y_\mu) = \max_{\eta \in A} d(U^x_\lambda, U^y_\mu)(\eta),$$

for all $U^x_\lambda, U^y_\mu \in SP(\tilde{U})$. Then the following hold:

(1) m_d is a metric on $SP(\tilde{U})$.

(2) For any sequence $\{U^x_{\lambda,n}\}_n$ of soft points and a soft point U^y_{λ},:

(a) $\{U^x_{\lambda,n}\}_n$ is a Cauchy sequence in $(\tilde{U}, d, A)$ if and only if it is a Cauchy sequence in $(SP(\tilde{U}), m_d)$.

(b) $d(U^y_{\lambda}, U^x_{\lambda,n}) \to \overline{0}$ if and only if $m_d(U^y_{\lambda}, U^x_{\lambda,n}) \to 0$.

(3) $(\tilde{U}, d, A)$ is complete if and only if $(SP(\tilde{U}), m_d)$ is complete.

Proof. (1) Let $U^x_{\lambda}, U^y_{\mu}, U^z_{\upsilon} \in SP(\tilde{U})$. Then, we have:

(a) $m_d(U^x_{\lambda}, U^y_{\mu}) \geq 0$ by condition M1 of definition 5.8.

(b) $m_d(U^x_{\lambda}, U^y_{\mu}) = 0 \Leftrightarrow U^x_{\lambda} = U^y_{\mu}$ by condition M2 of definition 5.8.

(c) $m_d(U^x_{\lambda}, U^y_{\mu}) = m_d(U^y_{\mu}, U^x_{\lambda})$ by condition M3 of definition 5.8.

(d) $m_d(U^x_{\lambda}, U^y_{\mu}) \leq m_d(U^x_{\lambda}, U^z_{\upsilon}) + m_d(U^y_{\mu}, U^z_{\upsilon})$ by condition M4 of Definition 5.8.

Indeed, we have

$$\begin{aligned} m_d(U^x_{\lambda}, U^y_{\mu}) &= \max_{\eta\in A} d(U^x_{\lambda}, U^y_{\mu})(\eta) \leq \max_{\eta\in A} d(U^x_{\lambda}, U^z_{\upsilon})(\eta) + \max_{\eta\in A} d(U^z_{\upsilon}, U^y_{\mu})(\eta) \\ &= m_d(U^x_{\lambda}, U^z_{\upsilon}) + m_d(U^y_{\mu}, U^z_{\upsilon}). \end{aligned}$$

(2a) Let $\{U^x_{\lambda,n}\}_n$ be a Cauchy sequence in $(\tilde{U}, d, A)$. Given $\varepsilon > 0$, use the constant soft real number $\overline{\varepsilon}\tilde{>}\overline{0}$. Then $m \in \mathbb{N}$ such that $d(U^x_{\lambda,i}, U^x_{\lambda.j})\tilde{<}\overline{\varepsilon}$ for all $i, j \geq m$. Hence $d(U^x_{\lambda,i}, U^x_{\lambda.j})(\eta) < \varepsilon$ for all $\eta \in A$ and $i, j \geq m$ and $m_d(U^x_{\lambda,i}, U^x_{\lambda.j}) < \varepsilon$ for all $i, j \geq m$. We deduce that $\{U^x_{\lambda,n}\}_n$ is a Cauchy sequence in $(m_d, SP(\tilde{U}))$.

Conversely, let $\{U^x_{\lambda,n}\}_n$ be a Cauchy sequence in $(m_d, SP(\tilde{U}))$. Given $\widehat{\varepsilon}\tilde{>}\overline{0}$, there exists $\varepsilon = \min_{\eta\in A} \widehat{\varepsilon}(\eta) > 0$, because A is a finite set . Then $m \in \mathbb{N}$ such that $m_d(U^x_{\lambda,i}, U^x_{\lambda.j}) < \varepsilon$ for all $i, j \geq m$. Hence $d(U^x_{\lambda,i}, U^x_{\lambda.j})(\eta) < \varepsilon \leq \widehat{\varepsilon}(\eta)$ for all $\eta \in A$ and $i, j \geq m$. We deduce that $\{U^x_{\lambda,n}\}_n$ is a Cauchy sequence in $(\tilde{U}, d, A)$.

(2b) follows from (2a).

(3) is a consequence of (2a) and (2b).

In Section 5.1, we established a Banach contraction principle for complete soft metric spaces $(\tilde{U}, d, A)$ such that A is a (nonempty) finite set, and showed that the condition that A is finite cannot be omitted. Now we will see that we can utilize Theorem 5.3 with the help of the compatible metric constructed above.

Theorem 5.4 *Let $(\tilde{U}, d, A)$ be a complete soft metric space with A a finite set. Suppose that the soft mapping $f : \tilde{U} \tilde{\to} \tilde{U}$ satisfies*

$$d(f(U_\lambda^x), f(U_\mu^y)) \tilde{\leq} \overline{c} d(U_\lambda^x, U_\mu^y)), \tag{5.2}$$

for all $U_\lambda^x, U_\mu^y \in SP(\tilde{U})$ where $\overline{0} \tilde{\leq} \overline{c} \tilde{<} \overline{1}$. Then f has a unique fixed point, i.e., there is a unique soft point U_λ^x such that $f(U_\lambda^x) = U_\lambda^x$.

Proof. Consider the metric m_d on $SP(\tilde{U})$ as constructed in Theorem 5.3. Since $(\tilde{U}, d, A)$ is complete it follows from Theorem 5.3 (3) that $(SP(\tilde{U}), m_d)$ is a complete metric space.

Since for each $U_\lambda^x \in SP(\tilde{U})$ (or equivalently, $U_\lambda^x \tilde{\in} \tilde{U}$) there is a unique soft point U_μ^y such that $f(U_\lambda^x) = U_\mu^y$ (see Remark 5.2), the restriction of f to $SP(\tilde{U})$ is a self mapping on $SP(\tilde{U})$, also denoted by f. Note also the real number c generating the constant soft real number $\overline{c}$ satisfies that $0 \leq c < 1$. Finally, we obtain the following contraction condition, for each $U_\lambda^x, U_\mu^y \in SP(\tilde{U})$,

$$\begin{aligned} m_d(f(U_\lambda^x), f(U_\mu^y)) &= \max_{\eta \in A} d(f(U_\lambda^x), f(U_\mu^y))(\eta) \leq \max_{\eta \in A} c \left(d(U_\lambda^x, U_\mu^y)(\eta) \right) \\ &= c[\max_{\eta \in A} d(U_\lambda^x, U_\mu^y))(\eta)] = c m_d(U_\lambda^x, U_\mu^y). \end{aligned}$$

Hence f has a unique fixed point by the Banach contraction principle.

Our next result provides a soft metric generalization of the celebrated Kannan fixed point theorem [22].

Theorem 5.5 *Let $(\tilde{U}, d, A)$ be a complete soft metric space with A a finite set. Suppose that the soft mapping $f : \tilde{U} \tilde{\to} \tilde{U}$ satisfies*

$$d(f(U_\lambda^x), f(U_\mu^y)) \tilde{\leq} \overline{c} \left\{ d(U_\lambda^x, f(U_\lambda^x)) + d(U_\mu^y, f(U_\mu^y)) \right\}, \tag{5.3}$$

for all $U_\lambda^x, U_\mu^y \in SP(\tilde{U})$, where $\overline{0} \tilde{\leq} \overline{c} \tilde{<} \overline{1/2}$. Then f has a unique fixed point.

Proof. Since $(\tilde{U}, d, A)$ is complete it follows from Theorem 5.3 (3) that $(SP(\tilde{U}), m_d)$ is a complete metric space.

Moreover, the restriction of f to $SP(\tilde{U})$ is self mapping on $SP(\tilde{U})$, exactly as in the proof of Theorem 5.4. Note also that the real number c generating the constant soft real number $\overline{c}$ satisfies $0 \leq c < 1/2$. Finally, we obtain the following contraction condition for each $U_\lambda^x, U_\mu^y \in SP(\tilde{U})$,

$$\begin{aligned} m_d(f(U_\lambda^x), f(U_\mu^y)) &= \max_{\alpha \in A} d(f(U_\lambda^x), f(U_\mu^y))(\alpha) \\ &\leq \max_{\alpha \in A} c \left((d(U_\lambda^x, f(U_\lambda^x)) + d(U_\mu^y, f(U_\mu^y)))(\alpha) \right) \\ &\leq c \left\{ m_d(U_\lambda^x, f(U_\lambda^x)) + m_d(U_\mu^y, f(U_\mu^y)) \right\}. \end{aligned}$$

Hence f has a unique fixed point by Kannan's fixed point theorem.

The following example shows that the condition "A is a finite set" cannot be omitted in the preceding theorem.

Example 5.8 Let $U = A = \{1/n : n \in \mathbb{N}\}$. According to [8, Example 4.3] the mapping $d : SP(\tilde{U}) \times SP(\tilde{U}) \to \mathbb{R}(A)^*$ given by

$$d(U_\lambda^x, U_\mu^y) = |\overline{x} - \overline{y}| + \left|\overline{\lambda} - \overline{\mu}\right|,$$

for all $U_\lambda^x, U_\mu^y \in SP(\tilde{U})$ where $|.|$ denotes the modulus of soft real numbers is a soft metric on $\tilde{U}$. Furthermore, the soft metric space $(\tilde{U}, d)$ is complete.

Let $f : \tilde{U} \tilde{\to} \tilde{U}$ such that $f(U_\lambda^x) = U_1^{x/4}$ for all $x \in U$, $\lambda \in A$. We show that f satisfies the contraction condition (5.3) of Theorem 5.5 with $\overline{c} = \overline{1/3}$. In fact, given $x, y \in U$ and $\lambda, \mu \in A$, for each $\eta \in A$ we have

$$\begin{aligned} d(f(U_\lambda^x), f(U_\mu^y))(\eta) &= d(U_1^{x/4}, U_1^{y/4})(\eta) = \frac{1}{4}|x-y| \leq \frac{1}{4}(x+y) \\ &= \frac{1}{3}\left((x - \frac{x}{4}) + (y - \frac{y}{4})\right) \\ &\leq \frac{1}{3}\left(d(U_\lambda^x, f(U_\lambda^x)) + d(U_\mu^y, f(U_\mu^y))\right)(\eta). \end{aligned}$$

Therefore $d(f(U_\lambda^x), f(U_\mu^y)) \tilde{\leq} \overline{1/3}\left\{d(U_\lambda^x, f(U_\lambda^x)) + d(U_\mu^y, f(U_\mu^y))\right\}$. However f has no fixed point.

Now we present an example where we can apply Theorem 5.5 but not Theorem 5.4.

Example 5.9 Let $U = \mathbb{R}^+$ and $A = \{0, 1\}$. Again, according to [8, Example 4.3], the mapping $d : SP(\tilde{U}) \times SP(\tilde{U}) \to \mathbb{R}(A)^*$ given by

$$d(U_\lambda^x, U_\mu^y) = |\overline{x} - \overline{y}| + \left|\overline{\lambda} - \overline{\mu}\right|$$

for all $U_\lambda^x, U_\mu^y \in SP(\tilde{U})$, is a soft metric on $\tilde{U}$. Since $\mathbb{R}^+$ is complete for the Euclidean metric, we deduce that $(\tilde{U}, d)$ is a complete soft metric space.

Let $f : \tilde{U} \tilde{\to} \tilde{U}$ such that $f(U_0^x) = f(U_1^x) = U_0^0$ if $x \in [0, 2)$, and $f(U_0^x) = f(U_1^x) = U_0^{1/2}$ if $x \in [2, \infty)$.

Let $\overline{c}$ be a constant soft real number such that $\overline{0} \tilde{\leq} \overline{c} \tilde{<} \overline{1}$. Then, there is a real number $c \in [0, 1)$ such that $c = \overline{c}(\eta)$ for all $\eta \in A$. Choose $y \in [0, 2)$ such that $c(2 - y) < 1/2$. Then, for each $\eta \in A$ we have

$$d(f(U_0^2), f(U_0^y))(\eta) = d(U_0^{1/2}, U_0^0)(\eta) = \frac{1}{2} > c(2-y) = cd(U_0^2, U_0^y)(\eta).$$

Therefore f does not satisfy condition (5.2) of Theorem 5.4 for any $\overline{c}$ satisfying $\overline{0} \tilde{\leq} \overline{c} \tilde{<} \overline{1}$.

However, taking, without loss of generality, $x \in [0,2)$ and $y \in [2,\infty)$, we obtain, for $\lambda, \mu, \eta \in A$,

$$d(f(U^x_\lambda), f(U^y_\mu))(\eta) = d(U^0_0, U^{1/2}_0)(\eta) = \frac{1}{2} = \frac{1}{3}(2-\frac{1}{2}) \leq \frac{1}{3}(x+y-\frac{1}{2})$$
$$= \frac{1}{3}\left(d(U^x_\lambda, U^0_0) + d(U^y_\mu, U^{1/2}_\mu)\right)(\eta).$$

Therefore f satisfies condition (5.3) of Theorem 5.5 for $\overline{c} = \overline{1/3}$. In fact, U^0_0 is the unique fixed point of f.

Meir and Keeler proved [30] their well-known fixed point theorem that every Meir-Keeler contractive self mapping on a complete metric space has a unique fixed point, where a self mapping T on a metric space (X,d) is said to be a Meir-Keeler contractive mapping if for each $\varepsilon > 0$ there exists $\delta > 0$ such that for each $x, y \in X$,

$$\varepsilon \leq d(x,y) < \varepsilon + \delta \Rightarrow d(T(x), T(y)) < \varepsilon.$$

In a recent paper [7], Chen and Lin discussed the extension of the Meir-Keeler fixed point theorem to soft metric spaces and introduced the following notion [7, Definition 15]:

Let $(\tilde{U}, d, A)$ be a soft metric space and let $\varphi : A \to A$. A soft mapping $(f,\varphi) : \tilde{U} \tilde{\to} \tilde{U}$ is called a soft Keir-Meeler contractive mapping if for each soft real number $\widehat{\varepsilon} \tilde{>} \overline{0}$ there exists $\widehat{\delta} \tilde{>} \overline{0}$ such that for each $U^x_\lambda, U^y_\mu \in SP(\tilde{U})$,

$$\widehat{\varepsilon} \tilde{\leq} d(U^x_\lambda, U^y_\mu) \tilde{<} \widehat{\varepsilon} + \widehat{\delta} \Rightarrow d((f,\varphi)(U^x_\lambda), (f,\varphi)(U^y_\mu)) \tilde{<} \widehat{\varepsilon}.$$

Chen and Lin [7, Theorem 1] established that every soft Keir-Meeler contractive mapping on a complete soft metric space has a unique fixed point.

The following example shows that this result is not correct even if A is a finite set (the error in the proof seems occur on page 4, lines 18 and 19: compare Definition 5.7 above).

Example 5.10 Let $U = \{2\}$, $A = \{0,1\}$, and d be a soft metric on $\tilde{U}$ defined as:

$d(U^2_\lambda, U^2_\lambda) = \overline{0}$ for all $\lambda \in A$,

and

$$d(U^2_0, U^2_1)(0) = d(U^2_1, U^2_0)(0) = 0, \quad d(U^2_0, U^2_1)(1) = d(U^2_1, U^2_0)(1) = 1.$$

Since m_d is the discrete metric on $SP(\tilde{U})$ it follows from Theorem 5.3 (3) that $(\tilde{U}, d)$ is a complete soft metric space.

For $f : U \to U$, necessarily follows that $f(2) = 2$. Let $\varphi : A \to A$ given by $\varphi(0) = 1$ and $\varphi(1) = 0$. Then $(f,\varphi)(U^2_0) = U^2_1$ and $(f,\varphi)(U^2_1) = U^2_0$. For each $\widehat{\varepsilon} \tilde{>} \overline{0}$ we have $d(U^2_0, U^2_1)(0) = 0 < \widehat{\varepsilon}(0)$. It follows that condition $\widehat{\varepsilon} \tilde{\leq} d(U^2_\lambda, U^2_\mu)$ is not satisfied for any $\lambda, \mu \in A$, and thus (f,φ) is trivially a soft Keir-Meeler contractive mapping on $(\tilde{U}, d)$. However (f,φ) has no fixed point.

The above example suggests the following modification of [7, Definition 15]:

Definition 5.20 Let $(\tilde{U}, d, A)$ be a soft metric space and let $\varphi : A \to A$. A soft mapping $(f, \varphi) : \tilde{U} \tilde{\to} \tilde{U}$ is called a soft contraction of Meir-Keeler type if for each soft real number $\widehat{\varepsilon} \tilde{>} \bar{0}$ there exists $\widehat{\delta} \tilde{>} \bar{0}$ such that for each $U_\lambda^x, U_\mu^y \in SP(\tilde{U}), d(U_\lambda^x, U_\mu^y) \tilde{<} \widehat{\varepsilon} + \widehat{\delta}$, and $\widehat{\varepsilon}(\eta) \leq d(U_\lambda^x, U_\mu^y)(\eta)$ for some $\eta \in A \Rightarrow d((f,\varphi)(U_\lambda^x), (f,\varphi)(U_\mu^y)) \tilde{<} \widehat{\varepsilon}$.

Remark 5.6 Let $(\tilde{U}, d)$ be the complete soft metric space of Example 5.8. Define $f(x) = x/4$ for all $x \in U$, and $\varphi(\lambda) = 1$ for all $\lambda \in A$. Then $(f, \varphi)(U_\lambda^x) = U_1^{x/4}$ for all $x \in U$ and $\lambda \in A$. Although (f, φ) has no fixed point, it is easy to check that the conditions of Definition 5.20 hold. However, we can state the following positive result.

Theorem 5.6 *Let $(\tilde{U}, d, A)$ be a complete soft metric space with A a finite set. Then, every soft contraction of Meir-Keeler type on $(\tilde{U}, d, A)$ has a unique fixed point.*

Proof. We first note that, by Theorem 5.3 (3), the metric space $(SP(\tilde{U}), m_d)$ is complete.

Now let (f, φ) be a soft contraction of Meir-Keeler type on $(\tilde{U}, d, A)$. As in the proof of Theorem 5.4, the restriction of (f, φ) to $SP(\tilde{U})$ is a self mapping on $SP(\tilde{U})$, which is also denoted by (f, φ).

We want to show that (f, φ) is a Meir-Keeler contractive mapping on $(SP(\tilde{U}), m_d)$. Indeed, given $\varepsilon > 0$ consider the constant soft real number $\bar{\varepsilon}$. Since $\bar{\varepsilon} \tilde{>} \bar{0}$, there exists $\widehat{\delta} \tilde{>} \bar{0}$ for which the conditions of Definition 5.20 are satisfied. Also, $\delta = \min_{\eta \in A} \widehat{\delta}(\eta) > 0$ because A is finite.

Consider $U_\lambda^x, U_\mu^y \in SP(\tilde{U})$ satisfying $\varepsilon \leq m_d(U_\lambda^x, U_\mu^y) < \varepsilon + \delta$. Then

$$d((U_\lambda^x, U_\mu^y)(\eta) < \varepsilon + \delta \leq (\bar{\varepsilon} + \widehat{\delta})(\eta)$$

for all $\eta \in A$, so $d((U_\lambda^x, U_\mu^y) \tilde{<} \bar{\varepsilon} + \widehat{\delta}$. Furthermore, from $\varepsilon \leq m_d(U_\lambda^x, U_\mu^y)$ we deduce that

$$\bar{\varepsilon}(\eta_0) \leq d(U_\lambda^x, U_\mu^y)(\eta_0),$$

where $m_d(U_\lambda^x, U_\mu^y) = d(U_\lambda^x, U_\mu^y)(\eta_0)$, $\eta_0 \in A.$,

Since (f, φ) is a soft mapping of Meir-Keeler type,

$$d((f,\varphi)(U_\lambda^x), (f,\varphi)(U_\mu^y)) \tilde{<} \bar{\varepsilon}.$$

From this relation it follows that $m_d((f,\varphi)(U_\lambda^x), (f,\varphi)(U_\mu^y)) < \varepsilon$. We deduce that (f, φ) is a Meir-Keeler contractive mapping on $(SP(\tilde{U}), m_d)$. Hence (f, φ) has a unique fixed point.

We conclude this section by obtaining a soft metric extension of the celebrated Caristi-Kirk [6, 24] theorem that a metric space (X, d) is complete if and only if every Caristi mapping on (X, d) has a fixed point.

Let us recall that a self mapping T on a metric space (X, d) is a Caristi mapping provided that there exists a lower semicontinuous function $\phi : X \to \mathbb{R}^+$ such that $d(x, T(x)) + \phi(T(x)) \leq \phi(x)$ for all $x \in X$.

Caristi proved that every Caristi mapping on a complete metric space has a fixed point, while Kirk proved that Caristi's fixed point theorem characterizes metric completeness.

In Definition 5.22 below we propose a notion of soft Caristi mapping. We first generalize, in a natural way, Definition 5.9 to the case of a net. Thus, given a soft metric space $(\tilde{U}, d)$, we say that a net $\{U^x_{\lambda,\alpha}\}_{\alpha\in\Lambda}$ of soft points in $\tilde{U}$ is convergent in $(\tilde{U}, d)$ if there is a soft point U^y_μ such that for each $\widehat{\varepsilon}\tilde{>}\bar{0}$, there exists $\alpha_0 \in \Lambda$ satisfying $d(U^x_{\lambda,\alpha}, U^y_\mu)\tilde{<}\widehat{\varepsilon}$, whenever $\alpha \geq \alpha_0$.

Definition 5.21 Let $(\tilde{U}, d, A)$ be a soft metric space. A mapping $\phi : SP(\tilde{U}) \to \mathbb{R}(A)^*$ is called lower semicontinuous on $(\tilde{U}, d, A)$ if whenever $\{U^x_{\lambda,\alpha}\}_{\alpha\in\Lambda}$ is a net of soft points in $\tilde{U}$ that converges in $(\tilde{U}, d, A)$ to a soft point U^y_μ, the following holds: for each $\widehat{\varepsilon}\tilde{>}\bar{0}$ there exists $\widehat{\delta}\tilde{>}\bar{0}$ such that $\phi(U^y_\mu)\tilde{\leq}\phi(U^x_{\lambda,\alpha}) + \widehat{\varepsilon}$ whenever $d(U^x_{\lambda,\alpha}, U^y_\mu)\tilde{<}\widehat{\delta}$.

Definition 5.22 Let $(\tilde{U}, d, A)$ be a soft metric space. A soft mapping $f : \tilde{U}\tilde{\to}\tilde{U}$ is called a soft Caristi mapping if there exists a lower semicontinuous mapping $\phi : SP(\tilde{U}) \to \mathbb{R}(A)^*$ such that $d(U^x_\lambda, f(U^x_\lambda)) + \phi(f(U^x_\lambda))\tilde{\leq}\phi(U^x_\lambda)$ for all $U^x_\lambda \in SP(\tilde{U})$.

Theorem 5.7 *Let $(\tilde{U}, d, A)$ be a soft metric space with A a finite set. $(\tilde{U}, d, A)$ is complete if and only if every soft Caristi mapping on $(\tilde{U}, d, A)$has a fixed point.*

Proof. Suppose that $(\tilde{U}, d, A)$ is complete and let f be a soft Caristi mapping on $(\tilde{U}, d, A)$. Then, there exists a lower semicontinuous mapping $\phi : SP(\tilde{U}) \to \mathbb{R}(A)^*$ such that $d(U^x_\lambda, f(U^x_\lambda))+\phi(f(U^x_\lambda))\tilde{\leq}\phi(U^x_\lambda)$ for all $U^x_\lambda \in SP(\tilde{U})$. Exactly as in Theorem 5.4, the restriction of f to $SP(\tilde{U})$ is a self mapping on $SP(\tilde{U})$ denoted by f. Now define $\Phi : SP(\tilde{U}) \to \mathbb{R}^+$ as

$$\Phi(U^x_\lambda) - \sum_{\eta\in A}(\phi(U^x_\lambda)(\eta)),$$

for all $U^x_\lambda \in SP(\tilde{U})$. It is not difficult to see that Φ is lower semicontinuous on the complete metric space $(SP(\tilde{U}), m_d)$. Furthermore, given $U^x_\lambda \in SP(\tilde{U})$ let $\eta_{x,\lambda} \in A$ such that $m_d(U^x_\lambda, f(U^x_\lambda)) = d(U^x_\lambda, f(U^x_\lambda))(\eta_{x,\lambda})$. Then

$$\begin{aligned} m_d(U^x_\lambda, f(U^x_\lambda)) + \Phi(f(U^x_\lambda)) &= d(U^x_\lambda, f(U^x_\lambda))(\eta_{x,\lambda}) + \sum_{\eta\in A}(\phi(f(U^x_\lambda)(\eta)) \\ &\leq \phi(U^x_\lambda)(\eta_{x,\lambda}) + \sum_{\eta\in A\setminus\{\eta_{x,\lambda}\}}(\phi(f(U^x_\lambda)(\eta)) \\ &= \Phi(U^x_\lambda). \end{aligned}$$

We deduce that f is a Caristi mapping on $(SP(\tilde{U}), m_d)$, and hence has a fixed point.

Conversely, suppose that every soft Caristi mapping on $(\tilde{U}, d, A)$ has a fixed point, and let T be a Caristi mapping on the complete metric space $(SP(\tilde{U}), m_d)$. Then, there exists a lower semicontinuous function $\phi : SP(\tilde{U}) \to \mathbb{R}^+$ such that $m_d(U_\lambda^x, T(U_\lambda^x)) + \phi(T(U_\lambda^x)) \leq \phi(U_\lambda^x)$ for all $U_\lambda^x \in SP(\tilde{U})$.

Let $f_T : \tilde{U} \to \tilde{U}$ such that $f_T(U_\lambda^x) = T(U_\lambda^x)$, and define $\varphi : SP(\tilde{U}) \to \mathbb{R}(A)^*$ such that $\varphi(U_\lambda^x) = \overline{\phi(U_\lambda^x)}$ for all $U_\lambda^x \tilde{\in} \tilde{U}$. Clearly φ is lower semicontinuous on $(\tilde{U}, d, A)$ and for each $U_\lambda^x \tilde{\in} \tilde{U}$ and $\eta \in A$ we obtain

$$\begin{aligned}(d(U_\lambda^x, f_T(U_\lambda^x)) + \varphi(f_T(U_\lambda^x)))\,(\eta) &\leq m_d(U_\lambda^x, T(U_\lambda^x)) + \phi(T(U_\lambda^x)) \\ &\leq \phi(U_\lambda^x) = \varphi(U_\lambda^x)(\eta).\end{aligned}$$

We deduce that $d(U_\lambda^x, f(U_\lambda^x)) + \varphi(f_T(U_\lambda^x)) \tilde{\leq} \varphi(U_\lambda^x)$, and, consequently, f_T is a soft Caristi mapping on $(\tilde{U}, d, A)$. Therefore f_T, and hence T, has a fixed point. $(SP(\tilde{U}), m_d)$ is complete by the Caristi-Kirk theorem. Completeness of $(\tilde{U}, d, A)$ is now a consequence of Theorem 5.3 (3).

5.4 Conclusion

In an attempt to reverse the trend of obtaining soft metric extensions of existing fixed point results in ordinary metric spaces, we showed that a soft metric space under the restriction that a set of parameters is finite gives rise to a compatible metric on the collection of all soft points of an absolute soft set. We also studied some essential properties of the induced metric thus obtained. We proved that recently obtained soft fixed point results can be deduced from comparable results in ordinary metric spaces. We presented examples to show that restriction of finite parameter sets is unavoidable. It will be interesting to study the limits to which soft fixed point theory may be extended.

References

[1] Abbas M., Murtaza G., Romaguera S., Soft contraction theorem. J. Nonlinear Conv. Anal. 2015; 16:423-435.

[2] Abbas M., Murtaza G., Romaguera S., On the fixed point theory of soft metric spaces. Fixed Point Theory Appl. 2016; 1:1-11.

[3] Aktaş H., Çağman N., Soft sets and soft groups. Inform. Sci. 2007; 177:2726–2735.

[4] Ali M.I., Feng F., Liu X., Min W.K., Shabir M., On some new operations in soft set theory. Comput. Math. Appl. 2009; 57:1547–1553.

[5] Cagman N., Karatas S., Enginoglu S., Soft topology. Comput. Math. Appl. 2011; 62:351–358.

[6] Caristi J., Fixed point theorems for mappings satisfying inwardness conditions. Trans. Amer. Math. Soc. 1976; 215:241-251.

[7] Chen C.M., Lin I.J., Fixed point theory of the soft Meir-Keeler type contractive mappings on a complete soft metric space. Fixed Point Theory Appl. 2015, Art. 184.

[8] Das S., Samanta S.K., Soft metric. Ann. Fuzzy Math. Inform. 2013; 6:77-94.

[9] Das S., Samanta S.K., Soft real sets, soft real numbers and their properties. J. Fuzzy Math. 2012; 20:551–576.

[10] Feng F., Liu X., Leoreanu-Fotea V., Jun Y.B., Soft sets and soft rough sets. Inf. Sci. 2011; 181:1125–1137.

[11] Feng F., Li C.X., Davvaz B.. Ali M.I., Soft sets combined with fuzzy sets and rough sets: a tentative approach. Soft Computing. 2010; 14:8999–9911.

[12] Feng F., Jun Y.B., Liu X.Y., Li L.F., An adjustable approach to fuzzy soft set-based decision making. J. Comput. Appl. Math. 2009; 234:10–20.

[13] Feng F., Liu., Soft rough sets with applications to demand analysis. In Proceedings of International Workshop on Intelligent System Applications. May 2009, Wuhan, China, IEEE, pp. 1-4 .

[14] Feng F., Jun Y.B., Zhao X., Soft semirings. Comput. Math. Appl. 2008; 56:2621–2628.

[15] Herawan T, Deris M.M., On multi-soft set construction in information systems. In Proceedings of Fifth International Conference on Intelligent Computing, September 2009, Ulsan, South Korea, Springer, pp. 101-110.

[16] Herawan T., Rose A.N.M., Deris M.M., Soft set theoretic approach for dimensionality reduction. In Proceedings of International Conference on Database Theory and Applications, December 2009, Jeju Island, South Korea, Springer, pp. 171-178.

[17] Jiang Y., Tang Y., Chen Q., Wang J., Tang S., Extending soft sets with description logics. Comput. Math. Appl. 2009; 59:2087–2096.

[18] Jun Y.B., Lee K.J., Khan A., Soft ordered semigroups. Math. Logic Q. 2010; 56:42–50.

[19] Jun Y.B., Lee K.J., Park C.H., Soft set theory applied to ideals in d-algebras. Comput. Math. Appl. 2009; 57:367–378.

[20] Jun Y.B., Soft BCK/BCI-algebras. Comput. Math. Appl. 2008; 56:1408–1413.

[21] Jun Y.B., Park C.H., Applications of soft sets in ideal theory of BCK/BCI-algebras. Inform. Sci. 2008; 178:2466–2475.

[22] Kannan R., Some results on fixed points II. Am. Math. Mon. 1969; 76: 405-408.

[23] Kim Y.K., Min W.K., Full soft sets and full soft decision systems. J. Int. Fuzzy Syst. 2014; 26(2):925-933.

[24] Kirk W.A., Caristi's fixed-point theorem and metric convexity. Colloq. Math. 1976; 36:81-86.

[25] Kong Z., Gao L., Wang L., Li S., The normal parameter reduction of soft sets and its algorithm. Comput. Math. Appl. 2008; 56:3029–3037.

[26] Li F., Notes on the soft operations. ARPN J. Syst. Software. 2011; 1:205-208.

[27] Maji P.K., Biswas R., Roy A.R., Soft set theory. Comput. Math. Appl. 2003; 45:555–562.

[28] Maji P.K., Roy A.R., Biswas R., An application of soft sets in a decision making problem. Comput. Math. Appl. 2002; 44:1077–1083.

[29] Majumdar P., Samanta S.K., Generalized fuzzy soft sets. Comput. Math. Appl. 2010; 59:1425–1432.

[30] Meir A., Keeler E., A theorem on contraction mappings. J. Math. Anal. Appl. 1969; 28:326-329.

[31] Molodtsov D., Soft set theory: First results. Comput. Math. Appl. 1999; 37:19–31.

[32] Yazar M.I., Aras C.G., Bayramov S., Fixed point theorems of soft contractive mappings. arXiv-arxiv.org/pdf/1308.4657.

[33] Mushrif M.M., Sengupta S., Ray A.K., Texture classification using a novel, soft-set theory based classification algorithm. Lect. Notes Comput. Sci. 2006; 3851:246–254.

[34] Qin K., Hong Z., On soft equality. J. Comput. Appl. Math. 2010; 234:1347–1355.

[35] Roy A.R., Maji P.K., A fuzzy soft set theoretic approach to decision making problems. J. Comput. Appl. Math. 2007; 203:412–418.

[36] Wardowski D., On a soft mapping and its fixed points, Fixed Point Theory Appl. 2013: Art. 182.

[37] Xiao Z., Gong K., Xia S., Zou Y., Exclusive disjunctive soft sets. Comput. Math. Appl. 2009; 59:2128–2137.

[38] Xiao Z., Gong K., Zou Y., A combined forecasting approach based on fuzzy soft sets. J. Comput. Appl. Math. 2009; 228:326–333.

[39] Xu W., Ma J., Wang S., Hao G., Vague soft sets and their properties. Comput. Math. Appl. 2010; 59:787–794.

[40] Yang C.F., A note on soft set theory. Comput. Math. Appl. 2008; 56:1899-1900.

[41] Yang X., Lin T.Y., Yang J., Li Y., Yu D., Combination of interval-valued fuzzy set and soft set. Comput. Math. Appl. 2009; 58:521–527.

[42] Zadeh L.A., Fuzzy Sets. Inform. control. 1965; 8:103–112.

[43] Zhu P., Wen Q., Operations on soft sets revisited. arXiv:1205.2857v1, 2012.

[44] Zhu P., Wen Q., Probabilistic soft sets, In Proceedings of IEEE International Conference on Granular Computing, 2010: 635–638.

[45] Zou Y., Xiao Z., Data analysis approaches of soft sets under incomplete information. Knowl. Based Syst. 2008; 21:941–945.

Chapter 6

Best Proximity Point Theorems for Cyclic Contractions Mappings

Chirasak Mongkolkeha and Poom Kumam

6.1 Introduction

Many problems in mathematics such as equilibrium and variational inequalities, mathematical economics, game theory and optimization can be formulated as equations of the form $Tx = x$, where T is a self mapping in some suitable framework, because of its ability to solve ordinary differential equations, integral equations, matrix equations and others. However, given nonempty subsets A and B of X. A mapping $T : A \to B$ (or $T : A \cup B \to A \cup B$) using the equation $Tx = x$ does not necessarily have a solution. It is worthwhile to find an approximate solution x under mapping T so that the error $d(x, Tx)$ is minimal. This is the basis of the optimal approximation theory that includes a generalized fixed point. Whenever A coincides with B, the optimization problem known as a best proximity point of the mapping T reduces to a fixed point problem. Fan [7] introduced a classical best approximation theorem: if A is a nonempty compact convex subset of a Hausdorff locally convex topological vector space B and $T : A \to B$ is a continuous mapping, there exists an element $x \in A$ such that $d(Tx, x) = d(Tx, A) := \inf\{d(Tx, y) : y \in A\}$. Several authors, including Prolla [15], Reich [16], Sehgal [20, 21], derived extensions of Fan's theorem in many directions.

Let A and B be nonempty subsets of a metric space (X, d). The following notations and notions will be used in what follows.

$$d(A, B) := \inf\{d(x, y) : x \in A \text{ and } y \in B\},$$

$$A_0 := \{x \in A : d(x, y) = d(A, B) \text{ for some } y \in B\},$$
$$B_0 := \{y \in B : d(x, y) = d(A, B) \text{ for some } x \in A\}.$$

If $A \cap B \neq \emptyset$, then A_0 and B_0 are nonempty. It is interesting that A_0 and B_0 are contained in the boundaries of A and B, respectively, provided A and B are closed subsets of a normed linear space such that $d(A, B) > 0$ (see [18]).

6.2 Best proximity point theorems for cyclic contractions

In 2003, Kirk *et al.* [9] extended Banach's contraction principle to cyclic mappings. This was show that a fixed point solution of such a mapping implies that A must intersect B, and hence this result implies Banach's contraction principle. In 2006, Eldred and Veeramani [4] introduced cyclic contraction.

Definition 6.1 $T : A \cup B \to A \cup B$ is called a *cyclic mapping* if and only if $T(A) \subset B$ and $T(B) \subset A$.

Definition 6.2 [4] $T : A \cup B \to A \cup B$ is called a *cyclic contraction* if T is a cyclic mapping and $k \in [0, 1)$ such that $d(Tx, Ty) \leq kd(x, y) + (1 - k)d(A, B)$ for all $x \in A$ and $y \in B$.

It is easy to see that, if $d(A, B) = 0$, the cyclic contraction mapping is reduced to a contraction mapping.

Definition 6.3 A point $x \in A$ is called a *best proximity point* of T in A if $d(x, Tx) = d(A, B)$.

Lemma 6.1 *[4] Let A and B be nonempty subsets of a complete metric space (X, d) and $T : A \cup B \to A \cup B$ be a cyclic contraction. Then, for any $x_0 \in A \cup B$, $d(x_n, x_{n+1}) \to d(A, B)$ as $n \to \infty$, where $\{x_n\}$ is a sequence in $A \cup B$ defined by $x_{n+1} = Tx_n$ for each $n \geq 0$.*

Proof. For each $n \in \mathbb{N}$, we have

$$\begin{aligned}
d(x_n, x_{n+1}) &\leq kd(x_{n-1}, x_n) + (1 - k)d(A, B) \\
&\leq k(kd(x_{n-2}, x_{n-1}) + (1 - k)d(A, B)) + (1 - k)d(A, B) \\
&= k^2 d(x_{n-2}, x_n) + (1 - k^2)d(A, B).
\end{aligned}$$

Inductively, we have

$$d(x_n, x_{n+1}) \leq k^n d(x_0, x_1) + (1-k)d(A, B)$$

Therefore $d(x_n, x_{n+1}) \to d(A, B)$ as $n \to \infty$.

Lemma 6.2 *[4] Let A and B be nonempty closed subsets of a complete metric space (X, d) and $T : A \cup B \to A \cup B$ be a cyclic contraction. For any $x_0 \in A \cup B$, define a sequence $\{x_n\}$ in $A \cup B$ by $x_{n+1} := Tx_n$ for each $n \geq 0$. If $\{x_{2n}\}$ has a convergence subsequence of $\{x_n\}$ in A, then $x \in A$ such that $d(x, Tx) = d(A, B)$.*

Proof. Let $\{x_{2n_k}\}$ be a subsequence $\{x_{2n}\}$ such that $\lim_{k\to\infty} x_{2n_k} = x$ for some $x \in A$. Since

$$d(A, B) \leq d(x, x_{2n_k-1}) \leq d(x, x_{2n_k}) + d(x_{2n_k}, x_{2n_k-1}) \tag{6.1}$$

for each $k \geq 1$. Letting $k \to \infty$ in (6.1), we obtain $d(x, x_{2n_k-1}) \to d(A, B)$. Thus, by Lemma 6.1, we have $d(A, B) \leq d(x_{2n_k}, Tx) \leq d(x_{2n_k-1}, x)$ for each $k \geq 1$, which implies that $d(x, Tx) = d(A, B)$.

Lemma 6.3 *[4] Let A and B be nonempty subsets of a complete metric space (X, d) and $T : A \cup B \to A \cup B$ be a cyclic contraction. If, for any $x_0 \in A \cup B$, we define a sequence $\{x_n\}$ in $A \cup B$ by $x_{n+1} := Tx_n$ for each $n \geq 0$, the sequences $\{x_{2n}\}$ and $\{x_{2n+1}\}$ are bounded.*

Proof. Let $x_0 \in A$ (the proof when $x_0 \in B$ is similar). Since by Lemma 6.1, $d(x_{2n}, x_{2n+1})$ converges to $d(A, B)$, it is enough to prove that $\{x_2n + 1\}$ is bounded. If $\{x_2n + 1\}$ is not bounded, N_0 such that

$$d(T^2x_0, T^{2N_0+1}x_0) > M \quad \text{and} \quad d(T^2x_0, T^{2N_0-1}x_0) \leq M,$$

where $M > \max\left\{\frac{2d(x_0,Tx_0)}{1/k^2-1} + d(A, B), d(T^2x_0, Tx_0)\right\}$. By the cyclic contraction property of T,

$$\begin{aligned}
\frac{M - d(A, B)}{k^2} + d(A, B) &< d(x_0, T^{2N_0-1}x_0) \\
&\leq d(x_0, T^2x_0) + d(T^2x_0, T^{2N_0-1}x_0) \\
&\leq 2d(x_0, Tx_0) + M.
\end{aligned}$$

Thus, $M\frac{2d(x_0,Tx_0)}{1/k^2-1} + d(A, B)$, which is a contradiction.

Definition 6.4 A subset K of a metric space X is said to be *bounded compact* if every bounded sequence in K has a subsequence converging to a point in K.

Theorem 6.1 *[4] Let A and B be nonempty closed subsets of a complete metric space (X, d) and $T : A \cup B \to A \cup B$ be a cyclic contraction. If either A or B is bounded compact, $x \in A \cup B$ such that $d(x, Tx) = d(A, B)$.*

Proof. Suppose that A is bounded compact. For any $x_0 \in A \cup B$, define a sequence $\{x_n\}$ in $A \cup B$ by $x_{n+1} := Tx_n$ for each $n \geq 0$. Then, by Lemma 6.3, a sequence $\{x_{2n}\}$ is bounded. Since A is bounded compact, there exists a subsequence of $\{x_{2n}\}$ converging to some point in A. Thus, by Lemma 6.2, $x \in A$ such that $d(x, Tx) = d(A, B)$.

Recall that a Banach space X is said to be:

(1) *uniformly convex* if there exists a strictly increasing function $\delta : (0, 2] \to [0, 1]$ such that, for all $x, y, p \in X, R > 0$ and $r \in [0, 2R]$,

$$\left.\begin{array}{l} \|x-p\| \leq R \\ \|y-p\| \leq R \\ \|x-y\| \geq r \end{array}\right\} \implies \left\|\frac{x+y}{2} - p\right\| \leq \left(1 - \delta\left(\frac{r}{R}\right)\right) R;$$

(2) *strictly convex* if, for all $x, y, p \in X$ and $R > 0$,

$$\left.\begin{array}{l} \|x-p\| \leq R \\ \|y-p\| \leq R \\ x \neq y \end{array}\right\} \implies \left\|\frac{x+y}{2} - p\right\| < R.$$

It is well known that every uniformly convex Banach space is strictly convex.

Lemma 6.4 *[4] Let A be a nonempty closed and convex subset of a uniformly convex Banach space X and B be a nonempty closed subset of X. Let $\{x_n\}$ and $\{z_n\}$ be two sequences in A and $\{y_n\}$ be a sequence in B satisfying the following conditions:*

(a) $\|z_n - y_n\| \to d(A, B)$ as $n \to \infty$;

(b) for any $\varepsilon > 0$, there exists a positive integer N_0 such that, for all $m > n \geq N_0$, $\|x_m - y_n\| \leq d(A, B) + \varepsilon$.

Then, for any $\varepsilon > 0$, there exists a positive integer N_1 such that, for all $m > n \geq N_1$, $\|x_m - z_n\| \leq \varepsilon$.

Proof. Assume the contrary, that $\epsilon_0 > 0$ such that for every $k \in \mathbb{N}$, there exists $m_k > n_k \geq k$, for which $\|x_{m_k} - z_{n_k}\| \geq \epsilon_0$. Choose $0 < \gamma < 1$ such that $\frac{\epsilon_0}{\gamma} > d(A, B)$ and ϵ such that $0 < \epsilon < \min\{\frac{\epsilon_0}{\gamma} - d(A, B), \frac{d(A,B), \delta(\gamma)}{1-\delta(\gamma)}\}$. For $\epsilon > 0$ there exists N_0 such that for all $m_k > n_k \geq N_0$, $\|x_{m_k} - y_{n_k}\| \geq d(A, B) + \epsilon_0$. Also, there exists N_2 such that for all $n_k \geq N_2$, $\|z_{n_k} - y_{n_k}\| \geq d(A, B) + \epsilon_0$. Choose $N_1 = \max\{N_0, N_2\}$. By uniform convexity, for all $m_k > n_k \geq N_1$,

$$\left\|\frac{x_{m_k} + z_{n_k}}{2} - y\right\| \leq \left(1 - \delta\left(\frac{\epsilon_0}{d(A, B) + \epsilon}\right)\right)(d(A, B) + \epsilon).$$

Since δ is strictly increasing and by the choice of ϵ, we have

$$\left\| \frac{x_{m_k} + z_{n_k}}{2} - y \right\| \leq d(A, B),$$

for all $m_k > n_k \geq N_1$, which is a contradiction and hence the proof is complete.

In a similar way we can prove the following lemma.

Lemma 6.5 *[4] Let A be a nonempty closed and convex subset of a uniformly convex Banach space X and B be a nonempty closed subset of X. Let $\{x_n\}$ and $\{z_n\}$ be two sequences in A and $\{y_n\}$ be a sequence in B satisfying the following conditions:*

(a) $\|x_n - y_n\| \to d(A, B)$ *as* $n \to \infty$;

(b) $\|z_n - y_n\| \to d(A, B)$ *as* $n \to \infty$.

Then $\|x_n - z_n\| \to 0$.

Theorem 6.2 *[4] Let A and B be nonempty closed and convex subsets of a uniformly convex Banach space X. Suppose that $T : A \cup B \to A \cup B$ is a cyclic contraction. Then there exists a unique best proximity point $x \in A$. Moreover, for any $x_0 \in A$, if we define a sequence $\{x_n\}$ in $A \cup B$ by $x_{n+1} := Tx_n$ for each $n \geq 0$, a subsequence $\{x_{2n}\}$ of $\{x_n\}$ converges to the point x.*

Proof. Suppose that $d(A, B) = 0$, then $A \cap B \neq \emptyset$ and the theorem follows from Banach contraction theorem, as T is a contraction map on $A \cap B$. We assume $d(A, B) > 0$. For any $x_0 \in A \cup B$, define a sequence $\{x_n\}$ in $A \cup B$ by $x_{n+1} := Tx_n$ for each $n \geq 0$. Then, by Lemma 6.1,

$$\|x_{2n} - x_{2n+1}\| \to d(A, B), \quad \|x_{2n+2} - x_{2n+1}\| \to d(A, B)$$

as $n \to \infty$. By Lemma 6.5, we get

$$\|x_{2n} - x_{2n+2}\| \to 0 \tag{6.2}$$

as $n \to \infty$. Similarly, we can prove that $\|x_{2n+1} - x_{2n+3}\| \to 0$ as $n \to \infty$.

Next, we show that, for any $\varepsilon > 0$, there exists a positive integer N_0 such that, for all $m > n \geq N_0$, $\|x_{2m} - x_{2n+1}\| \leq d(A, B) + \varepsilon$. Suppose the contrary. Then $\varepsilon_0 > 0$ such that, for all $k \geq 1$, there exists $m_k > n_k \geq k$ satisfying $\|x_{2m_k} - x_{2n_k+1}\| > d(A, B) + \varepsilon_0$, $\|x_{2(m_k-1)} - x_{2n_k+1}\| \leq d(A, B) + \varepsilon_0$. Observe that

$$\begin{aligned} d(A, B) + \varepsilon_0 &< \|x_{2m_k} - x_{2n_k+1}\| \\ &\leq \|x_{2m_k} - x_{2(m_k-1)}\| + \|x_{2(m_k-1)} - x_{2n_k+1}\| \\ &\leq \|x_{2m_k} - x_{2(m_k-1)}\| + d(A, B) + \varepsilon_0. \end{aligned} \tag{6.3}$$

Thus, by using (6.2) and (6.3), we have $\lim\limits_{k \to \infty} \|x_{2m_k} - x_{2n_k+1}\| = d(A, B) + \varepsilon_0$.

Now, using the triangle inequality and T as a cyclic contraction,

$$\begin{aligned}
&\|x_{2m_k} - x_{2n_k+1}\| \\
&\le \|x_{2m_k} - x_{2m_k+2}\| + \|x_{2m_k+2} - x_{2n_k+3}\| + \|x_{2n_k+3} - x_{2n_k+1}\| \\
&\le \|x_{2m_k} - x_{2m_k+2}\| + \|x_{2m_k+1} - x_{2n_k+2}\| + \|x_{2n_k+3} - x_{2n_k+1}\| \\
&\le \|x_{2m_k} - x_{2m_k+2}\| + \|x_{2n_k+3} - x_{2n_k+1}\| + k^2\|x_{2m_k} - x_{2n_k+1}\| \\
&\quad + (1-k^2)d(A,B).
\end{aligned}$$

Hence

$$d(A,B) + \varepsilon_0 \le k^2(d(A,B) + \varepsilon_0) + (1-k^2)d(A,B) = k^2 d(A,B) + k^2\varepsilon_0 \quad (6.4)$$

which is a contradiction. Therefore, by Lemma 6.4, we conclude that $\{x_{2n}\}$ is a Cauchy sequence and so converges to some $x \in A$. Using Lemma 6.2, it follows that $\|x - Tx\| = d(A,B)$ and hence $T^2x = x$.

Finally, we show that A has a unique best proximity point. Suppose that there exists another $y \in A$ with $y \neq x$ such that $\|y - Ty\| = d(A,B)$ and thus $T^2y = y$. Therefore $\|Tx - y\| = \|Tx - T^2y\| \le \|x - Ty\|$, $\|Ty - x\| = \|Ty - T^2x\| \le \|y - Tx\|$ and so $\|Tx - y\| = \|x - Ty\|$. On the other hand, by the definition of $d(A,B)$, we have $\|Tx - y\| \ge d(A,B)$. If $\|Tx - y\| > d(A,B)$ and

$$\begin{aligned}
\|x - Ty\| &\le k\|Tx - y\| + (1-k)d(A,B) \\
&< k\|Tx - y\| + (1-k)\|Tx - y\| \\
&= \|x - Ty\|,
\end{aligned}$$

which is a contradiction. Hence $\|y - Tx\| = \|x - Tx\|$. Since A is convex and X is strictly convex, it follows that

$$\|x - Tx\| = \left\|\frac{x - Tx}{2} + \frac{y - Tx}{2}\right\| = \left\|\frac{x+y}{2} - Tx\right\| < d(A,B),$$

which is a contradiction. Therefore, $x = y$.

6.3 Best proximity point with property UC

In 2009, Suzuki, Kikkawa and Vetro [22] introduced the property UC which extends useful Lemma 6.5 of Eldred and Veeramani in metric spaces as follows:

Definition 6.5 [22] Let (X,d) be a metric space and let A and B be nonempty subsets of X. Then (A,B) is said to satisfy the *property* (UC) if and only if $\{x_n\}$ and $\{\acute{x}_n\}$ are the sequences in A and $\{y_n\}$ is a sequence in B such that

$$\lim_{n\to\infty} d(x_n, y_n) = d(A,B) \text{ and } \lim_{n\to\infty} d(\acute{x}_n, y_n) = d(A,B),$$

then $\lim_{n\to\infty} d(x_n, \acute{x}_n) = 0$.

Proposition 6.1 *Let A and B be nonempty subsets of a metric space (X,d) such that $d(A,B) = 0$. Then (A,B) satisfies the property UC.*

Proposition 6.2 *Let A and B be nonempty subsets of a uniformly convex Banach space X. Assume that A is convex. Then (A,B) has the property UC.*

Example 6.1 [22] Let c be a positive real number and let f, g and h be functions from $\mathbb{R}$ into $(0,\infty)$ such that

$$f(t) \le 1,\ \ g(t) \le 1,\ \ 0 \le c + f(t)h(t) - h(t) \ \text{ and } \ 0 \le c + g(t)h(t) - h(t).$$

Using $A' = \{a(t) \mid t \in \mathbb{R}\}$, $B' = \{b(t) \mid t \in \mathbb{R}\}$, $X = A' \cup B'$ and define a metric d on X by $d(a(s), a(s')) = |s - s'|$,

$$d(a(s), b(t)) = \begin{cases} c + f(t)h(t) - h(t) + |s-t| & \text{if } s \le t - h(t), \\ c + f(t)|s-t| & \text{if } t - h(t) \le s \le t, \\ c + g(t)|s-t| & \text{if } t \le s \le t + h(t), \\ c + g(t)h(t) - h(t) + |s-t| & \text{if } t + h(t) \le s, \end{cases}$$

$$d(b(t), b(t')) = \begin{cases} 0 & \text{if } t = t', \\ \inf\{d(a(s), b(t)) + d(a(s), b(t'))\ ;\ s \in \mathbb{R}\} & \text{if } t \ne t'. \end{cases}$$

Then A' is complete, $d(A', B') = c$ and (A', B') has the property UC.

Lemma 6.6 *[22] Let (X,d) be a metric space and A and B be nonempty subsets of X. Assume that (A,B) has the property UC. Let $\{x_n\}$ and $\{y_n\}$ be the sequences in A and B, respectively, such that either of the following holds:*

$$\lim_{m\to\infty} \sup_{n\ge m} d(x_m, y_n) = d(A,B)$$

or

$$\lim_{n\to\infty} \sup_{m\ge n} d(x_m, y_n) = d(A,B).$$

Then $\{x_n\}$ is a Cauchy sequence.

Lemma 6.7 *[22] Let (X,d) be a metric space and A and B be nonempty subsets of X. Assume that (A,B) has the property (UC). Let T be a cyclic mapping on $A \cup B$ such that*

$$d(T^2x, Tx) \le d(x, Tx) \ \ \textit{for all } x \in A \cup B \tag{6.5}$$

and

$$d(T^2x, Tx) < d(x, Tx) \text{ for all } x \in A \cup B \text{ with } d(A, B) < d(x, Tx). \quad (6.6)$$

Let $z \in A$. Then the following are equivalent:

(i) z is a best proximity point,

(ii) z is a fixed point of T^2 and Tz is a best proximity point in B.

Lemma 6.8 *[22] Let (X, d) be a metric space and A and B be nonempty subsets of X. Let T be a cyclic mapping on $A \cap B$. Assume that $r \in [0, 1)$such that*

$$d^*(T^2x, Tx) < rd^*(x, Tx) \text{ for all } x \in A \cup B \quad (6.7)$$

where $d^*(x, y) = d(x, y) - d(A, B)$ for all $(x, y) \in X \times X$. Then (6.5) and (6.6) hold.

Theorem 6.3 *Let (X, d) be a metric space and A and B be nonempty subsets of X such that (A, B) satisfies the property UC. Assume that A is complete. Let T be a cyclic mapping on $A \cap B$. Assume that $r \in [0, 1)$ such that*

$$d(Ty, Tx) < r \max\{d(x, y), d(x, Tx), d(Ty, y)\} + (1 - r)d(A, B) \quad (6.8)$$

for all $x \in A$ and $y \in B$. Then the following hold:

(i) T has a unique best proximity point z in A.

(ii) z is a unique fixed point of T^2 in A.

(iii) $\{T^{2n}x\}$ converges to z for every $x \in A$.

(iv) T has at least one best proximity point in B.

(v) If (B, A) satisfies the property UC; then Tz is a unique best proximity point in B and $\{T^{2n}y\}$ converges to Tz for every $y \in B$.

Proof. For each $x \in A$ and $y \in B$, we can write

$$d^*(Ty, Tx) < r \max\{d^*(x, y), d^*(x, Tx), d^*(Ty, y)\} + (1 - r)d(A, B).$$

So we have

$$d^*(T^2x, Tx) < rd^*(x, Tx) \text{ and } d^*(T^2y, Ty) < rd^*(y, Ty)$$

for $x \in A$ and $y \in B$ and (6.7) holds. We note that (6.5) and (6.6) hold by Lemma 6.8. Fix $x \in A$. We have

$$\lim_{n\to\infty} d^*(T^{2n}x, T^{2n+1}x)) \leq \lim_{n\to\infty} r^{2n}d^*(x, Tx) = 0,$$

and

$$\lim_{n\to\infty} d^*(T^{2n+2}x, T^{2n+1}x)) \leq \lim_{n\to\infty} r^{2n} d^*(T^2x, Tx) = 0.$$

From the property UC, $\lim_{m\to\infty} d^*(T^{2m}x, T^{2m+2}x)) = 0$ holds. Fix $\varepsilon > 0$ and choose $L \in \mathbb{N}$ such that

$$d(T^{2m}x, T^{2m+2}x)) < (1-r)\varepsilon, \ \ d^*(T^{2m}x, T^{2m+1}x)) < \varepsilon$$

$$\text{and} \ \ d^*(T^{2m+2}x, T^{2m+1}x)) < \varepsilon,$$

for all $m \geq L$. Fix $m \in \mathbb{N}$ with $m \geq L$. We shall show by induction

$$d^*(T^{2n}x, T^{2n+1}x)) \leq \varepsilon \tag{6.9}$$

for all $n \in \mathbb{N}$ with $n \geq m$. In case $n = m$, it is obvious that (6.14) holds. We assume that (6.14) holds for some $n \geq m$. Then we have

$$\begin{aligned} d^*(T^{2m}x, T^{2n+3}x)) &\leq d^*(T^{2m}x, T^{2m+2}x)) + d^*(T^{2m+2}x, T^{2n+3}x)) \\ &\leq (1-r)\varepsilon + r\max\{d^*(T^{2n+2}x, T^{2m+1}x)), \\ &\qquad d^*(T^{2n+2}x, T^{2n+3}x)), d^*(T^{2m+2}x, T^{2m+1}x))\} \\ &\leq (1-r)\varepsilon + r\max\{rd^*(T^{2n+1}x, T^{2m}x)), r\varepsilon, \varepsilon\} \\ &\leq (1-r)\varepsilon + r\max\{r\varepsilon, \varepsilon\} \\ &\leq \varepsilon \end{aligned} \tag{6.10}$$

and hence (6.14) holds when $n := n+1$ and by induction (6.14) also holds for all $n \in \mathbb{N}$ with $n \geq m$. Thus we obtain

$$\lim_{n\to\infty} \sup_{m\geq n} d^*(T^{2m}x, T^{2n+1}x)) = 0.$$

By Lemma 6.6, $\{T^{2n}x\}$ is a Cauchy sequence in A. Since A is complete, $\{T^{2n}x\}$ converges to some point $z \in A$. We have

$$\begin{aligned} d^*(z, Tz) &= \lim_{n\to\infty} d^*(T^{2n}x, Tz)) \\ &\leq \lim_{n\to\infty} r\max\{d^*(T^{2n-1}x, z)), d^*(T^{2n-1}x, T^{2n}x)), d^*(z, Tz))\} \\ &\leq \lim_{n\to\infty} r\max\{d^*(T^{2n-1}x, z)), r^{2n-2}d^*(Tx, T^2x)), d^*(z, Tz))\} \\ &\leq rd^*(z, Tz) \end{aligned} \tag{6.11}$$

which implies $d^*(z, Tz) = 0$ and thus z is a best proximity point of T.

Next, to show (i), we let z' be a best proximity point in A. By Lemma 6.7, we note that z and z' are fixed points of T^2, and Tz and Tz' are best proximity points in B. So we have

$$\begin{aligned} d^*(z, Tz') &= d^*(T^2 z, Tz') \\ &\le r \max\{d^*(Tz, z'), d^*(T^2 z, Tz), d^*(z', Tz'))\} \\ &= rd^*(z', Tz))\} \\ &= rd^*(T^2 z', Tz)) \\ &\le r^2 \max\{d^*(z, Tz'), d^*(z, Tz), d^*(T^2 z', Tz'))\} \\ &= r^2 d^*(z, Tz')) \end{aligned} \tag{6.12}$$

and hence $d^*(z, Tz') = 0$. It follows from $d^*(z, Tz') = d^*(z', Tz') = 0$ and the property UC implication that $z = z'$, we obtain (i). Also, we have shown (ii) through (iv).

Let us prove (v). We assume that (B, A) satisfies the property UC. Then by the same argument as above, we can prove that T has at most one proximity point in B since Tz is the unique proximity point in B. Fix $y \in B$, then we have

$$\lim_{n\to\infty} d^*(T^{2n}y, T^{2n+1}y)) \le \lim_{n\to\infty} r^{2n} d^*(y, Ty) = 0,$$

and

$$\begin{aligned} d^*(T^{2n+1}y, Tz) &\le \lim_{n\to\infty} r \max\{d^*(T^{2n}y, z)), d^*(T^{2n+1}y, T^{2n}y)), d^*(z, Tz))\} \\ &\le \lim_{n\to\infty} r \max\{d^*(T^{2n}y, T^{2n+1}y)) + d(T^{2n+1}y, z))\} \\ &= 0. \end{aligned} \tag{6.13}$$

By the property UC, we have $\lim_{n\to\infty} d^*(T^{2n}y, Tz)) = 0$. That is, $\{T^{2n}y\}$ converges to Tz.

Corollary 6.1 *Let (X, d) be a metric space and A and B be nonempty subsets of X such that (A, B) satisfies the property UC. Assume that A is complete. Let T be a cyclic contraction mapping on $A \cap B$. Then T has a unique best proximity point z in A and $\{T^{2n}x\}$ converges to z for every $x \in A$.*

Remark 6.1 It is interesting to compare Corollary 6.1 with Theorem 6.2. It is obvious that Corollary 6.1 is one of generalizations of the Banach contraction principle.

In 2011, Abka and Gabeleh [1] proved best proximity point theorems in partially ordered metric spaces with the property UC as follows:

$(\mathcal{A})$: if a nondecreasing sequence $\{x_n\}$ in X, $x_n \to x \in X$ as $n \to \infty$, then $x_n \preceq x$ for all $n \in \mathbb{N}$

Theorem 6.4 *Let A and B be nonempty closed subsets of a partially ordered metric space $(X, \preceq)$ and d be a metric on X. Let A and B be nonempty closed* *subsets of X such that (A, B) satisfies the property UC, and A is complete. Let the condition $(\mathcal{A})$ hold, and let $T : A \cup B \to A \cup B$ be a cyclic mapping such that T and T^2 are nondecreasing on A such that*

$$d(T\acute{x}, T^2x) \leq \alpha d(\acute{x}, Tx) + (1-\alpha)d(A, B), \tag{6.14}$$

and

$$d(T\acute{y}, T^2y) \leq \alpha d(\acute{y}, Ty) + (1-\alpha)d(A, B) \tag{6.15}$$

for some $\alpha \in [0, 1)$ and for all $(x, \acute{x}) \in A \times A$, $(y, \acute{y}) \in B \times B$ with $x \preceq \acute{x}$, $y \preceq \acute{y}$. Assume that $x_0 \in A$ with $x_0 \preceq T^2x_0$ and define $x_{n+1} = Tx_n$ for all $n \geq 1$. If $T|_A$ is continuous and $\{x_{2n}\}$ has convergent subsequence in A, then T has the best proximity point $p \in A$.

In 2012, Motivated by (6.14) and (6.15), Mongkolkeha and Kumam [13] proved new best proximity point theorems in partially ordered metric space as follows:

Theorem 6.5 *Let A and B be nonempty closed subsets of a partially ordered metric space $(X, \preceq)$ and d be a metric on X. Let $T : A \cup B \to A \cup B$ be a cyclic mapping such that T and T^2 are nondecreasing on A such that*

$$d(T\acute{x}, T^2x) \leq \alpha d(\acute{x}, Tx) + \beta d(x, T\acute{x}) + (1-\alpha-\beta)d(A, B),$$

and

$$d(T\acute{y}, T^2y) \leq \alpha d(\acute{y}, Ty) + \beta d(y, T\acute{y}) + (1-\alpha-\beta)d(A, B)$$

for some $\alpha, \beta \in [0, 1)$ with $\alpha + \beta < 1$ and for all $(x, \acute{x}) \in A \times A$, $(y, \acute{y}) \in B \times B$ with $x \preceq \acute{x}$, $y \preceq \acute{y}$. Assume that $x_0 \in A$ with $x_0 \preceq T^2x_0$ and define $x_{n+1} = Tx_n$ for all $n \geq 1$. If $T|_A$ is continuous and $\{x_{2n}\}$ has a convergent subsequence in A, T has best proximity point $p \in A$.

Proof. Let $\{x_{2n_k}\}$ be a subsequence of $\{x_{2n}\}$ converging to some $p \in A$. By the continuity of T, we get $x_{2n_k+1} = Tx_{2n_k} \to Tp$ as $n \to \infty$. Since T and T^2 are nondecreasing on A and $x_0 \preceq T^2x_0$, it follows that $\{T^{2n}x_0\}$ and $\{T^{2n-1}x_0\}$ are nondecreasing. Indeed,

$$\begin{aligned}
d(A, B) &\leq d(x_{2n_k}, x_{2n_k+1}) \\
&< \alpha d(Tx_{2n_k-2}, T^2x_{2n_k-2}) + \beta d(Tx_{2n_k-2}, T^2x_{2n_k-2}) \\
&\qquad + (1-\alpha-\beta)d(A, B) \\
&\leq \alpha^2 d(x_{2n_k-2}, Tx_{2n_k-2}) + 2\alpha\beta d(x_{2n_k-2}, Tx_{2n_k-2}) \\
&\qquad + \beta^2 d(x_{2n_k-2}, Tx_{2n_k-2}) + (1-\alpha^2-2\alpha\beta-\beta^2)d(A, B) \\
&\leq \alpha^3 d(x_{2n_k-3}, Tx_{2n_k-3}) + 3\alpha^2\beta d(x_{2n_k-3}, Tx_{2n_k-3}) \\
&\qquad + 3\alpha\beta^2 d(x_{2n_k-3}, Tx_{2n_k-3}) + \beta^3 d(x_{2n_k-3}, Tx_{2n_k-3}) \\
&\qquad\quad + (1-\alpha^3-3\alpha^2\beta-3\alpha\beta^2-\beta^3)d(A, B) \\
&\ \ \vdots
\end{aligned}$$

$$\leq \alpha^{2n_k} d(x_0, Tx_0) + \binom{2n_k}{1}\alpha^{2n_k-1}\beta d(x_0, Tx_0) + \cdots$$
$$+ \beta^{2n_k} d(x_0, Tx_0)\left(1 - \alpha^{2n_k} - \binom{2n_k}{1}\alpha^{2n_k-1}\beta - \cdots\right.$$
$$\left. - \binom{2n_k}{2n_k-1}\alpha\beta^{2n_k-1} - \beta^{2n_k}\right) d(A, B).$$

Taking $k \to \infty$ in the above equality, we obtain

$$d(p, Tp) = \lim_{n\to\infty} d(x_{2n_k}, Tx_{2n_k}) = d(A, B).$$

Corollary 6.2 *Let A and B be nonempty closed subsets of a partially ordered metric space $(X, \preceq)$ and d be a metric on X. Let $T : A \cup B \to A \cup B$ be a cyclic mapping such that T and T^2 are nondecreasing on A such that*

$$d(T\acute{x}, T^2x) \leq \alpha d(\acute{x}, Tx) + (1 - \alpha)d(A, B),$$

and

$$d(T\acute{y}, T^2y) \leq \alpha d(\acute{y}, Ty) + (1 - \alpha)d(A, B)$$

for some $\alpha \in [0, 1)$ and for all $(x, \acute{x}) \in A \times A$, $(y, \acute{y}) \in B \times B$ with $x \preceq \acute{x}$, $y \preceq \acute{y}$. Assume that $x_0 \in A$ with $x_0 \preceq T^2x_0$ and define $x_{n+1} = Tx_n$ for all $n \geq 1$. If $T|_A$ is continuous and $\{x_{2n}\}$ has convergent subsequence in A, T has best proximity point $p \in A$.

Corollary 6.3 *Let A and B be nonempty closed subsets of a partially ordered metric space $(X, \preceq)$ and d be a metric on X. Let $T : A \cup B \to A \cup B$ be a cyclic mapping such that T and T^2 are nondecreasing on A such that*

$$d(T\acute{x}, T^2x) \leq \beta d(x, T\acute{x}) + (1 - \beta)d(A, B),$$

and

$$d(T\acute{y}, T^2y) \leq \beta d(y, T\acute{y}) + (1 - \beta)d(A, B)$$

for some $\beta \in [0, 1)$ and for all $(x, \acute{x}) \in A \times A$, $(y, \acute{y}) \in B \times B$ with $x \preceq \acute{x}$, $y \preceq \acute{y}$. Assume that $x_0 \in A$ with $x_0 \preceq T^2x_0$ and define $x_{n+1} = Tx_n$ for all $n \geq 1$. If $T|_A$ is continuous and $\{x_{2n}\}$ has a convergent subsequence in A, T has best proximity point $p \in A$.

Lemma 6.9 *Let A and B be nonempty subsets of a partially ordered metric space $(X, \preceq)$ and d be a metric on X. Let $T : A \cup B \to A \cup B$ be a cyclic mapping such that*

$$d(T\acute{x}, T^2x) \leq \alpha d(\acute{x}, Tx) + \beta d(x, T\acute{x}) + (1 - \alpha - \beta)d(A, B), \tag{6.16}$$

and

$$d(T\acute{y}, T^2y) \leq \alpha d(\acute{y}, Ty) + \beta d(y, T\acute{y}) + (1 - \alpha - \beta)d(A, B) \tag{6.17}$$

for some $\alpha, \beta \in [0, 1)$ with $\alpha + \beta < 1$ and for all $(x, \acute{x}) \in A \times A$, $(y, \acute{y}) \in B \times B$ with $x \preceq \acute{x}$, $y \preceq \acute{y}$. Then

$$d^*(T\acute{x}, T^2x) \leq \alpha d^*(\acute{x}, Tx) + \beta d^*(x, T\acute{x})$$

and

$$d^*(T\acute{y}, T^2y) \leq \alpha d^*(\acute{y}, Ty) + \beta d^*(y, T\acute{y}),$$

where $d^*(a,b) = d(a,b) - d(A,B)$ *for* $(a,b) \in A \times B$.

Proof. By the definition of d^* and (6.16), we have

$$\begin{aligned} d^*(T\acute{x}, T^2x) &= d(T\acute{x}, T^2x) - d(A,B) \\ &\leq \alpha d(\acute{x}, Tx) + \beta d(x, T\acute{x}) + (1-\alpha-\beta)d(A,B) - d(A,B) \\ &= \alpha d(\acute{x}, Tx) + \beta d(x, T\acute{x}) - \alpha d(A,B) - \beta d(A,B) \\ &= \alpha\big(d(\acute{x}, Tx) - d(A,B)\big) + \beta\big(d(x, T\acute{x}) - d(A,B)\big) \\ &= \alpha d^*(\acute{x}, Tx) + \beta d^*(x, T\acute{x}). \end{aligned} \tag{6.18}$$

Similarly, we see that $d^*(T\acute{y}, T^2y) \leq \alpha d^*(\acute{y}, Ty) + \beta d^*(y, T\acute{y})$.

Theorem 6.6 *Let* $(X, \preceq)$ *be a partially ordered set and* d *be a metric on* X. *Let* A *and* B *be two nonempty subsets of* X *such that* (A,B) *satisfies the property UC and* A *is complete. Let* $T : A \cup B \to A \cup B$ *be a cyclic mapping such that* T *and* T^2 *are nondecreasing on* A. *Suppose that*

$$d(T\acute{x}, T^2x) \leq \alpha d(\acute{x}, Tx) + \beta d(x, T\acute{x}) + (1-\alpha-\beta)d(A,B),$$

and

$$d(T\acute{y}, T^2y) \leq \alpha d(\acute{y}, Ty) + \beta d(y, T\acute{y}) + (1-\alpha-\beta)d(A,B)$$

for some $\alpha, \beta \in [0,1)$ *with* $\alpha + \beta < 1$ *and for all* $(x, \acute{x}) \in A \times A$, $(y, \acute{y}) \in B \times B$ *with* $x \preceq \acute{x}$, $y \preceq \acute{y}$. *If* $T|_A$ *is continuous and that* $x_0 \in A$ *such that* $x_0 \preceq T^2x_0$ *and* $x_{n+1} = Tx_n$ *for all* $n \geq 1$, T *has a best proximity point* $p \in A$ *and* $x_{2n} \to p$.

Proof. Since T and T^2 are nondecreasing on A and $x_0 \preceq T^2x_0$, it follows that $\{T^{2n}x_0\}$ and $\{T^{2n-1}x_0\}$ are nondecreasing. Let $n \geq 1$ with $n \geq m$. By Lemma 6.9, we have

$$\begin{aligned} &d^*(T^{2m}x_0, T^{2n+1}x_0) \\ &\quad = d^*(T(T^{2n}x_0), T^2(T^{2m-2}x_0)) \\ &\quad \leq \alpha d^*(T(T^{2n-1}x_0), T^2(T^{2m-3}x_0)) + \beta d^*(T(T^{2n}x_0), T^2(T^{2m-4}x_0)) \\ &\quad \leq \alpha^2 d^*(T(T^{2n-2}x_0), T^2(T^{2m-4}x_0)) \\ &\qquad + 2\alpha\beta d^*(T(T^{2n-1}x_0), T^2(T^{2m-5}x_0)) + \beta^2 d^*(T(T^{2n}x_0), T^2(T^{2m-6}x_0)) \\ &\quad \leq \alpha^3 d^*(T^{2m-3}x_0, T^{2n-2}x_0) + 3\alpha^2\beta d^*(T^{2m-4}x_0, T^{2n-1}x_0) \\ &\qquad + 3\alpha\beta^2 d^*(T^{2m-5}x_0, T^{2n}x_0) + \beta^3 d^*(T^{2m-6}x_0, T^{2n+1}x_0) \end{aligned}$$

$$\vdots$$

$$\leq \beta^m d^*(x_0, T^{2n+1}x_0) + \binom{m}{1}\alpha\beta^{m-1}d^*(Tx_0, T^{2n}x_0) + \binom{m}{2}\alpha^2\beta^{m-2}d^*(T^2x_0, T^{2n-2}x_0) + \cdots + \alpha^m d^*(T^{2m-m}x_0, T^{(2n+1)-(m)}x_0).$$

Since $\alpha, \beta \in [0, 1)$, it follows from the above inequality that

$$\lim_{m\to\infty} \sup_{n\geq m} d^*(T^{2m}x_0, T^{2n+1}x_0) = 0. \tag{6.19}$$

Since (A, B) satisfies the property UC, it follows from Lemma 6.6 that $\{x_{2n}\}$ is a Cauchy sequence and since A is complete, $p \in A$ such that

$$T^{2n}x_0 = x_{2n} \to p.$$

By the continuity of T on A, we get $T^{2n+1}x_0 = T(T^{2n}x_0) \to Tp$ as $n \to \infty$. Since $\{T^{2n}x_0\}$ and $\{T^{2n-1}x_0\}$ are nondecreasing, we have

$$\begin{aligned}
d(A, B) &\leq d(T(T^{2n-1}x_0), T^2(T^{2n-1}x_0)) \\
&\leq \alpha d(T^{2n-1}x_0, T^{2n}x_0) + \beta d(T^{2n-1}x_0, T^{2n}x_0) + (1 - \alpha - \beta)d(A, B) \\
&\leq \alpha^2 d(T^{2n-2}x_0, T(T^{2n-2}x_0)) + 2\alpha\beta d(T^{2n-2}x_0, T(T^{2n-2}x_0)) \\
&\quad + \beta^2 d(T^{2n-2}x_0, T(T^{2n-2}x_0)) + (1 - \alpha^2 - 2\alpha\beta - \beta^2)d(A, B) \\
&\leq \alpha^3 d(T^{2n-3}x_0, T(T^{2n-3}x_0)) + 3\alpha^2\beta d(T^{2n-3}x_0, T(T^{2n-3}x_0)) \\
&\quad + 3\alpha\beta^2 d(T^{2n-3}x_0, T(T^{2n-3}x_0)) + \beta^3 d(T^{2n-3}x_0, T(T^{2n-3}x_0)) \\
&\quad + (1 - \alpha^3 - 3\alpha^2\beta - 3\alpha\beta^2 - \beta^3)d(A, B) \\
&\;\;\vdots \\
&\leq \alpha^{2n} d(x_0, Tx_0) + \binom{2n}{1}\alpha^{2n-1}\beta d(x_0, Tx_0) + \cdots \\
&\quad + \binom{2n}{2n-1}\alpha\beta^{2n-1}d(x_0, Tx_0) + \beta^{2n}d(x_0, Tx_0) + \Big(1 - \alpha^{2n} - \\
&\quad \binom{2n}{1}\alpha^{2n-1}\beta - \cdots - \binom{2n}{2n-1}\alpha\beta^{2n-1} - \beta^{2n}\Big)\, d(A, B).
\end{aligned}$$

Since $\alpha, \beta \in [0, 1)$, letting $n \to \infty$ in the above inequality, we obtain

$$d(p, Tp) = \lim_{n\to\infty} d((T^{2n}x_0, T^{2n+1}x_0) = d(A, B).$$

This completes the proof.

Corollary 6.4 *Let $(X, \preceq)$ be a partially ordered set and d be a metric on X. A and B are two nonempty subsets of X satisfying the property UC, and A is complete. Let $T : A \cup B \to A \cup B$ be a cyclic mapping such that T and T^2 are nondecreasing on A. Suppose that*

$$d(T\acute{x}, T^2x) \leq \alpha d(\acute{x}, Tx) + (1-\alpha)d(A, B),$$

and

$$d(T\acute{y}, T^2y) \leq \alpha d(\acute{y}, Ty) + (1-\alpha)d(A, B)$$

for some $\alpha \in [0, 1)$ and for all $(x, \acute{x}) \in A \times A$, $(y, \acute{y}) \in B \times B$ with $x \preceq \acute{x}$, $y \preceq \acute{y}$. If $T|_A$ is continuous and $x_0 \in A$ such that $x_0 \preceq T^2x_0$ and $x_{n+1} = Tx_n$ for all $n \geq 1$. T has best proximity point $p \in A$ and $x_{2n} \to p$.

Corollary 6.5 *Let $(X, \preceq)$ be a partially ordered set and d be a metric on X. Let A and B be two nonempty subsets of X such that (A, B) satisfies the property UC, and A is complete. Let $T : A \cup B \to A \cup B$ be a cyclic mapping such that T and T^2 are nondecreasing on A. Suppose that*

$$d(T\acute{x}, T^2x) \leq \beta d(x, T\acute{x}) + (1-\beta)d(A, B),$$

and

$$d(T\acute{y}, T^2y) \leq \beta d(y, T\acute{y}) + (1-\beta)d(A, B)$$

for some $\beta \in [0, 1)$ and for all $(x, \acute{x}) \in A \times A$, $(y, \acute{y}) \in B \times B$ with $x \preceq \acute{x}$, $y \preceq \acute{y}$. If T is continuous and $x_0 \in A$ such that $x_0 \preceq T^2x_0$ and $x_{n+1} = Tx_n$ for all $n \geq 1$, T has best proximity point $p \in A$ and $x_{2n} \to p$.

Now, we give an example to illustrate Theorem 6.6.

Example 6.2 Consider $X = \mathbb{R}^2$ with the usual metric and define the partial order $\preceq$ on $\mathbb{R}^2$ in the following way:

$$(x_1, y_1) \preceq (x_2, y_2) \Longleftrightarrow x_1 \leq x_2,\ y_1 \leq y_2$$

for all $(x_1, y_1), (x_2, y_2) \in \mathbb{R}^2$. Set $A = \{(1, a) : a \geq 0\}$, $B = \{(-1, b) : b \geq 0\}$ and define a mapping $T : A \cup B \to A \cup B$ by

$$T(1, a) = (-1, \tfrac{a}{2}),\ T(-1, b) = (1, \tfrac{b}{2})$$

for all $a, b \geq 0$. Then $d(A, B) = 2$. We show that A and B satisfy the property UC. Let $\{(1, a_n)\}, \{(1, a'_n)\}$ be two sequences in A and $\{(1, b_n)\}$ be a sequence in B such that

$$\lim_{n\to\infty} d((1, a_n), (-1, b_n)) = 2,\ \lim_{n\to\infty} d((1, a'_n), (-1, b_n)) = 2.$$

Thus $\lim_{n\to\infty} |a_n - b_n| = 0$ and $\lim_{n\to\infty} |a'_n - b_n| = 0$. Since

$$\lim_{n\to\infty} |a_n - a'_n| \leq \lim_{n\to\infty} |a_n - b_n| + \lim_{n\to\infty} |b_n - a'_n| = 0,$$

$$\lim_{n\to\infty} d((1, a_n), (1, a'_n)) = 0 \Longleftrightarrow \lim_{n\to\infty} |a_n - a'_n| = 0.$$

Hence A and B satisfy the property UC. Simple computations show that T satisfies the conditions of Theorem 6.6 for $\alpha = 2/3$, $\beta = 1/4$. Since $x_0 := (1, 0) \in A$, if we define $x_{n+1} = Tx_n$ for all $n \geq 1$, then $x_0 \preceq T^2 x_0$ and $\{x_{2n}\}$, $\{x_{2n-1}\}$ are nondecreasing. Therefore, T has a best proximity point. Clearly, this point is x_0 itself.

6.4 Best proximity point with proximally complete property

In 2011, Sankar Raj [19] introduced P-property to prove best proximity point results for various classes of non-self mappings as follows.

Definition 6.6 Let (A, B) be a pair of nonempty subsets of X with $A_0 \neq \emptyset$. The pair is said to have the *P-property* if and only if

$$\left.\begin{array}{l} d(x_1, y_1) = d(A, B) \\ d(x_2, y_2) = d(A, B) \end{array}\right\} \quad \Longrightarrow \quad d(x_1, x_2) = (y_1, y_2),$$

where $x_1, x_2 \in A_0$ and $y_1, y_2 \in B_0$.

Later, Abkar and Gabeleh [2] show that results with the P-property concerning best proximity points can be obtained from the same results in fixed point theory.

In 2010, Kosuru and Veeramani [10] (see also [11]) introduced the semi-sharp proximinal concept described below.

Definition 6.7 [10, 11] A pair (A, B) of nonempty subsets of metric space (X, d) is said to be sharp proximinal if and only if, for each x in A and y in B, there exist a unique x' in B and y' in A such that

$$d(x, x') = d(y', y) = d(A, B).$$

Definition 6.8 [10, 11] A pair (A, B) of nonempty subsets of metric space (X, d) is said to be semi-sharp proximinal if and only if, for each x in A and y in B, there exists at most one point x' in B and one point y' in A such that

$$d(x, x') = d(y', y) = d(A, B).$$

We can see immediately that property UC implies semi-sharp proximality. Moreover, it is easy to see that if (A, B) has the P-property, then both (A, B) and (B, A) are semi-sharp proximinal. Obviously a semi-sharp proximinal pair (A, B) may not necessarily have the P-property. Some examples of this concept follow.

Example 6.3 [6] If A and B are closed and convex subsets of a strictly convex Banach space, the pair (A, B) has a semi-sharp proximinal property.

Example 6.4 [3] Let $X = \mathbb{R}^{k}$ with the norm defined by $\|(x, y)\| = |x| + |y|$ for all $(x, y) \in X$. Let $A = \{(x, y) : 0 \leq x \leq 1, y = 0\}$, $B_1 = \{(x, y) : 0 \leq -x + y = 1, -1 \leq x \leq 0\}$ $B_2 = \{(x, y) : 0 \leq x \leq 1, y = 0\}$ and $B = B_1 \cup B_2$. Then the pair (A, B) have semi-sharp proximinal property.

In 2015, Espínola, Kosuru and Veeramani [6] introduced the concept of proximally complete.

Definition 6.9 [6] Let X be a metric space, and A and B nonempty subsets of X. A sequence $\{x_n\}$ in $A \cup B$, with $x_{2n} \in A$ and $x_{2n+1} \in B$ for $n \geq 0$, is said to be a cyclical Cauchy sequence iff for every $\epsilon > 0$ there exists an $N \in \mathbb{N}$ such that $d(x_n, x_m) < d(A, B) + \varepsilon$ whenever n is even, m is odd and $n, m \geq N$.

Note that if $d(A, B) = 0$, cyclical Cauchy sequences in $A \cup B$ are Cauchy sequences.

Example 6.5 [6] Let $X = (\mathbb{R}^3, \|\cdot\|_2)$, $A = \{(x, y, z) \in X : -1 \leq x \leq 1, y^2 + z^2 = 1\}$ and $B = \{(0, 0, 0\}$. For $n \in \mathbb{N}$, define $x_{2n+1} = (0, 0, 0)$ and

$$x_{2n} \begin{cases} (\frac{-1}{n}, 1, 0) & \text{if } n \text{ is odd;} \\ (\frac{1}{n}, -1, 0) & \text{if } n \text{ is even.} \end{cases}$$

Then $\{x_n\}$ is cyclically Cauchy, while $\{x_{2n}\}$ has $(0, 1, 0)$ and $(0, -1, 0)$ as accumulation points.

Remark 6.2 In Example 6.5, we see that

1. The cyclical Cauchy sequences may have more than one accumulation point.
2. The pair (A, B) does not have property UC even though A and B are both compact subsets of a uniformly convex Banach space.

Lemma 6.10 *[6] Every cyclical Cauchy sequence is bounded.*

Definition 6.10 [6] Let X be a metric space. A pair (A, B) of subsets of a metric space X is proximally complete iff, for every cyclically Cauchy sequence $\{x_n\}$ in $A \cup B$, the sequences $\{x_{2n}\}$ and $\{x_{2n+1}\}$ have convergent subsequences in A and B, respectively.

Example 6.6 Recalling the result in [6] and a trivial example, we can see that

1. If A and B are closed subsets of a complete metric space with $dist(A, B) = 0$, then (A, B) is a proximally complete pair.

2. Every boundedly compact pair of subsets of a metric space is proximally complete.

3. Any nonempty, closed, and convex pair (A, B) in a uniformly convex Banach space is proximally complete.

Example 6.7 [6] Let (X, d) be a metric space and Y be a complete subspace of X. For $x \neq y$ in X, define $A = Y \times \{x\}$ and $B = Y \times \{y\}$. The pair (A, B) is proximally complete in the metric space $(X \times X, d_1)$ with

$$d_1((u_1, v_1), (u_2, v_2)) = d(u_1, u_2) + d(v_1, v_2).$$

This applies, for instance, where X stands for the set of all complex valued continuous functions on $[0, 1]$ with norm

$$\|f_1 + f_2 i\| = \int_0^1 |f_1(t)|dt + \int_0^1 |f_2(t)|dt.$$

Let $A := \sin nt : n \in \mathbb{N}$ and $B := \sin nt + i : n \in \mathbb{N}$. Notice that X is not complete but that A and B are complete and the pair (A, B) is proximally complete.

Now, we present the best proximity points of Espínola, Kosuru and Veeramani [6] for cyclic contractions under proximally complete conditions.

Theorem 6.7 *Let (A, B) be a proximally complete pair semi-sharp proximinal in a metric space X. If $\{x_n\}$ is a cyclical Cauchy sequence in $A \cup B$, then x_{2n} tends to $x \in A$, and x_{2n+1} tends to $y \in B$. Moreover, $d(x, y) = (A, B)$.*

Theorem 6.8 *Let (A, B) be a proximally complete pair in a metric space X. If T is a cyclic contraction mapping on $A \cup B$, then $(x, y) \in A \times B$ such that $d(x, Tx) = (A, B)$ and $d(y, Ty) = (A, B)$ with $d(x, y) = (A, B)$.*

Theorem 6.9 *Let (A, B) be a proximally complete semi-sharp proximinal pair in a metric space (X, d). Suppose T is a cyclic contraction on $A \cup B$; then the following holds:*

(i) There exists a a unique point in x of T in A.

(ii) The sequence $\{T^{2n}x_0\}$ converges to x and the sequence $\{T^{2n+1}x_0\}$ converges to Tx, for any $x_0 \in A$.

(iii) Tx is a unique point in B such that $d(x, Tx) = d(A, B)$.

(iv) x and Tx are the only fixed points of T^2 in A and B, respectively.

Proof. By Theorem 6.8, $x \in A$ such that $d(x, Tx) = (A, B)$. Notice that

$$d(A, B) \leq d(Tx, T^2x) \leq d(x, Tx) = d(A, B).$$

Since (A, B) is a semi-sharp proximinal pair, $T^2x = x$. Suppose that $z \neq x \in A$ is such that $d(z, Tz) = (A, B)$. In a similar fashion, $T^2z = z$, and therefore

$$d(Tz, x) = d(Tz, T^2x) \leq d(z, Tx) = d(T^2z, Tx) < d(Tz, x),$$

which is a contradiction, and uniqueness is proved.

The proof of convergence of iterates now follows. Fix $x_0 \in A$ and define $x_n = Tx_{n+1}$ for $n \in \mathbb{N}$. It has been proved in Theorem 6.8 that $\{x_n\}$ is a cyclical Cauchy sequence and hence, by Theorem 6.8, $\{x_{2n}\}$ and $\{x_{2n+1}\}$ are convergent sequences. Moreover, from the claim in the proof of Theorem 6.8, $\{x_{2n}\}$ converges to the best proximity point x in A and $\{x_{2n+1}\}$ converges to the best proximity point y in B, with $d(x, y) = d(A, B)$. The semi-sharp proximinality of (A, B) yields $y = Tx$. Let us see now that x is the unique fixed point of T^2 in A. It is known that $T^2x = x$. If $y \in A$ satisfies $T^2y = y$, $T^{2n}y = y$ for all $n \in \mathbb{N}$. Therefore, $T^{2n}y \to x$ and $y = x$. In a similar way, one can show that Tx is a unique fixed point of T^2 in B.

6.5 Common best proximity points for proximity commuting mappings

In 2012, Sadiq Basha [17] extended the notion of commuting mapping for the common best proximity point theorem and introduced some new concepts:

Definition 6.11 Let $S : A \to B$ and $T : A \to B$. An element $x^* \in A$ is said to be a common best proximity point if it satisfies

$$d(x^*, Sx^*) = d(x^*, Tx^*) = d(A, B).$$

Observe that a common best proximity point is an element at which the multi-objective functions $x \to d(x, Sx)$ and $x \to d(x, Tx)$ attain common global minimum, since $d(x, Sx) \geq d(A, B)$ and $d(x, Tx) \geq d(A, B)$ for all x.

Definition 6.12 [17] Mappings $S : A \to B$ and $T : A \to B$ are said to *commute proximally* if it satisfies

$$[d(u, Sx) = d(v, Tx) = d(A, B)] \Longrightarrow Sv = Tu$$

for all $u, v, x \in A$.

It is easy to see that proximal commutativity of self mappings become commutativity of the mappings.

Definition 6.13 [17] A mapping $S : A \to B$ and $T : A \to B$ is said to be a *swapped proximally* if they satisfy the following condition

$$[d(y,u) = d(y,v) = d(A,B) \text{ and } Su = Tv] \Longrightarrow Sv = Tu$$

for all $u, v, \in A$ and $y \in B$.

Definition 6.14 A is said to be *approximatively compact with respect to* B if every sequence $\{x_n\}$ in A satisfies the condition that $d(y, x_n) \to d(y, A)$ for some $y \in B$ that has a convergent subsequence.

We observe that every set is approximatively compact with respect to itself. Every compact set is approximatively compact with respect to any set. Moreover, A_0 and B_0 are nonempty sets if A is compact and B is approximatively compact with respect to A.

Mongkolkeha and Kumam [14] using the concept of weak contraction with the commute proximally property proved the following theorem.

Theorem 6.10 *Let A and B be nonempty closed subsets of a complete metric space X such that A is approximatively compact with respect to B. Also, assume that A_0 and B_0 are nonempty. Let $S : A \to B$, $T : A \to B$ be nonself mappings satisfying the following conditions:*

(i) *For each x and y that are elements in A,*

$$d(Sx, Sy) \leq d(Tx, Ty) - \varphi(d(Tx, Ty)),$$

where, $\varphi : [0, \infty) \to [0, \infty)$ is a continuous and nondecreasing function such that $\varphi(t) = 0$ if and only if $t = 0$.

(ii) *T is continuous,*

(iii) *S and T commute proximally.*

(iv) *S and T can be swapped proximally.*

(v) *$S(A_0) \subseteq B_0$ and $S(A_0) \subseteq T(A_0)$.*

Then there exists an element $x \in A$ such that

$$d(x, Tx) = d(A, B), \; d(x, Sx) = d(A, B).$$

Moreover, if x^* is another common best proximity point of the mappings S and T,

$$d(x, x^*) \leq 2d(A, B).$$

Proof. Let x_0 be a fixed element in A_0. Since $S(A_0) \subseteq T(A_0)$, it follows that $x_1 \in A_0$ such that $Sx_0 = Tx_1$. Again, since $S(A_0) \subseteq T(A_0)$, $x_2 \in A_0$ such that $Sx_1 = Tx_2$. Similarly, we can find $x_n \in A_0$ such that

$$Sx_{n-1} = Tx_n \tag{6.20}$$

for all $n \geq 1$. It follows that

$$\begin{aligned} d(Sx_n, Sx_{n+1}) &\leq d(Tx_n, Tx_{n+1}) - \varphi(d(Tx_n, Tx_{n+1})) \\ &= d(Sx_{n-1}, Sx_n) - \varphi(d(Sx_{n-1}, Sx_n)) \\ &\leq d(Sx_{n-1}, Sx_n), \end{aligned} \tag{6.21}$$

which means that the sequence $\{d(Sx_{n-1}, Sx_n)\}$ is nonincreasing and bounded below. Hence $r \geq 0$ such that

$$\lim_{n\to\infty} d(Sx_{n-1}, Sx_n) = r. \tag{6.22}$$

If $r > 0$, then

$$d(Sx_n, Sx_{n+1}) \leq d(Sx_{n-1}, Sx_n) - \varphi(d(Sx_{n-1}, Sx_n)). \tag{6.23}$$

Taking $n \to \infty$ in (6.23), by the continuities of φ, we get $r \leq r - \varphi(r) < r$, which is a contradiction and hence $r = 0$. Therefore,

$$\lim_{n\to\infty} d(Sx_{n-1}, Sx_n) = 0. \tag{6.24}$$

Next, we will prove that $\{Sx_n\}$ is a Cauchy sequence.

Case I: Consider a positive integer n such that $Sx_n = Sx_{n+1}$. Observe that

$$\begin{aligned} d(Sx_{n+1}, Sx_{n+2}) &\leq d(Tx_{n+1}, Tx_{n+2}) - \varphi(d(Tx_{n+1}, Tx_{n+2})) \\ &= d(Sx_n, Sx_{n+1}) - \varphi(d(Sx_n, Sx_{n+1})) \\ &= 0, \end{aligned}$$

which implies that $Sx_{n+1} = Sx_{n+2}$. For all $m > n$, we conclude that $Sx_m = Sx_n$. Hence $\{Sx_n\}$ is a Cauchy sequence in B.

Case II: The successive terms of $\{Sx_n\}$ are different. Suppose that $\{Sx_n\}$ is not a Cauchy sequence. Then $\varepsilon > 0$ and subsequences $\{Sx_{m_k}\}, \{Sx_{n_k}\}$ of $\{Sx_n\}$ with $n_k > m_k \geq k$ such that

$$d(Sx_{m_k}, Sx_{n_k}) \geq \varepsilon, \quad d(Sx_{m_k}, Sx_{n_k-1}) < \varepsilon. \tag{6.25}$$

By using (6.25) and the triangular inequality, we get

$$\begin{aligned} \varepsilon &\leq d(Sx_{m_k}, Sx_{n_k}) \\ &\leq d(Sx_{m_k}, Sx_{n_k-1}) + d(Sx_{n_k-1}, Sx_{n_k}) \\ &< \varepsilon + d(Sx_{n_k-1}, Sx_{n_k}). \end{aligned} \tag{6.26}$$

Using (6.26) and (6.24), we have

$$d(Sx_{m_k}, Sx_{n_k}) \to \varepsilon \tag{6.27}$$

as $k \to \infty$. Again, by the triangular inequality, we get

$$d(Sx_{m_k}, Sx_{n_k}) \leq d(Sx_{m_k}, Sx_{m_k+1}) + d(Sx_{m_k+1}, Sx_{n_k+1}) + d(Sx_{n_k+1}, Sx_{n_k}) \tag{6.28}$$

and

$$d(Sx_{m_k+1}, Sx_{n_k+1}) \leq d(Sx_{m_k+1}, Sx_{m_k}) + d(Sx_{m_k}, Sx_{n_k}) + d(Sx_{n_k}, Sx_{n_k+1}). \tag{6.29}$$

From (6.24), (6.27), (6.28) and (6.29), we obtain

$$d(Sx_{m_k+1}, Sx_{n_k+1}) \to \varepsilon \tag{6.30}$$

as $k \to \infty$. In view of

$$\begin{aligned} d(Sx_{m_k+1}, Sx_{n_k+1}) &\leq d(Tx_{m_k+1}, Tx_{n_k+1}) - \varphi(d(Tx_{m_k+1}, Tx_{n_k+1})) \\ &= d(Sx_{m_k}, Sx_{n_k}) - \varphi(d(Sx_{m_k}, Sx_{n_k})), \end{aligned} \tag{6.31}$$

and letting, $k \to \infty$ in (6.31), we obtain

$$\varepsilon \leq \varepsilon - \varphi(\varepsilon),$$

which is a contradiction by the property of φ. Then we deduce that $\{Sx_n\}$ is a Cauchy sequence in B. Since B is a closed subset of a complete metric space X, then $y \in B$ such that $Sx_n \to y$ as $n \to \infty$. Consequently, the sequence $\{Tx_n\}$ also converges to y. From $S(A_0) \subseteq B_0$, there exists $u_n \in A$ such that

$$d(Sx_n, u_n) = d(A, B) \tag{6.32}$$

for all $n \geq 1$ and it follows from (6.20) and (6.32) that

$$d(Tx_n, u_{n-1}) = d(Sx_{n-1}, u_{n-1}) = d(A, B) \tag{6.33}$$

for all $n \geq 1$. By (6.32), (6.33) and the fact that the mappings S and T are commuting proximally, we obtain

$$Tu_n = Su_{n-1} \tag{6.34}$$

for all $n \geq 1$. Moreover, we have

$$\begin{aligned} d(y, A) &\leq d(y, u_n) \\ &\leq d(y, Sx_n) + d(Sx_n, u_n) \\ &= d(y, Sx_n) + d(A, B) \\ &\leq d(y, Sx_n) + d(y, A). \end{aligned} \tag{6.35}$$

Therefore, $d(y, u_n) \to d(y, A)$ as $n \to \infty$. Since A is approximatively compact with respect to B, there exists a subsequence $\{u_{n_k}\}$ of the sequence $\{u_n\}$ such that $\{u_{n_k}\}$ converges to some element $u \in A$. Further, since $d(y, u_{n_k-1}) \to d(y, A)$ and A is approximatively compact with respect to B, there exists a subsequence $\{u_{n_{k_j}-1}\}$ of the sequence $\{u_{n_k-1}\}$ such that $\{u_{n_{k_j}-1}\}$ converges to some element $v \in A$. By the continuity of the mappings S and T, we have

$$Tu = \lim_{j\to\infty} Tu_{n_{k_j}} = \lim_{j\to\infty} Su_{n_{k_j}-1} = Sv \tag{6.36}$$

and

$$\begin{aligned} d(y,u) &= \lim_{k\to\infty} d(Sx_{n_k}, u_{n_k}) = d(A,B), \\ d(y,v) &= \lim_{j\to\infty} d(Tx_{n_{k_j}}, u_{n_{k_j}-1}) = d(A,B). \end{aligned} \tag{6.37}$$

Since S and T can be swapped proximally, we get

$$Tv = Su. \tag{6.38}$$

Next, we prove that $Su = Sv$. Suppose the reverse. Then, by (6.36), (6.37), (6.38) and the property of φ, we have

$$\begin{aligned} d(Su,Sv) &\leq d(Tu,Tv) - \varphi(d(Tu,Tv)) \\ &= d(Sv,Su) - \varphi(d(Sv,Su)) \\ &< d(Sv,Su), \end{aligned}$$

which is a contradiction. Thus $Su = Sv$ and $Tu = Su$. Since $S(A_0)$ is contained in B_0, $x \in A$ such that

$$d(x,Tu) = d(A,B),\ d(x,Su) = d(A,B).$$

Since S and T are commuting proximally, $Sx = Tx$. Consequently,

$$\begin{aligned} d(Su,Sx) &\leq d(Tu,Tx) - \varphi(d(Tu,Tx)) \\ &= d(Su,Sx) - \varphi(d(Su,Sx)). \end{aligned} \tag{6.39}$$

In (6.39), if $Su \neq Sx$, then

$$d(Su,Sx) \leq d(Su,Sx) - \varphi(d(Su,Sx)) < d(Su,Sx),$$

which is impossible. $Su = Sx$ and hence $Tu = Tx$. It follows that

$$d(x,Tx) = d(x,Tu) = d(A,B)$$

and

$$d(x,Sx) = d(x,Su) = d(A,B).$$

Therefore, x is a common best proximity point of S and T.

Suppose that x^* is another common best proximity point of the mappings S and T. Then we have

$$d(x^*,Tx^*) = d(A,B)$$

and

$$d(x^*,Sx^*) = d(A,B).$$

Since S and T are commuting proximally, $Sx = Tx$ and $Sx^* = Tx^*$. Consequently,

$$\begin{aligned} d(Sx^*,Sx) &\leq d(Tx^*,Tx) - \varphi(d(Tx^*,Tx)) \\ &= d(Sx^*,Sx) - \varphi(d(Sx^*,Sx)). \end{aligned} \tag{6.40}$$

In (6.40), if $Sx^* \neq Sx$, then

$$d(Sx^*, Sx) \leq d(Sx^*, Sx) - \varphi(d(Sx^*, Sx)) < d(Sx^*, Sx),$$

which is impossible. We have $Sx = Sx^*$. Moreover, it follows that

$$\begin{aligned} d(x, x^*) &\leq d(x, Sx) + d(Sx, Sx^*) + d(Sx^*, x^*) \\ &= d(A, B) + d(A, B) \\ &= 2d(A, B). \end{aligned}$$

This completes the proof.

Using $\varphi(t) = (1 - \alpha)t$, where $0 \leq \alpha < 1$ in Theorem 6.10, we obtain following.

Corollary 6.6 [17] *Let A and B be nonempty closed subsets of a complete metric space X such that A is approximatively compact with respect to B. Also, assume that A_0 and B_0 are nonempty. Let $S : A \to B$, $T : A \to B$ be the nonself mapping satisfying the following conditions.*

(i) *There exists a non-negative real number $\alpha < 1$ such that*

$$d(Sx_1, Sx_2) \leq \alpha d(Tx_1, Tx_2) \quad \textit{for all } x_1, x_2 \in A.$$

(ii) *T is continuous.*

(iii) *S and T commute proximally.*

(iv) *S and T can be swapped proximally.*

(v) *$S(A_0) \subseteq B_0$ and $S(A_0) \subseteq T(A_0)$.*

Then $x \in A$ such that

$$d(x, Tx) = d(A, B), \quad d(x, Sx) = d(A, B).$$

Further, if x^* is another common best proximity point of the mappings S and T,

$$d(x, x^*) \leq 2d(A, B).$$

For a self mapping, Theorem 6.10 contains common fixed point theorems of Jungck [8] for commuting self mappings, which in turn generalize Banach's contraction principle.

Corollary 6.7 [8] *Let (X, d) be a complete metric space. Let S and T be self mappings on X satisfying the following conditions:*

(i) *T is continuous.*

(ii) *$S(X) \subseteq T(X)$.*

(iii) *S and T commute.*

Suppose $\alpha \in [0,1)$ such that

$$d(Sx, Sy) \leq \alpha d(Tx, Ty)$$

for all $x, y \in X$. Then there exists a unique common fixed point of S and T.

Now, we illustrate Theorem 6.10.

Example 6.8 Consider the complete metric space $\mathbb{R}^2$ with Euclidean metric. Let

$$A = \{(x,1) : 0 \leq x \leq 1\}$$

and

$$B = \{(x,-1) : 0 \leq x \leq 1\}.$$

Define two mappings $S : A \to B$, $T : A \to B$ as follows:

$$S(x,1) = \left(x - \frac{x^2}{2}, -1\right)$$

and

$$T((x,1)) = (x,-1).$$

It is easy to see that $d(A,B) = 2$, $A_0 = A$ and $B_0 = B$. Further, S and T are continuous and A is approximatively compact with respect to B.

First, we show that S and T satisfy the condition (i) of Theorem 6.10 with a function $\varphi : [0,\infty) \to [0,\infty)$ defined by $\varphi(t) = \frac{t^2}{2}$ for all $t \in [0,\infty)$. Let $(x,1),(y,1) \in A$. Without a loss generality, we see that $x > y$. Then we have

$$\begin{aligned}
d(S(x,1), S(y,1)) &= \left|\left(x - \frac{x^2}{2}\right) - \left(y - \frac{y^2}{2}\right)\right| \\
&= (x-y) - \frac{1}{2}(x^2 - y^2) \\
&= (x-y) - \frac{1}{2}((x-y)(x+y)) \\
&\leq (x-y) - \frac{1}{2}(x-y)^2 \\
&= d(T(x,1), T(y,1)) - \varphi(d(T(x,1), T(y,1))).
\end{aligned}$$

Next we show that S and T are commuting proximally. Let $(u,1),(v,1),(x,1) \in A$

$$d((u,1), S(x,1)) = d(A,B) = 2, \quad d((v,1), T(x,1)) = d(A,B) = 2.$$

It follows that

$$u = x - \frac{x^2}{2}, \; v = x$$

and hence

$$S(v,1) \ = \left(v - \frac{v^2}{2}, -1\right) \ = \left(x - \frac{x^2}{2}, -1\right) = (u,-1) = T(u,1).$$

Finally, we show that S and T can be swapped proximally. If it is true that

$$d((u,1),(y,-1)) = d((v,1),(y,-1)) = d(A,B) = 2, \quad S(u,1) = T(v,1)$$

for some $(u,1),(v,1) \in A$ and $(y,-1) \in B$, we get $u = v = 0$ and thus

$$S(v,1) = T(u,1).$$

Therefore, all the hypotheses of Theorem 6.10 are satisfied.

Furthermore, $(0,1) \in A$ is a common best proximity point of S and T because

$$d((0,1),S(0,1)) = d((0,1),(0,-1)) = d((0,1),T((0,1)) = d(A,B).$$

On the other hand, suppose that there exists $k \in [0,1)$ such that

$$d(S(x,1),S(y,1)) \le kd(T(x,1),T(y,1)),$$

that is,

$$\left|\left(x - \frac{x^2}{2}\right) - \left(y - \frac{y^2}{2}\right)\right| \le k|x-y|.$$

Using $y = 0$ and $x > 0$, it follows that

$$1 = \lim_{x \to 0+} \left|\left(1 - \frac{x}{2}\right)\right| \le k < 1,$$

which is a contradiction. Therefore, the results of Sadiq Basha in [17] cannot be applied to this example or our Theorem 6.10 result.

References

[1] Abkar A., Gabeleh M., Best proximity points for cyclic mappings in ordered metric spaces, J. Optim. Theory Appl. 2011; 150 (1):188–19.

[2] Abkar A, Gabeleh M., A note on some best proximity point theorems proved under P-property. Abstr. Appl. Anal. 2013: Art. 189567.

[3] Chen J., Xiao S., Wang H.i, Deng S., Best proximity point for the proximal nonexpansive mapping on the star-shaped sets. Fixed Point Theory Appl. 2015: Art. 19.

[4] Eldred A.A., Veeramani, P., Existence and convergence of best proximity points. J. Math. Anal. Appl. 2006; 323:1001–1006.

[5] Espínola R., Kosuru, G.S.R., Veeramani P., Pythagorean and best proximity point theorems. J. Optim. Theory Appl. 2015; 164:534–550.

[6] Espínola R., Kosuru G.S.R., Veeramani P., Pythagorean property and best-proximity point theorems. J Optim Theory Appl. 2015; 164:534–550.

[7] Fan K., Extensions of two fixed point theorems of F.E. Browder. Math. Z. 1969; 112:, 234–240.

[8] Jungck G., Commuting mappings and fixed points, Amer. Math. Monthly, 1976; 83(4):261–263.

[9] Kirk W.A., Srinivasan P.S., Veeramani P., Fixed points for mappings satisfying cyclical contractive conditions. Fixed Point Theory. 2003; 4:79–89.

[10] Kosuru GSR., Veeramani P., On existence of best proximity pair theorems for relatively nonexpansive mappings. J. Nonlinear Convex Anal. 2010; 11:71–77

[11] Kosuru GSR., P. Veeramani, A note on existence and convergence of best proximity points for pointwise cyclic contractions. Numer. Funct. Anal. Optim. 2011; 32(7), 821-830.

[12] Mongkolkeha C., Cho Y.J., Kumam P., Best proximity points for Geraghty's proximal contraction mappings. Fixed Point Theory Appl. 2013: Art. 180.

[13] Mongkolkeha C., Kumam P., Best proximity point theorems for generalized cyclic contractions in ordered metric spaces. J. Optim. Theory Appl. 2012; 155(1):215–226.

[14] Mongkolkeha C., Kumam P., Some common best proximity points for proximity commuting mappings. Optim. Lett. 2013; 7:1825–1836.

[15] Prolla J.B., Fixed point theorems for set valued mappings and existence of best approximations. Numer. Funct. Anal. Optim. 1983; 5(4):449–455.

[16] Reich S., Approximate selections, best approximations, fixed points and invariant sets. J. Math. Anal. Appl. 1978; 62:104–113.

[17] Sadiq Basha S., Common best proximity points: global minimization of multi-objective functions. J. Glob. Optim. 2012; 54(2):367–373.

[18] Sadiq Basha S., Veeramani P., Best proximity pair theorems for multifunctions with open fibres. J. Approx. Theory. 2000; 103:119–129.

[19] Sankar Raj V., A best proximity point theorem for weakly contractive non-self mappings. Nonlinear Anal. 2011; 74:4804–4808.

[20] Sehgal V.M., Singh S.P., A generalization to multifunctions of Fan's best approximation theorem. Proc. Amer. Math. Soc. 1988; 102:534–537.

[21] Sehgal V.M., Singh S.P., A theorem on best approximations. Numer. Funct. Anal. Optim. 1989; 10:181–184.

[22] Suzuki T., Kikkawab M., Vetro C., The existence of best proximity points in metric spaces with the property UC. Nonlinear Anal. 2009; 71:2918–2926.

Chapter 7

Applications of Metric Fixed Point Theory

Satish Shukla

This chapter is devoted to some elementary applications of fixed point theorems in root finding, establishing the existence and uniqueness of solutions of systems of linear algebraic equation, steady states of Markov processes, integral equations, initial value and boundary value problems, periodic and impulsive boundary value problems, equilibrium of difference equations and periodic solutions of cyclic systems.

7.1 Root finding problems

Let $T\colon I \to I$, I be any closed interval of real numbers. We want to find the solution of the equation $Tx = x$. It is clear that the solution is a fixed point of T.

Theorem 7.1 *Let $T\colon I \to I$ be a differentiable function and $\lambda > 0$ such that*

$$|T'x| \leq \lambda < 1$$

for all $x \in I$, where T' stands for the first derivative of T with respect to x. The equation $Tx = x$ has a unique solution in I and for any $x_0 \in X$ the sequence $\{x_n\}$ defined by $x_n = Tx_{n-1}$, $n \in \mathbb{N}$ converges to the solution of the equation.

Proof. Suppose, $X = I$ and d is the metric on I defined by $d(x, y) = |x - y|$ for all $x, y \in I$. Then (X, d) is a complete metric space. By Lagrange's mean value theorem for any $x, y \in I$, $x < y$ there exists $w \in (x, y)$ such that

$$Tx - Ty = (x - y)T'w.$$

Therefore,

$$\begin{aligned} d(Tx, Ty) &= |Tx - Ty| = |(x - y)T'w| \\ &= |x - y||T'w| \\ &\leq \lambda|x - y| \\ &= \lambda d(x, y). \end{aligned}$$

Since $\lambda < 1$, T is a contraction mapping on I and by Banach Contraction Principle (BCP), T has a unique fixed point in I; and the sequence $\{x_n\}$ converges to this fixed point. Thus, there exists a unique $u \in I$ such that $Tu = u$, which is the unique solution of T in I.

Remark 7.1 Suppose, we want to find a root of the equation $fx = 0$. Then, we rewrite the equation in the form $Tx = x$, in such a way that $|T'x| \leq \lambda < 1$ for some $\lambda > 0$ and for all $x \in I$. Now applying Theorem 7.1 we can find the solution of $Tx = x$ which is a root of $fx = 0$. For example, consider the equation $fx = \cos^2(x) - 3x = 0$. We rewrite it in the following form:

$$Tx = x = \frac{\cos^2(x)}{3}.$$

It is obvious that there is a root of equation $fx = 0$ in the interval $I = [0, 1]$. We can find $\lambda > 0$ such that $|T'x| \leq \lambda < 1$ for all $x \in I$. Therefore, by the above theorem there is a unique solution of the equation $Tx = x$ (i.e., the root of equation $fx = 0$). Now we can find the root as a limit of sequence $\{x_n\}$, $x_n = \dfrac{\cos^2(x_{n-1})}{3}$ with arbitrary $x_0 \in I$.

7.2 Solution of system of linear algebraic equations

We consider the following system of linear algebraic equations:

$$\left.\begin{aligned} a_{11}x_1 + a_{12}x_2 + \cdots + a_{1n}x_n &= b_1 \\ a_{21}x_1 + a_{22}x_2 + \cdots + a_{2n}x_n &= b_2 \\ &\vdots \\ a_{n1}x_1 + a_{n2}x_2 + \cdots + a_{nn}x_n &= b_n. \end{aligned}\right\} \tag{7.1}$$

A rearrangement of the above system is:

$$\begin{aligned} x_1 &= x_1 - [a_{11}x_1 + a_{12}x_2 + \cdots + a_{1n}x_n] + b_1 \\ x_2 &= x_2 - [a_{21}x_1 + a_{22}x_2 + \cdots + a_{2n}x_n] + b_2 \\ &\vdots \\ x_n &= x_n - [a_{n1}x_1 + a_{n2}x_2 + \cdots + a_{nn}x_n] + b_n. \end{aligned}$$

In matrix form, the above system can be written as

$$x = x - xA + b \tag{7.2}$$

where $A^t = (a_{ij})_{n\times n}$ is the coefficient matrix, $x = (x_1, x_2, \ldots, x_n)$, $b = (b_1, b_2, \ldots, b_n)$ and A^t denotes the transpose of the matrix A. Denote $\alpha_{ij} = -a_{ij} + \delta_{ij}$, $1 \leq i, j \leq n$, where δ_{ij} is the Kronecker delta defined by

$$\delta_{ij} = \begin{cases} 1, & \text{if } i = j; \\ 0, & \text{otherwise.} \end{cases}$$

Then (7.2) can be written as

$$x = x\mathcal{A} + b \tag{7.3}$$

where $\mathcal{A}^t = (\alpha_{ij})_{n\times n}$. We first convert the above into a fixed point problem. Define a mapping $T\colon \mathbb{R}^n \to \mathbb{R}^n$ by:

$$Tx = x\mathcal{A} + b \quad \text{for all} \quad x \in \mathbb{R}^n.$$

It is obvious that $x \in \mathbb{R}^n$ is a solution of (7.1) if and only if it is a fixed point of T.

Theorem 7.2 *If* $\sum\limits_{j=1}^{n} |\alpha_{ij}| \leq \lambda < 1$, $1 \leq i \leq n$ *for some positive number* λ, *system* (7.1) *has a unique solution. Furthermore, for any* $y_0 \in \mathbb{R}^n$ *the sequence* $\{y_m\}$ *defined by* $y_m = x_{m-1}\mathcal{A} + b$, $m \in \mathbb{N}$ *converges to the solution of the system.*

Proof. Consider the space $X = \mathbb{R}^n$ with the metric defined by:

$$d(x, y) = \max_{1\leq i\leq n} |x_i - y_i|$$

for all $x = (x_1, x_2, \ldots, x_n)$, $x = (y_1, y_2, \ldots, y_n) \in X$. (X, d) is a complete metric space. We show that the mapping T is a contraction on X. Then, for all $x, y \in X$ we have

$$\begin{aligned} d(Tx, Ty) &= d(x\mathcal{A} + b, y\mathcal{A} + b) \\ &= \max_{1 \le i \le n} \left| \sum_{j=1}^{n} \alpha_{ij}(x_j - y_j) \right| \\ &\le \max_{1 \le i \le n} \sum_{j=1}^{n} |\alpha_{ij}| \, |x_j - y_j| \\ &\le \max_{1 \le i \le n} \sum_{j=1}^{n} |\alpha_{ij}| \max_{1 \le p \le n} |x_p - y_p| \\ &= \max_{1 \le i \le n} \sum_{j=1}^{n} |\alpha_{ij}| \, d(x, y) \\ &\le \lambda d(x, y). \end{aligned}$$

We see that T is a contraction, and the existence and uniqueness of the solution of system (7.1) follows from the BCP.

Example 7.1 Consider the following system of equations:

$$\left. \begin{aligned} -20x_1 + 4x_2 + 3x_3 &= -32 \\ 12x_1 - 45x_2 + 20x_3 &= 300 \\ 3x_1 - 4x_2 - 16x_3 &= -36. \end{aligned} \right\} \tag{7.4}$$

The system can be written as:

$$\left. \begin{aligned} x_1 &= -\tfrac{1}{4}x_1 + \tfrac{1}{4}x_2 + \tfrac{3}{16}x_3 + 2 \\ x_2 &= \tfrac{1}{5}x_1 + \tfrac{1}{4}x_2 + \tfrac{1}{3}x_3 - 5 \\ x_3 &= \tfrac{1}{4}x_1 - \tfrac{1}{3}x_2 - \tfrac{1}{3}x_3 + 3. \end{aligned} \right\} \tag{7.5}$$

$\mathcal{A}^t = \begin{pmatrix} -\frac{1}{4} & \frac{1}{4} & \frac{3}{16} \\ \frac{1}{5} & \frac{1}{4} & \frac{1}{3} \\ \frac{1}{4} & -\frac{1}{3} & -\frac{1}{3} \end{pmatrix}$, $x = (x_1, x_2, x_3)$, $b = (2, -5, 3)$ and $\sum_{j=1}^{3} |\alpha_{1j}| = \frac{11}{16} < 1$, $\sum_{j=1}^{3} |\alpha_{2j}| = \frac{47}{60} < 1$, $\sum_{j=1}^{3} |\alpha_{3j}| = \frac{11}{12} < 1$. Therefore, $\sum_{j=1}^{3} |\alpha_{ij}| \le \frac{11}{12} < 1$ for $1 \le i \le 3$ and by Theorem 7.2, system (7.4) has a unique solution which is the limit of the sequence $\{y_n\}$ defined by $y_n = y_{n-1}\mathcal{A} + b$, $n \in \mathbb{N}$ and $y_0 \in \mathbb{R}^3$ is arbitrary.

Corollary 7.1 *Let* $\alpha_{ij} = -a_{ij} + \delta_{ij}$, $1 \le i, j \le n$, $a_{ij} \in \mathbb{R}$, *where* δ_{ij} *is the Kronecker delta. If* $\sum_{j=1}^{n} |\alpha_{ij}| \le \lambda < 1$, $1 \le i \le n$ *for some positive number* λ, *the following homogeneous system of linear algebraic equations has no* *nontrivial solution:*

$$\begin{aligned} a_{11}x_1 + a_{12}x_2 + \cdots + a_{1n}x_n &= 0 \\ a_{21}x_1 + a_{22}x_2 + \cdots + a_{2n}x_n &= 0 \\ &\vdots \\ a_{n1}x_1 + a_{n2}x_2 + \cdots + a_{nn}x_n &= 0. \end{aligned}$$

Proof. Assuming $b = 0$ in Theorem 7.2, and using the fact that the given system always has a trivial solution, the proof follows.

7.3 Markov process and steady state vector

A stochastic process with an finite number of states (say n) such that the state of system at any stage depends only on its previous stage is called a Markov process. Suppose $x_0 = (x_0^1, x_0^2, \ldots, x_0^n) \in \mathbb{R}^n$ and $0 \leq x_i^0 \leq 1$, $1 \leq i \leq n$ are the initial probability vectors of a system which gives the initial probability distribution of each state of the system. Let $P = (p_{ij})_{n\times n}$ be a matrix such that $p_{ij} \geq 0$ is the probability that the system transits to state j from state i, $\sum_{j=1}^{n} p_{ij} = 1$ for $i = 1, 2, \ldots, n$. After one period the probability vector of the system, i.e., the probability distribution of each state is $x_0P = x_1$. Then P is called the transition matrix of the of the Markov process.

In general, $x_m = x_{m-1}P = x_0P^m$ gives the probability distribution of each state after m period, i.e., the the probability vector of the system after the mth period.

Definition 7.1 A vector $x = (x_1, x_2, \ldots, x_n) \in \mathbb{R}^n$ is called a steady state vector or stationary distribution of a system with transition matrix $P = (p_{ij})_{n\times n}$ if:

(i) $x_i \geq 0$, $1 \leq i \leq n$;

(ii) $\sum_{i=1}^{n} x_i = 1$;

(iii) $xP = x$.

For a Markov process after a long run, i.e., as the period $m \to \infty$, the probability vector of system after m^{th} period is equal to the probability vector of system after the $(m+1)^{th}$ period, i.e., $x_0P^{m+1} = x_mP = x_{m+1} = x_m$, then x_m is the steady state vector of the Markov process. We shall show that for $p_{ij} > 0$, $1 \leq i, j \leq n$ there exists a unique steady state vector for every Markov process.

Theorem 7.3 *If $p_{ij} > 0$, $1 \leq i, j \leq n$, then there exists a unique steady state vector for the Markov process with transition matrix $P = (p_{ij})_{n\times n}$.*

Proof. Let $X = \Pi^n = \left\{(x_1, x_2, \ldots, x_n) \colon x_i \geq 0, \sum\limits_{i=1}^{n} x_i = 1\right\}$, and define a mapping T on X by:

$$Tx = xP \quad \text{for all} \quad x \in X.$$

We shall show that T is a self mapping of X. IF $x = (x_1, x_2, \ldots, x_n) \in X$, we have

$$\sum_{i=1}^{n} x_i = 1.$$

By definition, $Tx = xP = y = (y_1, y_2, \ldots, y_n)$, where $y_j = \sum_{i=1}^{n} p_{ij} x_i$. Therefore, $y_j \geq 0$ for $1 \leq j \leq n$. Since $\sum_{j=1}^{n} p_{ij} = 1$, $1 \leq i \leq n$ and $\sum_{i=1}^{n} x_i = 1$,

$$\begin{aligned} \sum_{j=1}^{n} y_j &= \sum_{j=1}^{n}\sum_{i=1}^{n} p_{ij} x_i = \sum_{i=1}^{n}\left(\sum_{j=1}^{n} p_{ij}\right) x_i \\ &= \sum_{i=1}^{n} 1 \cdot x_i \\ &= 1. \end{aligned}$$

Thus, T is a self mapping of X.

Define $d \colon X \times X \to \mathbb{R}$ by:

$$d(x, y) = \sum_{i=1}^{n} |x_i - y_i|$$

for all $x = (x_1, x_2, \ldots, x_n), y = (y_1, y_2, \ldots, y_n) \in X$. Then (X, d) is a complete metric space. We shall show that T is a contraction on X. Suppose, $Tx = xP = u = (u_1, u_2, \ldots, u_n)$, $Ty = yP = v = (v_1, v_2, \ldots, v_n)$, $m_j = \min\limits_{1 \leq i \leq n} p_{ij}$ and $m = \sum\limits_{j=1}^{n} m_j$; then $p_{ij} \geq m_j, 0 < m \leq 1$ and we have

$$\begin{aligned} d(Tx, Ty) &= d(xP, yP) \\ &= \sum_{j=1}^{n} |u_j - v_j| = \sum_{j=1}^{n}\left|\sum_{i=1}^{n} p_{ij} x_i - \sum_{i=1}^{n} p_{ij} y_i\right| \\ &= \sum_{j=1}^{n}\left|\sum_{i=1}^{n} p_{ij}(x_i - y_i)\right| \\ &= \sum_{j=1}^{n}\left|\sum_{i=1}^{n} [(p_{ij} - m_j)(x_i - y_i) + m_j(x_i - y_i)]\right| \\ &= \sum_{j=1}^{n}\left|\sum_{i=1}^{n} (p_{ij} - m_j)(x_i - y_i) + \sum_{i=1}^{n} m_j(x_i - y_i)\right| \\ &= \sum_{j=1}^{n}\left|\sum_{i=1}^{n} (p_{ij} - m_j)(x_i - y_i)\right|. \end{aligned}$$

Since $\sum_{j=1}^{n} p_{ij} = 1$, using the definition of m_j and m in the above inequality, we obtain

$$\begin{aligned} d(Tx, Ty) &\leq \sum_{j=1}^{n}\sum_{i=1}^{n}(p_{ij} - m_j)\,|x_i - y_i| \\ &= \sum_{i=1}^{n}\left[\sum_{j=1}^{n} p_{ij} - \sum_{j=1}^{n} m_j\right]|x_i - y_i| \\ &= \sum_{i=1}^{n}(1 - m)\,|x_i - y_i| \\ &= (1 - m)d(x, y). \end{aligned}$$

T is a contraction on X. Therefore, by the BCP, there exists a unique vector $x^* \in X$ such that $x^*P = Tx^* = x^*$. Also, the sequence $\{x_m\}$ defined by $x_m = Tx_{m-1} = x_{m-1}P$, where $x_0 \in X$ is some initial probability vector, converges to the vector x^*, i.e.,

$$\lim_{m\to\infty} x_m = x^*.$$

After a long run, the system settles down and exhibits steady state behavior.

7.4 Solutions of integral equations

Suppose, $K_1, K_2 \colon [a, b] \times [a, b] \times \mathbb{R} \to \mathbb{R}$ and $g \colon [a, b] \to \mathbb{R}$ are given functions. We consider the following integral equation:

$$u(t) = \beta A(t, u(t)) + \gamma B(t, u(t)) + g(t),\ t \in [a, b], \tag{7.6}$$

where

$$A(t, u(t)) = \int_a^t K_1(t, s, u(s))ds, B(t, u(t)) = \int_a^t K_2(t, s, u(s))ds \ \text{ and } \ \beta, \gamma \in \mathbb{R}.$$

We assume that $|\beta| + |\gamma| > 0$; otherwise, the above equation reduced into a trivial form.

To study the existence and uniqueness of the solution of integral equation (7.6), we first reduce it to a fixed point problem.

Let $C(I, \mathbb{R})$ be the space of all real-valued continuous functions on I, where $I = [a, b]$ with the Bielecki norm given by $\|u\|_B := \sup_{t\in I} |u(t)|e^{-\theta t}$, $u \in C(I, \mathbb{R})$, where $\theta > 0$ is arbitrary but fixed. The induced metric d on $C(I, \mathbb{R})$ is given by:

$$d(u, v) = \sup_{t\in I} |u(t) - v(t)|e^{-\theta t} \quad \text{for all} \quad u, v \in C(I, \mathbb{R}).$$

We now define a function $T\colon C(I,\mathbb{R}) \to C(I,\mathbb{R})$ by:

$$[T(u)](t) = \beta A(t,u(t)) + \gamma B(t,u(t)) + g(t),\ t \in I. \tag{7.7}$$

A function $u^* = u^*(t)$ is called a solution of (7.6) if $u^* \in C(I,\mathbb{R})$ and it satisfies (7.6). It is obvious that the problem of finding the solution of (7.6) is equivalent to the problem of finding the fixed point of T.

We now ensure the existence and uniqueness of the fixed point of function T by applying some constraints on the functions K_1 and K_2.

Theorem 7.4 *Suppose the functions $K_1, K_2\colon I \times I \times \mathbb{R} \to \mathbb{R}$ and $g\colon I \to \mathbb{R}$ are continuous and the following condition is satisfied: for all $t, s \in I, u, v \in \mathbb{R}$ there exist two continuous functions $L_1, L_2\colon I \times I \times \mathbb{R} \times \mathbb{R} \to \mathbb{R}^+$ such that:*

$$| K_i(t,s,u(s)) - K_i(t,s,v(s)) | \le L_i(t,s,u(s),v(s)) \mid u(s) - v(s) \mid,\ i = 1,2. \tag{7.8}$$

If $\alpha > 0$ such that $L_i(t,s,u(s),v(s)) \le \alpha$, $i = 1,2$ for all $s,t \in I$, the integral equation (7.6) has a unique solution in $C(I,\mathbb{R})$.

Proof. It is well-known that $(C(I,\mathbb{R}),d)$ is a complete metric space. Define $T\colon C(I,\mathbb{R}) \to C(I,\mathbb{R})$ by Equation (7.7). We shall show that the function T is a contraction on $C(I,\mathbb{R})$. If $u,v \in C(I,\mathbb{R})$, by the definition of T we have

$$\begin{aligned}
d(Tu,Tv) &= \sup_{t\in I} | \beta A(t,u(t)) + \gamma B(t,u(t)) - \beta A(t,v(t)) - \gamma B(t,v(t)) | e^{-\theta t} \\
&= \sup_{t\in I} \left| \beta \int_a^t [K_1(t,s,u(s)) - K_1(t,s,v(s))]ds \right. \\
&\quad \left. + \gamma \int_a^t [K_2(t,s,u(s)) - K_2(t,s,v(s))]ds \right| e^{-\theta t} \\
&\le \sup_{t\in I} \left\{ |\beta| \int_a^t | K_1(t,s,u(s)) - K_1(t,s,v(s)) | ds \right. \\
&\quad \left. + |\gamma| \int_a^t | K_2(t,s,u(s)) - K_2(t,s,v(s)) | ds \right\} e^{-\theta t} \\
&\le \sup_{t\in I} \left\{ |\beta| \int_a^t L_1(t,s,u(s),v(s)) \mid u(s) - v(s) \mid ds \right. \\
&\quad \left. + |\gamma| \int_a^t L_2(t,s,u(s),v(s)) \mid u(s) - v(s) \mid ds \right\} e^{-\theta t} \\
&\le d(u,v) \sup_{t\in I} \int_a^t \Big\{ |\beta| L_1(t,s,u(s),v(s))ds \\
&\quad + |\gamma| L_2(t,s,u(s),v(s)) \Big\} e^{\theta(s-t)} ds \\
&\le d(u,v)\alpha(|\beta| + |\gamma|) \sup_{t\in I} \int_a^t e^{\theta(s-t)} ds \\
&\le d(u,v) \frac{\alpha(|\beta| + |\gamma|)}{\theta} \left[1 - e^{-\theta(b-a)}\right].
\end{aligned}$$

Because $\theta > 0$ was arbitrary, we choose $\theta = \alpha(|\beta| + |\gamma|)$. The above equation yields:

$$d(Tu, Tv) \leq \left[1 - e^{-\theta(b-a)}\right] d(u, v) = \lambda d(u, v)$$

where $\lambda = \left[1 - e^{-\theta(b-a)}\right] < 1$.

Thus, T is a contraction on $C(I, \mathbb{R})$. By the BCP, a unique fixed point u^* of T in $C(I, \mathbb{R})$ is the unique solution of integral equation (7.14) in $C(I, \mathbb{R})$.

Example 7.2 Let $I = [0, \pi/2]$. We want to find a function $u\colon I \to \mathbb{R}$ such that:

$$u(t) = \sin(t) + \int_0^t \ln(1 + s + |u(s)|)ds, \; t \in I. \tag{7.9}$$

Then, we consider the the space $C(I, \mathbb{R})$ with the metric d induced by the Bielecki norm. Define a function $T\colon C(I, \mathbb{R}) \to C(I, \mathbb{R})$ by:

$$[T(u)](t) = \sin(t) + \int_0^t \ln(1 + s + |u(s)|)ds.$$

By the Lagrange mean value theorem, we have

$$|\ln(1 + s + |u(s)|) - \ln(1 + s + |v(s)|)| \leq |u(s) - v(s)| \text{ for all } s \in I, u, v \in \mathbb{R}.$$

Therefore, all the conditions of Theorem 7.4 are satisfied with $L_1(t, s, u(s), v(s)) = L_2(t, s, u(s), v(s)) = \alpha = \beta = 1$, $\gamma \in \mathbb{R}$, $g(t) = \sin(t)$, $K_1(t, s, u(s)) = \ln(1+s+|u(s)|)$, $K_2(t, s, u(s)) = 0$ for all $s, t \in I$ and $u \in \mathbb{R}$. Thus, there exists a unique solution of integral equation (7.9) which is the limit of sequence $\{u_n\}$ defined by $u_n = Tu_{n-1}$, $n \in \mathbb{N}$ and $u_0 \in C(I, \mathbb{R})$ is arbitrary.

Next, we consider the following integral equation:

$$u(t) = u(a) + \int_a^t K(s, u)ds \tag{7.10}$$

where $K\colon [a, b] \times \mathbb{R} \to \mathbb{R}$ is a continuous function.

Corollary 7.2 *Suppose the following condition is satisfied: for all $t, s \in I = [a, b], u, v \in \mathbb{R}$ there exists a continuous function $L\colon I \times \mathbb{R} \times \mathbb{R} \to \mathbb{R}^+$ such that:*

$$| K(s, u(s)) - K(s, v(s)) | \leq L(s, u(s), v(s)) | u(s) - v(s) | .$$

If $\alpha > 0$ such that $L(s, u(s), v(s)) \leq \alpha$ for all $s \in I$, the integral equation (7.10) has a unique solution in $C(I, \mathbb{R})$.

Remark 7.2 In the above corollary, one can take any $t_0 \in [a, b]$ instead of a as lower limit of the integral in Equation (7.10); the conclusion remains the same.

We now consider the following Fredholm type integral equation:

$$u(t) = \beta A(t, u(t)) + \gamma B(t, u(t)) + g(t),\ t \in [a, b], \tag{7.11}$$

where

$$A(t, u(t)) = \int_a^b K_1(t, s, u(s))ds, B(u(t), t) = \int_a^b K_2(t, s, u(s))ds \text{ and } \beta, \gamma \in \mathbb{R}$$

and $K_1, K_2\colon [a, b] \times [a, b] \times \mathbb{R} \to \mathbb{R}$ and $g\colon [a, b] \to \mathbb{R}$ are given functions. Again, we assume that $|\beta| + |\gamma| > 0$, and d is the induced metric on $C(I, \mathbb{R})$ by the Bielecki norm. Define $T\colon C(I, \mathbb{R}) \to C(I, \mathbb{R})$ by:

$$[T(u)]t = \beta A(t, u(t)) + \gamma B(t, u(t)) + g(t),\ t \in I \tag{7.12}$$

where $I = [a, b]$. A function $u^* = u^*(t)$ is called a solution of (7.11), if $u^* \in C(I, \mathbb{R})$ and it satisfies (7.11).Then finding the solution of (7.11) is reduced to the problem of finding the fixed point of T, i.e., u^* is a solution of (7.11) if and only if it is a fixed point of T.

Theorem 7.5 *Suppose, the functions $K_1, K_2\colon I \times I \times \mathbb{R} \to \mathbb{R}$ and $g\colon I \to \mathbb{R}$ are continuous and the following condition is satisfied: for all $t, s \in I, u, v \in \mathbb{R}$ there exist two continuous functions $L_1, L_2\colon I \times I \times \mathbb{R} \times \mathbb{R} \to \mathbb{R}^+$ such that:*

$$| K_i(t, s, u(s)) - K_i(t, s, v(s)) | \leq L_i(t, s, u(s), v(s)) \mid u(s) - v(s) \mid,\ i = 1, 2. \tag{7.13}$$

If $L_i(t, s, u(s), v(s)) \leq e^{(t-b)\theta_1}$, $i = 1, 2$ for all $s, t \in I$, where $\theta_1 = |\beta| + |\gamma|$, the integral Equation (7.11) *has a unique solution in* $C(I, \mathbb{R})$.

Proof. Following the same line of proof of Theorem 7.4,

$$\begin{aligned} d(Tu, Tv) &\leq \sup_{t \in I} \left\{ |\beta| \int_a^b L_1(t, s, u(s), v(s)) \mid u(s) - v(s) \mid ds \right. \\ &\quad \left. + |\gamma| \int_a^b L_2(t, s, u(s), v(s)) \mid u(s) - v(s) \mid ds \right\} e^{-\theta t} \\ &\leq d(u, v) \sup_{t \in I} \int_a^t \Big\{ |\beta| L_1(t, s, u(s), v(s))ds \\ &\quad + |\gamma| L_2(t, s, u(s), v(s)) \Big\} e^{\theta(s-t)} ds \\ &\leq d(u, v)(|\beta| + |\gamma|) \sup_{t \in I} \int_a^b e^{(t-b)\theta_1} e^{\theta(s-t)} ds. \end{aligned}$$

Because $\theta > 0$ was arbitrary, we choose $\theta = \theta_1$. The above equation yields:

$$d(Tu, Tv) \leq \left[1 - e^{-\theta(b-a)}\right] d(u, v) = \lambda d(u, v)$$

where $\lambda = \left[1 - e^{-\theta(b-a)}\right] < 1$.

Thus, T is a contraction on $C(I, \mathbb{R})$. Now the existence of unique fixed point, i.e., the unique solution of (7.11), follows from the BCP.

7.5 Solutions of initial value and boundary value problems

We consider the Cauchy initial value problem. Suppose, $K\colon [a,b]\times\mathbb{R}\to\mathbb{R}$ is a continuous function and we want to find a continuous differentiable function u on $[a,b]$ such that:

$$\frac{du}{dt}=K(t,u);\ u(t_0)=u_0. \tag{7.14}$$

This is called the Cauchy initial value problem.

Theorem 7.6 *Suppose, the following condition is satisfied: for all $t,s\in I=[a,b]$, $u,v\in\mathbb{R}$ there exists a continuous function $L\colon I\times\mathbb{R}\times\mathbb{R}\to\mathbb{R}^+$ such that:*

$$\mid K(s,u(s))-K(s,v(s))\mid\leq L(s,u(s),v(s))\mid u(s)-v(s)\mid.$$

If $\alpha>0$ such that $L(s,u(s),v(s))\leq\alpha$ for all $s\in I$, the problem (7.14) *has a unique solution.*

Proof. To determine the existence and uniqueness of the initial value problem (7.14), first we convert it into an integral equation. Integrating (7.14) from t_0 to $t\in[a,b]$:

$$u(t)=u_0+\int_{t_0}^{t}K(s,u)ds. \tag{7.15}$$

Therefore, finding a solution of (7.14) is equivalent to finding a solution of the integral equation (7.15).

Let $X=C([a,b],\mathbb{R})$. Then (X,d) is a complete metric space, where d is the metric induced by the Bielecki norm on $C(I,\mathbb{R})$. The existence and uniqueness of solution follows from Corollary 7.2.

Example 7.3 Consider the following initial value problem:

$$\frac{du}{dt}=u,\ u(0)=1,\ t\in[0,T],\ T>0.$$

For this problem, we have $K(t,u)=u$, and

$$|K(s,u)-K(s,v)|=|u-v|.$$

Therefore, all the conditions of Theorem 7.6 are satisfied with $L(s,u(s),v(s))=\alpha=1$, and the initial value problem has a unique solution. The solution is the limit of sequence $\{z_n\}$, $z_n=Tz_{n-1}$, $n\in\mathbb{N}$ and $z_0\in C([0,T]$ and $\mathbb{R})$ is arbitrary with

$$[T(u)](t)=u(0)+\int_0^t u(s)ds.$$

Suppose we start with $z_0 = u(0) = 1$, and obtain the following terms of the sequence:

$$\begin{aligned} z_1(t) &= [T(z_0)](t) = 1 + \int_0^t 1 \cdot ds = 1 + t; \\ z_2(t) &= [T(z_1)](t) = 1 + \int_0^t z_1(s)ds = 1 + t + \frac{t^2}{2}; \\ &\vdots \\ z_n(t) &= [T(z_{n-1})](t) = 1 + \int_0^t z_{n-1}(s)ds = \sum_{i=0}^{n} \frac{t^i}{i!}. \end{aligned}$$

The limit of this sequence, i.e., $\lim_{n\to\infty} z_n(t) = e^t$ for all $t \in [0, T]$, is the unique solution of the initial value problem.

Example 7.4 Consider the following initial value problem:

$$\frac{du}{dt} = t + \arctan(u), \; u(0) = 0, \; t \in [0, \pi/2].$$

$K(t, u) = t + \arctan(u)$, and so, by the Lagrange mean value theorem,

$$|K(s, u) - K(s, v)| = |\arctan(u) - \arctan(v)| \leq |u - v| .$$

Therefore, all the conditions of Theorem 7.6 are satisfied with $L(s, u(s), v(s)) = \alpha = 1$, and the initial value problem has a unique solution. It is the limit of sequence $\{z_n\}$, $z_n = Tz_{n-1}$, $n \in \mathbb{N}$ and $z_0 \in C([0, T]$ and $\mathbb{R})$ is arbitrary with

$$[T(u)](t) = \int_0^t [s + \arctan(u(s))]ds.$$

Note that the calculation of the terms of sequence is not as easy as it was in the previous example, even when, the BCP ensures the existence and the uniqueness of the solution of the problem.

We next prove that the existence of the solution of the problem (7.14) can be established under some weaker conditions on the function K. Here, we use Theorem 3.13 of Chapter 3 and apply it to establish the existence of the solution of the initial value problem (7.14).

Theorem 7.7 *Let $r > 0$ be a fixed real value and the following conditions are satisfied:*

(i) $K \colon [a, b] \times \mathbb{R} \to \mathbb{R}$ is continuous.

(ii) $y_0 \in C(I,\mathbb{R})$ *such that* $y_0 + \int_{t_0}^{t} K(s,y_0)ds \in B(y_0,r)$.

(iii) If there exists a continuous function $L\colon I \times \mathbb{R} \times \mathbb{R} \to \mathbb{R}^+$ *such that for all* $s \in I$, $z \in B(y,r)$ *the following condition is satisfied:*

$$|K(s,y(s)) - K(s,z(s))| \leq L(s,y(s),z(s))|y(s) - z(s)|.$$

If $k > 0$ *such that* $L(s,y(s),z(s)) \leq k$ *for all* $s \in I$, *problem* (7.14) *has a solution.*

Proof. It is obvious that the initial value problem (7.14) is equivalent to the following integral equation:

$$u(t) = u_0 + \int_{t_0}^{t} K(s,u)ds. \tag{7.16}$$

Therefore, finding a solution of problem (7.14) is equivalent to finding a solution of the integral equation (7.16).

Define a function $\alpha\colon C(I,\mathbb{R}) \times C(I,\mathbb{R}) \to \mathbb{R}^+$ by

$$\alpha(u,v) = \begin{cases} 1, & \text{if } v \in B(u,r); \\ 0, & \text{otherwise.} \end{cases}$$

Note that $(C(I,\mathbb{R}),d)$ is a complete metric space, where d is the metric induced by the Bielecki norm on $C(I,\mathbb{R})$. Define $T\colon C(I,\mathbb{R}) \to C(I,\mathbb{R})$ by

$$[T(u)]t = u_0 + \int_{t_0}^{t} K(s,u)ds, \ t \in I.$$

If $\alpha(y,z) = 1$, that is, $z \in B(y,r)$, by the definition of T and (iii),

$$\begin{aligned} d(Tu,Tv) &= \sup_{t\in I} \left| \int_{t_0}^{t} K(s,y)ds - \int_{t_0}^{t} K(s,z)ds \right| e^{-\theta t} \\ &\leq \sup_{t\in I} \int_{t_0}^{t} | K(s,y) - K(s,z) | e^{-\theta t} ds \\ &\leq \sup_{t\in I} \int_{t_0}^{t} L(s,y(s),z(s))|y(s) - z(s)| e^{-\theta t} ds \\ &\leq kd(y,z) \cdot \sup_{t\in I} \int_{t_0}^{t} e^{\theta(s-t)} ds \\ &\leq \frac{k}{\theta} \cdot d(y,z) \sup_{t\in I} [1 - e^{\theta(t_0 - t)}]. \end{aligned}$$

Using $\theta = k$, the above inequality yields

$$d(Tu, Tv) \leq \sup_{t \in I}[1 - e^{\theta(t_0 - t)}] \cdot d(y, z) = \lambda d(y, z) \tag{7.17}$$

where $\lambda = \sup_{t \in I}[1 - e^{\theta(t_0 - t)}] = 1 - \inf_{t \in I} e^{\theta(t_0 - t)} < 1$. Thus, T is an α-ψ-contraction with $\psi(t) = \lambda t$ for all $t \in [0, \infty)$.

Now we shall show that T is α-admissible. If $\alpha(y, z) = 1$, that is, $z \in B(y, r)$, by the inequality (7.17) we have

$$d(Ty, Tz) \leq \lambda d(y, z) < d(y, z) \leq r.$$

This implies that $Tz \in B(Ty, r)$, i.e., $\alpha(Ty, Tz) = 1$. Therefore, T is α-admissible.

Suppose $\{x_n\}$ is a sequence in X such that $x_n \to x$ as $n \to \infty$. Then $n_0 \in \mathbb{N}$ such that $d(x_n, x) < r$ for all $n > n_0$. Choosing $n(i) = n_0 + i$, we obtain a subsequence $\{x_{n(i)}\}$ such that $d(x_{n(i)}, x) < r$ for all $i \in \mathbb{N}$, i.e., $\alpha(x_{n(i)}, x) = 1$ for all $i \in \mathbb{N}$. By condition (ii), $y_0 \in C(I, \mathbb{R})$ such that $\alpha(y_0, [Ty_0](t)) = 1$. Thus, by Theorem 3.13, T has a fixed point which is a solution of integral Equation (7.15).

We next study the initial value problem for some classes of second order differential equations. We first consider the autonomous case: let $T > 0$, $I = [0, T]$ and consider the problem:

$$x''(t) = \xi(x(t)),\ t \in I, \qquad x(0) = \alpha,\ x'(0) = \beta, \tag{7.18}$$

where α, $\beta \in \mathbb{R}$ and $\xi \colon \mathbb{R} \to \mathbb{R}$ is a continuous function.

The Green's function associated with (7.18) is given by

$$G(t, \tau) = \begin{cases} t - \tau, & \text{for } t > \tau; \\ 0, & \text{for } 0 \leq t \leq \tau. \end{cases}$$

It can be easily seen that the solution to (7.18) is given by the solution of the integral equation of the following form:

$$\begin{aligned} x(t) &= \int_0^T G(t, \tau)\xi(x(\tau))d\tau + \zeta(t) \\ &= \int_0^t (t - \tau)\xi(x(\tau))d\tau + \zeta(t), \text{ for } t \in I, \end{aligned} \tag{7.19}$$

where $\zeta(t) = \alpha + \beta t$.

To define the concept of solution to (7.18), we consider $C^2(I, \mathbb{R})$, the space of all functions from I into $\mathbb{R}$ having a continuous second order derivative on I. A solution to (7.18) is a function $x \in C^2(I, \mathbb{R})$ which satisfies the conditions in (7.18). The procedure we follow to prove the solutions to problem (7.18) is to establish a connection between them and the solutions to the integral equation (7.19).

To study the existence of solutions to the integral equation (7.19), we consider $C(I,\mathbb{R})$, the Banach space of all continuous functions from $I = [0,T]$ into $\mathbb{R}$, with the Bielecki norm given by $\|u\|_B := \sup_{t\in I} |u(t)|e^{-\theta t}$, $u \in C(I,\mathbb{R})$, where $\theta > 0$ is arbitrary but fixed. The induced metric d on $C(I,\mathbb{R})$ is given by:

$$d(u,v) = \sup_{t\in I} |u(t) - v(t)|e^{-\theta t} \quad \text{for all} \quad u, v \in C(I,\mathbb{R}).$$

Define an operator $\Phi\colon C(I,\mathbb{R}) \to C(I,\mathbb{R})$ by

$$[\Phi(x)](t) = \int_0^t (t-\tau)\xi(x(\tau))d\tau + \zeta(t), \text{ for } t \in I, x \in C(I,\mathbb{R}).$$

It is obvious that the solutions to the integral equation (7.19) coincide with the fixed points of the operator Φ, i.e., $x \in C(I,\mathbb{R})$ such that $[\Phi(x)](t) = x(t)$, for every $t \in I$. Moreover, a solution to (7.18) trivially satisfies the integral Equation (7.19) and, if $x \in C(I,\mathbb{R})$ is a solution to the integral Equation (7.19); then we can prove that $x \in C^2(I,\mathbb{R})$ and the conditions in (7.18) are fulfilled. Hence, the solutions to the initial value problem (7.18) are the fixed points of the operator Φ. All these considerations allow us to prove the existence and uniqueness of solution to the initial value problem (7.18), as established in the following theorem.

Theorem 7.8 *Suppose that the following conditions are satisfied:* $L > 0$ *such that for all* $x, y \in \mathbb{R}$, *we have*

$$\mid \xi(x) - \xi(y) \mid \leq L \mid x - y \mid .$$

Then the initial value problem (7.18) *has a unique solution.*

Proof. We consider the metric d induced by the Bielecki norm. Since $\theta > 0$ can be selected arbitrarily, we choose $\theta = LT > 0$; then, for all $x, y \in C(I,\mathbb{R})$,

$$\begin{aligned} d(\Phi(x), \Phi(y)) &= \sup_{t\in I} \left| \int_0^t (t-\tau)[\xi(x(\tau)) - \xi(y(\tau))]e^{-\theta t} d\tau \right| \\ &\leq \sup_{t\in I} \int_0^t (t-\tau) \mid \xi(x(\tau)) - \xi(y(\tau)) \mid e^{-\theta t} d\tau \\ &\leq \sup_{t\in I} \int_0^t (t-\tau)\, L \mid x(\tau) - y(\tau) \mid e^{-\theta\tau} e^{\theta(\tau - t)} d\tau \\ &\leq L d(x,y) \sup_{t\in I} \left\{ e^{-\theta t} \int_0^t (t-\tau) e^{\theta\tau} d\tau \right\} \\ &\leq L d(x,y) \sup_{t\in I} \left\{ t e^{-\theta t} \int_0^t e^{\theta\tau} d\tau \right\} \\ &\leq \frac{LT}{\theta}(1 - e^{-\theta T}) d(x,y). \end{aligned}$$

Since $\theta = LT$, we have, for all $x, y \in C(I, \mathbb{R})$,

$$d(\Phi(x), \Phi(y)) \leq \lambda \cdot d(x, y),$$

where $0 < \lambda = \frac{LT}{\theta}(1-e^{-\theta T}) = 1-e^{-\theta T} < 1$. Thus, Φ is a contraction mapping on $C(I, \mathbb{R})$ and, in consequence, Φ has a unique fixed point in $C(I, \mathbb{R})$, which is the unique solution to the initial value problem (7.18).

Instead of problem (7.18), we could have considered a nonautonomous problem of the type

$$x''(t) = \xi(t, x(t)),\ t \in I = [0, T], \quad x(0) = \alpha,\ x'(0) = \beta, \tag{7.20}$$

where $\alpha, \beta \in \mathbb{R}$ and $\xi \colon I \times \mathbb{R} \to \mathbb{R}$ is continuous.

Taking $\zeta(t) = \alpha + \beta t$, $t \in I$, it is clear that the solutions to (7.20) coincide with those of the integral equation

$$\begin{aligned} x(t) &= \int_0^T G(t,\tau)\xi(\tau, x(\tau))d\tau + \zeta(t) \\ &= \int_0^t (t-\tau)\xi(\tau, x(\tau))d\tau + \zeta(t), \text{ for } t \in I, \end{aligned} \tag{7.21}$$

and also with the fixed points of the mapping $\widetilde{\Phi} \colon [C(I, \mathbb{R})] \to C(I, \mathbb{R})$ defined as

$$[\widetilde{\Phi}(x)](t) = \int_0^t (t-\tau)\xi(\tau, x(\tau))d\tau + \zeta(t), \text{ for } t \in I, x_1, \ldots, x_k \in C(I, \mathbb{R}).$$

Thus, the extension of Theorem 7.8 follows.

Theorem 7.9 *Suppose that the following conditions are satisfied:*

(a) $\xi \colon I \times \mathbb{R} \to \mathbb{R}$ *is a continuous function.*

(b) There exists a nonnegative and integrable function $L : I \to \mathbb{R}$ *such that for all* $t \in I$ *and* $x, y \in \mathbb{R}$,

$$\mid \xi(t, x) - \xi(t, y) \mid \leq L(t) \mid x - y \mid$$

and $\theta > 0$ *such that*

$$\sup_{t \in I} \left\{ e^{-\theta t} \int_0^t (t-\tau)L(\tau)e^{\theta\tau}\, d\tau \right\} < 1. \tag{7.22}$$

Then the initial value problem (7.20) has a unique solution.

Proof. We consider again $C(I, \mathbb{R})$ with the induced metric by the Bielecki norm, where $\theta > 0$ is given by the statement. Hence, similarly to the proof of Theorem 7.8, we have, for all $x, y \in C(I, \mathbb{R})$,

$$d_B(\widetilde{\Phi}(x), \widetilde{\Phi}(y)) \leq \sup_{t\in I} \int_0^t (t-\tau)\, L(\tau) \mid x(\tau) - y(\tau) \mid e^{-\theta\tau} e^{\theta(\tau - t)} d\tau$$
$$\leq \lambda\, d(x, y),$$

where $0 < \lambda := \sup_{t\in I} \left\{ e^{-bt} \int_0^t (t-\tau) L(\tau) e^{b\tau} d\tau \right\} < 1$. Therefore, $\widetilde{\Phi}$ is a contraction mapping on $C(I, \mathbb{R})$ and, in consequence, $\widetilde{\Phi}$ has a unique fixed point in $C(I, \mathbb{R})$, which is the unique solution to the initial value problem (7.20).

Remark 7.3 Condition (7.22) trivially holds if $\theta > 0$ such that

$$\sup_{t\in I} \left\{ e^{-\theta t} \int_0^t L(\tau) e^{\theta\tau}\, d\tau \right\} < \frac{1}{T},$$

or if

$$\lim_{\theta\to\infty} \left(\sup_{t\in I} \left\{ e^{-\theta t} \int_0^t L(\tau) e^{b\tau}\, d\tau \right\} \right) < \frac{1}{T},$$

since $\sup_{t\in I} \left\{ e^{-\theta t} \int_0^t (t-\tau) L(\tau) e^{\theta\tau}\, d\tau \right\} \leq T \sup_{t\in I} \left\{ e^{-\theta t} \int_0^t L(\tau) e^{\theta\tau}\, d\tau \right\}$.
In particular, if L is non-negative, integrable and bounded (there exists $\mathcal{L} > 0$ such that $L(t) \leq \mathcal{L}$, $t \in I$), then

$$\sup_{t\in I} \left\{ e^{-\theta t} \int_0^t (t-\tau) L(\tau) e^{\theta\tau}\, d\tau \right\} \leq T\mathcal{L} \sup_{t\in I} \left\{ e^{-\theta t} \int_0^t e^{\theta\tau}\, d\tau \right\} = T\mathcal{L} \frac{1 - e^{-\theta T}}{\theta},$$

so that it is enough to choose $\theta > T\mathcal{L}$.

Example 7.5 Let $I = [0, 1]$ to examine whether the following problem has a unique solution or not:

$$\left.\begin{array}{l} -x''(t) + \lambda \sin(x) = f(t),\ t \in I; \\ x(0) = x'(0) = 0 \end{array}\right\} \tag{7.23}$$

where $\lambda \in \mathbb{R}$ is nonzero and $f\colon I \to \mathbb{R}$ is a continuous function. Then, for $\xi(t, x) = \lambda \sin(x) - f(t)$ we have

$$|\xi(t, x) - \xi(t, y)| = |\lambda|\, |\sin(x) - \sin(y)| \leq |\lambda|\, |x - y|.$$

Also,

$$\sup_{t\in I} \left\{ e^{-|\lambda| t} \int_0^t (t-\tau) L(\tau) e^{|\lambda|\tau}\, d\tau \right\} \leq 1 - e^{-|\lambda| t} < 1.$$

Thus, all the conditions of Theorem 7.9 are satisfied with $L(t) = \theta = |\lambda|$, and (7.23) has a unique solution.

We now consider a problem of the type

$$\left.\begin{array}{ll} x''(t) = \xi(t, x(t)), & t \in I = [0, T],\ t \neq t_j,\ j = 1, \ldots, l, \\ x(t_j^+) = \alpha_j, \quad x'(t_j^+) = \beta_j, & j = 0, \ldots, l, \end{array}\right\} \quad (7.24)$$

where α_j, $\beta_j \in \mathbb{R}$, $j = 0, \ldots, l$, $0 = t_0 < t_1 < t_2 < \cdots < t_l < t_{l+1} = T$ and $\xi \colon I \times \mathbb{R} \to \mathbb{R}$ is such that, for every $j = 0, \ldots, l$, its restriction to $(t_j, t_{j+1}] \times \mathbb{R}$ is continuous and admits a continuous extension to the set $[t_j, t_{j+1}] \times \mathbb{R}$.

To define the solution to problem (7.24), we consider the spaces

$$\begin{aligned} PC(I, \mathbb{R}) &= \{x : I \to \mathbb{R} \,:\, x \text{ is continuous in } I \setminus \{t_1, \ldots, t_l\} \\ &\qquad\qquad \text{and } \exists\, x(t_j^+),\ x(t_j^-) = x(t_j),\ j = 1, \ldots, l\} \\ &= \{x : I \to \mathbb{R} \,:\, x \in C((t_j, t_{j+1}), \mathbb{R}),\ j = 0, \ldots, l, \text{and } \exists\, x(0^+) = x(0), \\ &\qquad\qquad x(T^-) = x(T),\ x(t_j^+),\ x(t_j^-) = x(t_j),\ j = 1, \ldots, l\}, \end{aligned}$$

$$\begin{aligned} E := \{x \in PC(I, \mathbb{R}) \,&:\, x \in C^2(I \setminus \{t_1, \ldots, t_l\}, \mathbb{R}) \\ &\qquad \text{and } \exists\, x'(t_j^+),\ x'(t_j^-),\ x''(t_j^+),\ x''(t_j^-),\ j = 1, \ldots, l\} \\ = \{x \in PC(I, \mathbb{R}) \,&:\, x \in C^2((t_j, t_{j+1}), \mathbb{R}),\ j = 0, \ldots, l, \text{ and } \exists\, x'(0^+) = x'(0), \\ &\qquad x''(0^+) = x''(0),\ x'(T^-) = x'(T),\ x''(T^-) = x''(T), \\ &\qquad x'(t_j^+),\ x'(t_j^-),\ x''(t_j^+),\ x''(t_j^-),\ j = 1, \ldots, l\}. \end{aligned}$$

Hence, a solution to (7.24) is a function $x \in E$ satisfying the conditions in (7.24). For problem (7.24), the Green's function $G : I \times I \to \mathbb{R}$ is given by

$$G(t, \tau) = \begin{cases} t - \tau, & t_j < \tau < t \leq t_{j+1}, \text{ for some } j = 0, \ldots, l, \\ 0, & \text{otherwise.} \end{cases}$$

We can also use the functions $G_j : [t_j, t_{j+1}] \times [t_j, t_{j+1}] \to \mathbb{R}$, $j = 0, \ldots, l$, defined by $G_j(t, \tau) = t - \tau$, if $t_j \leq \tau < t \leq t_{j+1}$, and $G_j(t, \tau) = 0$, if $t_j \leq t \leq \tau \leq t_{j+1}$, in such a way that $G(t, \tau) = G_j(t, \tau)$, for $(t, \tau) \in (t_j, t_{j+1}) \times (t_j, t_{j+1})$. Therefore, taking $\zeta_j(t) = \alpha_j + \beta_j(t - t_j)$, for $t \in (t_j, t_{j+1}]$, and $j = 0, \ldots, l$, the solutions to (7.24) are the solutions $x \in PC(I, \mathbb{R})$ to the family of integral equations:

$$\begin{aligned} x(t) &= \int_0^T G(t, \tau)\xi(\tau, x(\tau))d\tau + \zeta_j(t) \\ &= \int_{t_j}^{t_{j+1}} G_j(t, \tau)\xi(\tau, x(\tau))d\tau + \zeta_j(t) \\ &= \int_{t_j}^{t} (t - \tau)\xi(\tau, x(\tau))d\tau + \zeta_j(t),\ t \in (t_j, t_{j+1}]. \end{aligned}$$

$PC(I, \mathbb{R})$ is a Banach space with the norm

$$\|x\|_{PC} := \max_{0 \leq j \leq l} \sup_{t \in (t_j, t_{j+1}]} \left\{ |\, x(t) \,|\, e^{-\theta(t - t_j)} \right\},\ x \in PC(I, \mathbb{R}),$$

where $\theta > 0$ is arbitrary but fixed. The distance induced by $\|\cdot\|_{PC}$ is $d_{PC}(x,y) := \max_{0\le j\le l}\sup_{t\in(t_j,t_{j+1}]}\left\{| x(t)-y(t) | e^{-\theta(t-t_j)}\right\}$, for $x, y \in PC(I,\mathbb{R})$.

We consider the mapping $\widehat{\Phi}\colon PC(I,\mathbb{R}) \to PC(I,\mathbb{R})$, given by

$$[\widehat{\Phi}(x)](t) = \int_{t_j}^{t} (t-\tau)\xi(\tau, x(\tau))d\tau + \zeta_j(t),$$

for $t \in (t_j, t_{j+1}]$, $j = 0,\dots,l$, and $x \in PC(I,\mathbb{R})$, whose fixed points are the solutions sought.

We prove the following existence and uniqueness result for problem (7.24).

Theorem 7.10 *Suppose that the following conditions are satisfied:*

(a) $\xi\colon I\times\mathbb{R}\to\mathbb{R}$ is such that its restriction to $(t_j,t_{j+1}]\times\mathbb{R}$ is continuous and admits a continuous extension to the set $[t_j,t_{j+1}]\times\mathbb{R}$, for $j=0,\dots,l$.

(b) There exists a non-negative and integrable function $L : I \to \mathbb{R}$ such that for all $t \in I$ and $x, y \in \mathbb{R}$, we have

$$| \xi(t,x) - \xi(t,y) | \le L(t) \mid x - y \mid$$

and $\theta > 0$ such that

$$\max_{0\le j\le l}\sup_{t\in(t_j,t_{j+1}]}\left\{e^{-\theta t}\int_{t_j}^{t}(t-\tau)L(\tau)e^{\theta\tau}\,d\tau\right\} < 1. \tag{7.25}$$

Then the initial value problem (7.24) has a unique solution.

Proof. We take $PC(I,\mathbb{R})$ with the metric d_{PC}, where $\theta > 0$ is given in the statement. Analogous to the proof of Theorem 7.9, we have, for all $x, y \in PC(I,\mathbb{R})$,

$$\begin{aligned}
&d_{PC}(\widehat{\Phi}(x),\widehat{\Phi}(y))\\
&\le \max_{0\le j\le l}\sup_{t\in(t_j,t_{j+1}]}\int_{t_j}^{t}(t-\tau)\,L(\tau)\mid x(\tau)-y(\tau)\mid e^{-\theta(\tau-t_j)}e^{\theta(\tau-t)}d\tau\\
&\le \lambda d_{PC}(x,y),
\end{aligned}$$

where $0 < \lambda := \max_{0\le j\le l}\sup_{t\in(t_j,t_{j+1}]}\left\{e^{-\theta t}\int_{t_j}^{t}(t-\tau)L(\tau)e^{\theta\tau}d\tau\right\} < 1$. Thus, $\widehat{\Phi}$ is a contraction on $PC(I,\mathbb{R})$ and the the existence and uniqueness of solution follow.

Remark 7.4 If L is non-negative, integrable and bounded (with upper bound

$\mathcal{L} > 0$), condition (7.25) trivially holds since

$$\max_{0\le j\le l} \sup_{t\in(t_j,t_{j+1}]} \left\{ e^{-\theta t} \int_{t_j}^{t} (t-\tau)L(\tau)e^{\theta\tau}\, d\tau \right\}$$

$$\le T\mathcal{L} \max_{0\le j\le l} \sup_{t\in(t_j,t_{j+1}]} \left\{ e^{-\theta t} \int_{t_j}^{t} e^{\theta\tau}\, d\tau \right\} = \frac{T\mathcal{L}}{\theta} \max_{0\le j\le l} \left\{ 1 - e^{-\theta(t_{j+1}-t_j)} \right\}$$

and the same choice of $\theta > T\mathcal{L}$ is useful.

Next, we consider the solution of periodic boundary value problem when the lower or the upper solution of the problem exists. Consider the following periodic boundary value problem:

$$\frac{du}{dt} = K(t,u); u(0) = u(T) \tag{7.26}$$

where $K\colon [0,T]\times\mathbb{R}\to\mathbb{R}$ is a continuous function.

Definition 7.2 (i) A solution to (7.26) is a function $u^* \in C^1([0,T],\mathbb{R})$ satisfying conditions in (7.26).

(ii) A lower solution for (7.26) is a function $\alpha \in C^1([0,T],\mathbb{R})$ such that

$$\begin{aligned} \frac{d\alpha}{dt} &\le K(t,\alpha); \\ \alpha(0) &\le \alpha(T). \end{aligned}$$

An upper solution for (7.26) satisfies the reversed inequalities.

Definition 7.3 Let $(X,\sqsubseteq)$ be a partially ordered set and $T\colon X\to X$ be a mapping. T is said to be nondecreasing if $x\sqsubseteq y \implies Tx\sqsubseteq Ty$.

The space $C([a,b],\mathbb{R})$ is a partially ordered set with the order $\sqsubseteq$ defined by $x\sqsubseteq y$ if and only if $x(t)\le y(t)$ for all $t\in[a,b]$.

To prove the existence and uniqueness of the solution of problem (7.26), we need the following theorem due to Nieto and R-López [13]. The mapping in this theorem follows a weaker contractive condition than the Banach contraction.

Theorem 7.11 (Nieto and R-López [13]) *Let $(X,\sqsubseteq)$ be a partially ordered set and assume a metric d in X such that (X,d) is a complete metric space. Let $T\colon X\to X$ be a nondecreasing mapping such that $k\in[0,1)$ with*

$$d(Tx,Ty)\le kd(x,y),\ y\sqsubseteq x. \tag{7.27}$$

Suppose at least one of the following conditions is satisfied:

(a) T is continuous.

(b) If a nondecreasing sequence $\{x_n\}\to x\in X$, then $x_n\sqsubseteq x$, $\forall n$.

If $x_0 \in X$ with $x_0 \sqsubseteq Tx_0$, then T has a fixed point. In addition, if every pair of elements in X has a lower or upper bound, the fixed point of T is unique.

Theorem 7.12 (Nieto and R-López [13]) *Consider problem (7.26) in which K is continuous and $\lambda > 0, \mu > 0$ with $\mu < \lambda$ such that for all $t \in [0, T]$ and $x(t), y(t) \in \mathbb{R}$ with $x(t) \leq y(t)$ we have*

$$0 \leq K(t, y) + \lambda y - [K(t, x) + \lambda x] \leq \mu(y - x).$$

The existence of a lower solution for (7.26) provides unique solution of (7.26).

Proof. Problem (7.26) is written as:

$$u'(t) + \lambda u(t) = K(t, u(t)) + \lambda u(t);$$
$$u(0) = u(T).$$

This problem is equivalent to the integral equation

$$u(t) = \int_0^T G(t, s)[K(s, u(s)) + \lambda u(s)]ds$$

where

$$G(t, s) = \begin{cases} \dfrac{e^{\lambda(T+s-t)}}{e^{\lambda T} - 1}, & 0 \leq s < t \leq T; \\ \dfrac{e^{\lambda(s-t)}}{e^{\lambda T} - 1}, & 0 \leq t < s \leq T. \end{cases}$$

Define an operator $T\colon C(I, \mathbb{R}) \to C(I, \mathbb{R})$ by

$$T(u(t)) = \int_0^T G(t, s)[K(s, u(s)) + \lambda u(s)]ds, \ t \in I \tag{7.28}$$

where $I = [0, T]$. Note that a fixed point of T is a solution of (7.26).

We shall show that all the conditions of Theorem 7.11 are satisfied. $X = C(I, \mathbb{R})$ is a partially ordered set with the partial order $\sqsubseteq$ defined by

$$x \sqsubseteq y \text{ if and only if } x(t) \leq y(t) \text{ for all } t \in [a, b]$$

and it is a complete metric space with the metric induced by the supremum norm defined on X given by:

$$d(x, y) = \sup_{t \in I} |x(t) - y(t)|, \ x, y \in C(I, \mathbb{R}).$$

Suppose, $\{x_n\}$ is a monotone nondecreasing sequence (with respect to the partial order $\sqsubseteq$) in X. Then

$$x_1(t) \leq x_2(t) \leq \cdots \leq x_n(t) \leq \cdots \text{ for all } t \in I.$$

For every $t \in I$ the convergence of this sequence of real numbers to $x(t)$ implies

$$x_n(t) \leq x(t), \text{ for all } t \in I, \ n \in \mathbb{N}.$$

Therefore, $x_n \sqsubseteq x$ for all $n \in \mathbb{N}$ and the limit is an upper bound for all the terms x_n in the sequence.

The mapping T is nondecreasing. Indeed, since for all $v \sqsubseteq u$

$$K(t,u) + \lambda u \geq K(t,v) + \lambda v,$$

and for $v \sqsubseteq u$ we have

$$\begin{aligned} T(v(t)) &= \int_0^T G(t,s)[K(s,v(s)) + \lambda v(s)]ds \\ &\leq \int_0^T G(t,s)[K(s,u(s)) + \lambda u(s)]ds \\ &= T(u(t)) \end{aligned}$$

for all $t \in I$. Therefore, $Tv \sqsubseteq Tu$.

Further, if $v \sqsubseteq u$ we have

$$\begin{aligned} d(Tu,Tv) &= \sup_{t\in I} |T(u(t)) - T(v(t))| \\ &\leq \sup_{t\in I} \int_0^T G(t,s)\,|K(s,u(s)) + \lambda u(s) - K(s,v(s)) - \lambda v(s)|\,ds. \end{aligned}$$

Using the condition $0 \leq K(t,y) + \lambda y - [K(t,x) + \lambda x] \leq \mu(y - x)$, the above inequality yields

$$\begin{aligned} d(Tu,Tv) &\leq \sup_{t\in I} \int_0^T G(t,s)\mu(u(s) - v(s))ds \\ &\leq \mu d(u,v) \sup_{t\in I} \int_0^T G(t,s)ds \\ &= \mu d(u,v) \sup_{t\in I} \frac{1}{e^{\lambda T} - 1} \left(\left[\frac{e^{\lambda(T+s-t)}}{\lambda} \right]_0^t + \left[\frac{e^{\lambda(s-t)}}{\lambda} \right]_t^T \right) \\ &= \mu d(u,v) \frac{1}{\lambda(e^{\lambda T} - 1)} (e^{\lambda T} - 1) \\ &= \frac{\mu}{\lambda} d(u,v). \end{aligned}$$

Thus, T satisfies the condition (7.27) of Theorem 7.11 with constant $k = \frac{\mu}{\lambda}$.

We now show that $\alpha \in X$ such that $\alpha \sqsubseteq T\alpha$. Suppose α is a lower solution of problem (7.26), i.e.,

$$\begin{aligned} \frac{d\alpha(t)}{dt} &\leq K(t,\alpha); \\ \alpha(0) &\leq \alpha(T). \end{aligned}$$

Then, we have

$$\frac{d\alpha(t)}{dt} + \lambda\alpha(t) \leq K(t, \alpha(t)) + \lambda\alpha(t).$$

Multiplying by $e^{\lambda t}$ and rewriting we get

$$\frac{d}{dt}(\alpha e^{\lambda t}) \leq [K(t, \alpha(t)) + \lambda\alpha(t)]e^{\lambda t}.$$

By integrating from 0 to $t \in I$, we obtain

$$\alpha(t)e^{\lambda t} \leq \alpha(0) + \int_0^t [K(s, \alpha(s)) + \lambda\alpha(s)]e^{\lambda s}ds \text{ for all } t \in I. \tag{7.29}$$

Since $\alpha(0) \leq \alpha(T)$, using (7.29) we obtain

$$\begin{aligned} \alpha(0)e^{\lambda T} &\leq \alpha(T)e^{\lambda T} \\ &\leq \alpha(0) + \int_0^T [K(s, \alpha(s)) + \lambda\alpha(s)]e^{\lambda s}ds \end{aligned}$$

which yields

$$\alpha(0) \leq \int_0^T \frac{e^{\lambda s}}{e^{\lambda T} - 1}[K(s, \alpha(s)) + \lambda\alpha(s)]ds.$$

The above inequality with (7.29) gives

$$\begin{aligned} \alpha(t)e^{\lambda t} &\leq \int_0^T \frac{e^{\lambda s}}{e^{\lambda T} - 1}[K(s, \alpha(s)) + \lambda\alpha(s)]ds + \int_0^t [K(s, \alpha(s)) + \lambda\alpha(s)]e^{\lambda s}ds \\ &= \int_0^t \frac{e^{\lambda(T+s)}}{e^{\lambda T} - 1}[K(s, \alpha(s)) + \lambda\alpha(s)]ds \\ &\quad + \int_t^T \frac{e^{\lambda s}}{e^{\lambda T} - 1}[K(s, \alpha(s)) + \lambda\alpha(s)]ds. \end{aligned}$$

With the definition of the function G the above inequality gives

$$\alpha(t) \leq \int_0^T G(t, s)[K(s, \alpha(s)) + \lambda\alpha(s)]ds = T(\alpha(t)), \quad \text{for all } t \in I.$$

Thus, $\alpha(t) \sqsubseteq T(\alpha(t))$ and the solution of problem 7.26 follows from Theorem 7.11. It is easy to see that the condition for uniqueness in Theorem 7.11 is satisfied and therefore the solution is unique.

Remark 7.5 If any one of the upper solutions is known, the assumption "there exists $x_0 \in X$ with $x_0 \sqsubseteq Tx_0$", in Theorem 7.11 is replaced by the assumption "there exists $x_0 \in X$ with $Tx_0 \sqsubseteq x_0$". By repetition of the same process, one can find the solution of a boundary value problem.

7.6 Difference equations and cyclic systems

The study of nonlinear difference equations, which plays an important role in modelling problems in economics, biology, ecology, genetics, psychology, sociology, probability theory and others (see introduction from [14] and references) is a topic of great interest. The difference equations of order greater than one are studied by many authors. Here, we consider the equilibrium points of such difference equations. We need the following definitions.

Definition 7.4 Let I be an interval of real numbers and $T\colon I^k \to \mathbb{R}$ be a mapping. Consider the following k-th order nonlinear difference equation:

$$x_{n+k} = T(x_n, x_{n+1}, \ldots, x_{n+k-1}), \quad n = 1, 2, \ldots. \tag{7.30}$$

(i) A function $\psi\colon \mathbb{N} \to I$ is called a solution of difference equation (7.30) if $\psi(n+k) = \psi_{n+k} = T(\psi_1, \psi_2, \ldots, \psi_{n+k-1})$ for all $n \in \mathbb{N}$.

(ii) A point $v \in I$ is called an equilibrium point of difference equation (7.30) if $v = T(v, v, \ldots, v)$, i.e., if v is a fixed point of T.

(iii) The equilibrium point $u \in A$ is called a global attractor if, for arbitrary points $x_1, x_2, \ldots, x_k \in A$, we have $\lim\limits_{n\to\infty} x_n = u$.

Theorem 7.13 *Suppose, I is closed, $T\colon I^k \to I$ and there exist constants $\alpha_i \geq 0$, $1 \leq i \leq k$ such that*

$$\begin{aligned} &|T(x_1, x_2, \ldots, x_k) - T(x_2, x_3, \ldots, x_{k+1})| \\ &\leq \alpha_1|x_1 - x_2| + \alpha_2|x_2 - x_3| + \cdots + \alpha_k|x_k - x_{k+1}| \end{aligned}$$

for all $x_1, x_2, \ldots, x_k, x_{k+1} \in I$. Then for every set of initial conditions $\{x_0, x_1, \ldots, x_{k-1}\} \subset I$, the difference equation (7.30) *has a unique equilibrium point in I. Furthermore, the equilibrium point of* (7.30) *is a global attractor.*

Proof. Consider the metric space (I, d) where $d(x, y) = |x - y|$ for all $x, y \in I$. (I, d) is a complete metric space and T satisfies the condition (1.19) of Prešić's fixed point theorem, i.e., Theorem 1.11. Therefore, by Theorem 1.11 $u \in I$ such that

$$T(u, u, \ldots, u) = u.$$

Thus, u is a unique equilibrium point of (7.30) in I. Again, by Theorem 1.11 for every set of initial conditions $\{x_0, x_1, \ldots, x_{k-1}\} \subset I$, the sequence $\{x_n\}$ defined by

$$x_{n+k} = T(x_n, x_{n+1}, \ldots, x_{n+k-1}), \quad n = 1, 2, \ldots,$$

converges to the unique fixed point u and u is a global attractor.

Many authors studied the global behavior of the recursive sequence

$$x_{n+2} = \frac{a + bx_{n+1}}{c + dx_n}, \quad n = 0, 1, 2, \ldots, \tag{7.31}$$

where a, b, c and d are non-negative real numbers (see, [7, 9, 10]). Note that if $b, d \neq 0$, replacing x_n by $\left(\frac{b}{d}\right) x_n$, Equation (7.31) can be written as

$$x_{n+2} = \frac{\alpha + x_{n+1}}{\beta + x_n}, \quad n = 0, 1, 2, \ldots, \tag{7.32}$$

where α, β are non-negative real numbers. As an example, consider the second order difference equation (7.32) and prove the following corollary of Theorem 7.13.

Corollary 7.3 *If $\alpha \leq \beta$ and $\beta > 2$, the difference Equation (7.32) has a unique equilibrium point in the interval $I = [0, \infty)$ which is a global attractor.*

Proof. Note that I is complete metric space with the usual metric and the corresponding operator $T \colon I^2 \to I$ for Equation (7.32) is given by

$$T(x, y) = \frac{\alpha + y}{\beta + x} \quad \text{for all} \quad x, y \in I.$$

For any $x, y, z \in X$ we have

$$\begin{aligned}
|T(x,y) - T(y,z)| &= \left| \frac{\alpha + y}{\beta + x} - \frac{\alpha + z}{\beta + y} \right| \\
&= \frac{|\alpha(y - x) + \beta(y - z) + y^2 - xz|}{(\beta + x)(\beta + y)} \\
&= \frac{|(\alpha + y)(y - x) + (\beta + x)(y - z)|}{(\beta + x)(\beta + y)} \\
&\leq \frac{\alpha + y}{(\beta + x)(\beta + y)}|y - x| + \frac{1}{\beta + y}|y - z| \\
&\leq \frac{1}{\beta}[|y - x| + |y - z|] \quad (\text{as } \alpha \leq \beta \text{ and } x, y \in [0, \infty)).
\end{aligned}$$

Since, $\beta > 2$, T satisfies the contractive condition of Theorem 7.13 with $\alpha_1 = \alpha_2 = \frac{1}{\beta}$. Thus, by Theorem 7.13, Equation (7.32) has a unique equilibrium point in the interval $I = [0, \infty)$ which is a global attractor.

Next we consider the cyclic systems in n variables of the following form:

$$\left.\begin{aligned}
x_1 &= T(x_2, x_3, \ldots, x_{k+1}) \\
x_2 &= T(x_3, x_4, \ldots, x_{k+2}) \\
&\vdots \\
x_{n-1} &= T(x_n, x_1, \ldots, x_{k-1}) \\
x_n &= T(x_1, x_2, \ldots, x_k)
\end{aligned}\right\} \tag{7.33}$$

where $T\colon I^k \to I$ is a function, I is a closed interval and $k \in \mathbb{N}$ satisfies $1 \leq k < n$. If $m \in \mathbb{N}$ and $x \in I$, we define the iterates of T on x by $T^0(x, x, \ldots, x) = x$ and

$$T^m(x, x, \ldots, x) = T(T^{m-1}(x, x, \ldots, x), T^{m-1}(x, x, \ldots, x), \ldots, T^{m-1}(x, x, \ldots, x)).$$

A point $u \in I$ is said to be a periodic point of T of period m if

$$u = T^m(u, u, \ldots, u).$$

A point $(u_1, u_2, \ldots, u_k) \in I^k$ is said to be a solution of the system (7.33) if it satisfies (7.33) and is said to be an m-periodic solution of (7.33) if it is a solution and a periodic point of T of period m.

Theorem 7.14 *Let there exist non-negative constants $\alpha_1, \alpha_2, \ldots, \alpha_k$ such that $\sum_{i=1}^{k} \alpha_i < 1$ and*

$$|T(x_1, \ldots, x_k) - T(x_2, \ldots, x_{k+1})| \leq \alpha_1|x_1 - x_2| + \cdots + \alpha_k|x_k - x_{k+1}| \quad (7.34)$$

for all $x_1, x_2, \ldots, x_k, x_{k+1} \in I$. Then a periodic solution of system (7.33) *exists and it is the only solution of* (7.33).

Proof. We first show the existence of a solution. From (7.34) it is clear that the mapping T satisfies the condition (1.19) of Prešić's fixed point theorem, and so $u \in I$ of T such that $T(u, u, \ldots, u) = u$. Notice that

$$T^m(u, u, \ldots, u) = T(T^{m-1}(u, u, \ldots, u), \ldots, T^{m-1}(u, u, \ldots, u)) = u$$

for all $m \in \mathbb{N}$. Therefore, u is a periodic point of T of order $m \in \mathbb{N}$. Since $u = T(u, u, \ldots, u)$, the point $(u, u, \ldots, u)$ is a solution of (7.33). We shall show that it is the unique solution of (7.33).

Suppose, $(u_1, u_2, \ldots, u_k) \in I^k$ is another solution of system (7.33); then we have:

$$\begin{aligned} |u_1 - u_2| &= |T(u_2, u_3, \ldots, u_{k+1}) - T(u_3, u_4, \ldots, u_{k+2})| \\ &\leq \alpha_1|u_2 - u_3| + \alpha_2|u_3 - u_4| + \cdots + \alpha_k|u_{k+1} - u_{k+2}|. \end{aligned}$$

Similarly, we can show that

$$\begin{aligned} |u_2 - u_3| &\leq \alpha_1|u_3 - u_4| + \alpha_2|u_4 - u_5| + \cdots + \alpha_k|u_{k+2} - u_{k+3}| \\ &\vdots \\ |u_n - u_1| &\leq \alpha_1|u_1 - u_2| + \alpha_2|u_2 - u_3| + \cdots + \alpha_k|u_k - u_{k+1}|. \end{aligned}$$

By the above inequalities we obtain:

$$\begin{aligned}
& |u_1 - u_2| + |u_2 - u_3| + \cdots + |u_n - u_1| \\
\leq \; & \alpha_1 \left(|u_2 - u_3| + |u_3 - u_4| + \cdots + |u_{n-1} - u_n| + |u_n - u_1| + |u_1 - u_2|\right) \\
& + \alpha_2 \left(|u_3 - u_4| + |u_4 - u_5| + \cdots + |u_1 - u_2| + |u_2 - u_3|\right) \\
& \vdots \\
& + \alpha_k \left(|u_{k+1} - u_{k+2}| + |u_{k+2} - u_{k+3}| + \cdots + |u_{k-1} - u_k| + |u_k - u_{k+1}|\right) \\
= \; & \left(\sum_{i=1}^{k} \alpha_i\right) \left(|u_1 - u_2| + |u_2 - u_3| + \cdots + |u_n - u_1|\right).
\end{aligned}$$

It yields a contradiction since $0 \leq \sum_{i=1}^{k} \alpha_i < 1$. Therefore the only possibility is that $|u_1 - u_2| + |u_2 - u_3| + \cdots + |u_n - u_1| = 0$, i.e., $u_1 = u_2 = \cdots = u_n = v$ (say). Since $(u_1, u_2, \ldots, u_k) \in I^k$ is a solution of (7.33), we must have $v = T(v, v, \ldots, v)$, i.e., v is another fixed point of T. By the uniqueness of the fixed point of T we have $v = u$. Thus, the system (7.33) can have only one solution which is periodic.

Example 7.6 Suppose the following system is to be solved in $\mathbb{R}^3$:

$$\left.\begin{aligned}
x &= a\sin(y) + bz \\
y &= a\sin(z) + bx \\
z &= a\sin(x) + by
\end{aligned}\right\} \tag{7.35}$$

where $a, b \in \mathbb{R}$ are such that $0 \leq a + b < 1$.

Define $T\colon \mathbb{R}^2 \to \mathbb{R}$ by

$$T(x, y) = a\sin(x) + by \quad \text{for all} \quad x, y \in \mathbb{R}.$$

Then we have:

$$\begin{aligned}
|T(x, y) - T(y, z)| &= |a\sin(x) + by - a\sin(y) - bz| \\
&\leq a\,|\sin(x) - \sin(y)| + b\,|y - z| \\
&\leq a\,|x - y| + b\,|y - z|.
\end{aligned}$$

It shows that all the conditions of Theorem 7.14 are satisfied for $k = 2, n = 3$ and $\alpha_1 = a, \alpha_2 = b$ and I is taken as the whole real number system. Therefore, by Theorem 7.14 there exists a periodic solution of system (7.35) that is the only solution of (7.35). This solution can be obtained by solving the equation $x = f(x, x)$, i.e., $x = \dfrac{a}{1-b}\sin(x)$. Obviously, the solution of this equation is the limit of the sequence $\{x_n\}$ defined by $x_{n+2} = a\sin(x_n) + bx_{n+1}$, $n \in \mathbb{N}$, where x_1 and $x_2 \in \mathbb{R}$ are arbitrary.

For further and comparative study of applications of fixed point theorem, we refer the reader to Taleb and Hanebaly [20], Siddiqi [18], Dhage and Dhage [5], Kreyszig [11], Nieto and R-López [13], Benchohra et al. [3], Khan et al. [8], Murthy [12], George et al. [6], Agarwal et al. [1], Agarwal et al. [2], Singh et al. [19], Shukla et al. [16], Shukla and Radenović [17], Berinde and Păcurar [4], Pata [15] and the references therein.

References

[1] Agarwal R.P., Meehan M., O'Regan D., Fixed Point Theory and Applications. Cambridge University Press, 2001.

[2] Agarwal R.P., O'Regan D., Sahu D.R., Fixed Point Theory for Lipschitzian-type Mappings with Applications. Springer, 2009; p. 6.

[3] Benchohra M., Henderson, J., Ntouyas S., Impulsive Differential Equations and Inclusions. Contemporary Mathematics and Its Applications. 2006; p. 2.

[4] Berinde V., Păcurar M., Two elementary applications of some Prešić type fixed point theorems. Creat. Math. Inform. 2011; 20(1):32-42.

[5] Dhage B.C., Dhage S.B., Approximating positive solutions of PBVPs of nonlinear first order ordinary quadratic differential equations. Appl. Math. Lett. 2015; 46:133-142.

[6] George R., Reshma K.P., Rajagopalan R., A generalised fixed point theorem of Prešić type in cone metric spaces and application to Markov process. Fixed Point Theory Appl. 2011: 85.

[7] Jaroma J.H., On the global asymptotic stability of $x_{n+1} = \frac{a+bx_n}{A+x_{n-1}}$, In Proceeding of First International Conference on Difference Equations and Applications. San Antonio, May 1994. Gordon and Breach, 1995; 281-94.

[8] Khan M.S., Berzig M., Samet B., Some convergence results for iterative sequences of Prešić type and applications. Difference Equations. 2012; 2012:38.

[9] Kocić V.L., Ladas G., Global behavior of nonlinear difference equations of higher order with applications. Kluwer, 1993.

[10] Kocić V.L., Ladas G., Rodrigues I.W., On the rational recursive sequences. J. Math. Anal. Appl. 1993; 173:127-57.

[11] Kreyszig E., Introductory Functional Analysis with Applications, John Wiley, 1978.

[12] Murthy P.P., Important tools and possible applications of metric fixed point theory. Nonlinear Anal. 2001; 47:3479-3490.

[13] Nieto J.J., R-López R., Contractive mapping theorems in partially ordered sets and applications to ordinary differential equations. Order 2005; 22: 223-239.

[14] Păcurar M., Fixed points of almost Prešić operators by a k-step iterative method. An. Ştiinţ. Univ. "Al.I. Cuza" Iaşi. Mat. (N.S.). 2011; 57:199-210.

[15] Pata V., Fixed Point Theorems and Applications. Mimeo, 2014.

[16] Shukla S., Gopal D., R-López R., Fuzzy Prešić-Ćirić operators and applications to certain nonlinear differential equations. Math. Model. Anal. DOI: 10.1080/03081079.2016.1153084.

[17] Shukla S., Radenović S., Some generalizations of Prešić type mappings and applications. Ann. of the Alexandru Ioan Cuza Univ. Math. DOI: 10.1515/aicu-2015-0026.

[18] Sidiqui A.H., Functional Analysis with Applications, Tata McGraw-Hill, 1986.

[19] Singh S.L., Pathak H.K., Mishra S.N., On a Suzuki type general fixed point theorem with applications. Fixed Point Theory Appl. 2010: Art. 234717.

[20] Taleb A.A., Hanebaly E., A fixed point theorem and its application to integral equations in modular function spaces. Proc. Amer. Math. Soc. 127(8), 2335-2342.

Index